AF595015

# Advances in Plant Sulfur Research

# Advances in Plant Sulfur Research

Special Issue Editors

**Dimitris L. Bouranis**
**Mario Malagoli**
**Jean-Christophe Avice**
**Elke Bloem**

MDPI • Basel • Beijing • Wuhan • Barcelona • Belgrade • Manchester • Tokyo • Cluj • Tianjin

*Special Issue Editors*
Dimitris L. Bouranis
Agricultural University of Athens
Greece

Mario Malagoli
University of Padova
Italy

Jean-Christophe Avice
Université de Caen Normandie
France

Elke Bloem
Institute for Crop and Soil Science,
Julius Kühn-Institut (JKI),
Federal Research Centre for Cultivated Plants
Germany

*Editorial Office*
MDPI
St. Alban-Anlage 66
4052 Basel, Switzerland

This is a reprint of articles from the Special Issue published online in the open access journal *Plants* (ISSN 2223-7747) (available at: https://www.mdpi.com/journal/plants/special_issues/Sulfur_Metabolism).

For citation purposes, cite each article independently as indicated on the article page online and as indicated below:

LastName, A.A.; LastName, B.B.; LastName, C.C. Article Title. *Journal Name* **Year**, *Article Number*, Page Range.

**ISBN 978-3-03936-006-2 (Hbk)**
**ISBN 978-3-03936-007-9 (PDF)**

Cover image courtesy of Elke Bloem.

# Contents

# About the Special Issue Editors

**Dimitris L. Bouranis** is Professor of Plant Physiology at the Crop Science Department, Agricultural University of Athens (AUA), Greece. His research interests focus on plant nutrition physiology with a special focus on sulfur physiology and use efficiency, fertilization with sulfur-containing specialty fertilizers, and sulfur interactions with iron, nitrogen, and phosphorus in graminaceous species. He serves as Director of the PlanTerra Institute for Plant Nutrition and Soil Quality at AUA and as an elected member of the Hellenic Academy of Agricultural Sciences.

**Mario Malagoli** is Professor of Plant Biology at the Department of Agronomy Food Natural Resources Animals and Environment, University of Padova, Italy. His research interests focus on plant physiology, plant nutrition, sulfur nutrition, sulfur and abiotic stresses, sulfur fertilization, sulfur and tolerance to heavy metals, and sulfur in grapevine cultivation.

**Jean-Christophe Avice** is Professor at Université de Caen Normandie, UMR INRA/UCN 950 Ecophysiologie Végétale et Agronomie (EVA). His research interests focus on nutrient use efficiency, plant nutrition, nitrogen and sulfur fertilization, plant responses to abiotic stress, plant senescence, seed quality, remobilization of nutrients, and proteolytic mechanisms.

**Elke Bloem** is a senior researcher at the Institute for Crop and Soil Science, Julius Kühn-Institut (JKI), Federal Research Centre for Cultivated Plants. Her research interests focus on sulfur nutrition, glucosinolates, sulfur-containing metabolites, gaseous S compounds, hydrogensulfide, cysteine, glutathione, biotic stress, abiotic stress, sulfur and fungal pathogens, and medicinal plants.

*Editorial*

# Advances in Plant Sulfur Research

**Dimitris L. Bouranis [1,*], Mario Malagoli [2], Jean-Christophe Avice [3] and Elke Bloem [4]**

1 Plant Physiology Laboratory, Crop Science Department, Agricultural University of Athens, Iera Odos 75, 11855 Athens, Greece
2 Department of Agronomy, Food, Natural Resources, Animals and Environment, University of Padova, Agripolis, 35020 Legnaro Pd, Italy; mario.malagoli@unipd.it
3 UMR INRA-UCN 950 Ecophysiologie Végétale, Agronomie & Nutritions N.C.S., Normandie Université, UFR des Sciences, FED 4277 Normandie Végétal, Université de Caen Normandie, F-14032 Caen, France; jean-christophe.avice@unicaen.fr
4 Institute for Crop and Soil Science, Julius Kühn-Institut (JKI), Federal Research Centre for Cultivated Plants, Bundesallee 69 (Gebäude 250), D-38116 Braunschweig, Germany; elke.bloem@julius-kuehn.de
* Correspondence: bouranis@aua.gr

Received: 8 February 2020; Accepted: 10 February 2020; Published: 17 February 2020

**Abstract:** As an essential nutrient required for plant growth and development, sulfur (S) deficiency in productive systems limits yield and quality. This special issue hosts a collection of original research articles, mainly based on contributions from the 11th International Plant Sulfur Workshop held on 16–20 September 2018 in Conegliano, Italy, focusing on the following topics: (1) The germinative and post-germinative behaviour of *Brassica napus* seeds when severe S limitation is applied to the parent plants; (2) the independence of S deficiency from the mRNA degradation initiation enzyme PARN in Arabidopsis; (3) the glucosinolate distribution in the aerial parts of sel1-10, a disruption mutant of the sulfate transporter SULTR1;2, in mature *Arabidopsis thaliana* plants; (4) the accumulation of S-methylcysteine as its $\gamma$-glutamyl dipeptide in *Phaseolus vulgaris*; and (5) the role of ferric iron chelation-strategy components in the leaves and roots of maize, have provided new insights into the effect of S availability on plant functionality. Moreover, the role of S deficiency in root system functionality has been highlighted, focusing on (6) the contribution of root hair development to sulfate uptake in Arabidopsis, and (7) the modulation of lateral root development by the CLE-CLAVATA1 signaling pathway under S deficiency. The role of S in plants grown under drought conditions has been investigated in more detail focusing (8) on the relationship between S-induced stomata closure and the canonical ABA signal transduction machinery. Furthermore, (9) the assessment of S deficiency under field conditions by single measurements of sulfur, chloride, and phosphorus in mature leaves, (10) the effect of fertilizers enriched with elemental S on durum wheat yield, and (11,12) the impact of elemental S on the rhizospheric bacteria of durum wheat contributed to enhance the scientific knowledge on S nutrition under field conditions.

**Keywords:** sulfur deficiency; sulfur nutrition physiology; sulfur interactions

---

Sulfur (S) is an essential nutrient required for plant growth and development. The environmental protection measures put in place in the last decades and the reduction of $SO_2$ emission to the atmosphere paradoxically limited the availability of S as input in large scale agriculture with consequent crop S deficiency. Therefore, S containing fertilizers are nowadays used worldwide to support enhancing crop yield and quality. The emergence of S deficiency in such productive systems attracted the scientific attention and triggered significant research interest. In the last 25 years a tremendous boost has taken place in all aspects of plant S research. Milestones in this field include so far the central role of sulfate transporters in response to S availability, sulfur as a part of plant metabolic network, sulfotranferases activity, impact of S on $N_2$ fixation of legumes, role of S compounds in abiotic and biotic stress tolerance, S interactions in crop ecosystems, S-induced-resistance, molecular links between metals in the environment and plant S metabolism, role of sulfate and S-rich compounds in heavy metal tolerance and accumulation.

Despite the amazing amount of the rapidly accumulating information, there are still open questions and challenges on this fascinating field of research. Sulfur has unique characteristics in various aspects, some of which provide potential areas for practical applications, as summarized by Kopriva et al. (2019) [1], pointing out that still significant gaps in our knowledge remain in several aspects. For example, the effects of the availability of other mineral nutrients on S utilization need to be studied more in detail by focusing on their interactions. The functions of S metabolites in assisting mutualistic interactions between living organisms (i.e., microbes, animals, and plants) are not fully understood, while with the exception of a number of known cases characterized so far for essential biological or nutritional functions, the diversity of S metabolites appears to be an unexplored area worth further studies. Moreover, given the importance of the geochemical cycle of S in nature, the roles of S nutrition and S-containing metabolites in the environmental adaptation of plants need to be studied with more attention given to eco-physiological aspects.

This special issue hosts a collection of papers based mainly on contributions from the 11th International Plant Sulfur Workshop held on 16–20 September 2018 in Conegliano, Italy, and their contributions in plant S research is highlighted below.

## 1. New Insights into the Effect of Sulfur Availability on Plant Functionality

Sulfur limitation leads to a reduction of seed yield, nutritional quality, seed viability, and vigour. It is known that S metabolism is involved in the control of germination *sensu stricto* and seedling establishment, but it is largely unknown how the germination and the first steps of plant growth are impacted in seeds when the plants are subjected to sulfate limitation. D'Hooghe et al. (2019) [2] focused on the impact of various S-limited conditions applied to mother plants on the germination indexes and the rate of viable seedlings in a spring oilseed rape cultivar (cv. Yudal), as well as on the sulfate uptake capacity during development of the seedling. When seeds were produced under severe S limitation, viable seedlings from such seeds presented a higher dry biomass and were able to enhance the sulfate uptake by roots and the S translocation to shoots, although the rate of viable seedlings was significantly reduced along with the germination vigour and perturbations of post-germinative events were observed.

When plants are exposed to S limitation, the sulfate assimilation pathway is upregulated at the expense of growth-promoting measures, whilst after cessation of the stress, the protective measures are deactivated, and growth is restored. Indeed, transcripts of S deficiency marker genes are rapidly degraded when starved plants are resupplied with sulfur, but it remains unclear, which enzymes are responsible for the degradation of transcripts during the recovery from starvation. In eukaryotes, mRNA decay is often initiated by the cleavage of poly(A) tails via deadenylases, and mutations in the poly(A) ribonuclease PARN have been linked to altered abiotic stress responses in *Arabidopsis thaliana*. Moreover, the subcellular localization of PARN is currently disputed, with studies reporting both nuclear and cytosolic localization. Armbruster et al. (2019) [3] investigated the role of PARN in the recovery from S starvation. Despite the presence of putative PARN-recruiting AU-rich elements in S-responsive transcripts, S-depleted PARN hypomorphic mutants were able to reset their transcriptome to prestarvation conditions just as readily as wildtype plants. PARN was detected in cytoplasmic speckles, thus reconciling the diverging views in literature by identifying two PARN splice variants whose predicted localization is in agreement with those observations.

In *Arabidopsis thaliana*, a disruption mutant of SULTR1;2, sel1-10, has been characterized with phenotypes such as plants grown under S deficiency. Although the effects of S deficiency on S metabolism have been well investigated in seedlings, no studies have been performed on mature *A. thaliana* plants. Morikawa-Ichinose et al. (2019) [4] analyzed the accumulation and distribution of S-containing compounds in different parts of mature sel1-10, as well as wildtype (WT) plants grown under long-day conditions. While the levels of sulfate, cysteine, and glutathione were almost similar between sel1-10 and WT, levels of glucosinolates (GSLs) differed depending on plant part. GSLs levels in the leaves and stems were generally lower in sel1-10 than those in WT; however, sel1-10 seeds maintained similar levels of aliphatic GSLs to those in WT plants. GSL accumulation in reproductive tissues was likely to be prioritized due to its role in S storage and plant defense even when sulfate supply in sel1-10 was limited.

Seeds of common bean (*Phaseolus vulgaris*) on the one hand constitute an excellent source of vegetable dietary protein, but on the other hand present suboptimal levels of both the essential S-containing amino acids, methionine, and cysteine, whilst they accumulate large amounts of the $\gamma$-glutamyl dipeptide of S-methylcysteine, and lower levels of free S-methylcysteine and S-methylhomoglutathione. Previous findings suggested two distinct metabolite pools: Free S-methylcysteine levels were high at the beginning of seed development and declined at mid-maturation, whilst there was a biphasic accumulation of $\gamma$-glutamyl-*S*-methylcysteine, at early cotyledon and maturation stages. A possible model involves the formation of S-methylcysteine by cysteine synthase from O-acetylserine and methanethiol, whereas the majority of $\gamma$-glutamyl-*S*-methylcysteine may arise from S-methylhomoglutathione. According to Saboori-Robat et al. (2019) [5] metabolite profiling during development and in genotypes differing in total S-methylcysteine accumulation, showed that $\gamma$-glutamyl-*S*-methylcysteine accounts for most of the total S-methylcysteine in mature seed. Profiling of transcripts for candidate biosynthetic genes indicated that BSAS4;1 expression was correlated with both the developmental timing and levels of free S-methylcysteine accumulated, while homoglutathione synthetase (hGS) expression was correlated with the levels of $\gamma$-glutamyl-*S*-methylcysteine. Analysis of S-methylated phytochelatins revealed only small amounts of homophytochelatin-2 with a single S-methylcysteine. The mitochondrial localization of phytochelatin-2 synthase predominant in seed and its spatial separation from S-methylhomoglutathione may explain the lack of significant accumulation of S-methylated phytochelatins.

Plants have developed sophisticated mechanisms for acquiring iron (Fe) from the soil and in the graminaceous species, a chelation strategy is in charge, in order to take up ferric Fe from the rhizosphere. The ferric Fe chelation-strategy components may also be present in the aerial plant parts. Saridis et al. (2019) [6] searched for possible roles of those components in maize leaves, therefore the expression patterns of ferric Fe chelation-strategy components were monitored in the leaves and roots of mycorrhizal and nonmycorrhizal S-deprived maize plants, both before and after sulfate supply. Altering of S supply was chosen due to the strong impact of S on iron homeostasis, whilst mycorrhizal symbiosis was chosen as a treatment that forces the plant to optimize its photosynthetic efficiency, in order to feed the fungus. The results suggested a role for the aforementioned components in ferric chelation and/or unloading from the xylem vessels to the aerial plant parts and it was proposed that the gene expression of the DMA exporter ZmTOM1 can be used as an early indicator for the establishment of a mycorrhizal symbiotic relationship in maize.

## 2. The Role of S Deficiency in Root System Functionality

Root hairs contribute to nutrient uptake from the root environment, but the contribution varies among nutrients. In *Arabidopsis thaliana*, the two high-affinity sulfate transporters SULTR1;1 and SULTR1;2, are responsible for sulfate uptake by roots. Their increased expression under S deficiency (−S) stimulates sulfate uptake. Inspired by the higher and lower expression, respectively, of SULTR1;1 in mutants with more (werwolf [wer]) and fewer (caprice [cpc]) root hairs, Kimura et al. (2019) [7] examined the contribution of root hairs to sulfate uptake. Its uptake rates were similar among plant lines under both S sufficiency and deficiency. Under S deficiency, the expression of SULTR1;1 and SULTR1;2 was negatively correlated with the number of root hairs. These results suggested that both SULTR expression and sulfate uptake rates induced by S deficiency were independent of the number of root hairs. A negative correlation between primary root lengths and number of root hairs, along with a greater number of root hairs was observed under S deficiency than under S sufficiency, thus suggesting that under both S sufficiency and deficiency, sulfate uptake was influenced by the root biomass rather than the number of root hairs.

Architecture of the plant root system changes drastically in response to the availability of macronutrients in the soil environment. Despite the importance of root S uptake in plant growth and reproduction, molecular mechanisms underlying root development in response to S availability have not been fully characterized. Dong et al. (2019) [8] reported on the signaling module composed of the CLAVATA3 (CLV3)/EMBRYO SURROUNDING REGION (CLE) peptide and CLAVATA1 (CLV1) leucine-rich repeat receptor kinase, which regulate lateral root (LR) development in *Arabidopsis thaliana* upon changes in S availability. The wildtype seedlings exposed to prolonged S deficiency showed a phenotype with low LR density, which was restored upon sulfate supply. In contrast, under prolonged S deficiency the clv1 mutant showed a higher daily increase rate of LR density relative to the wildtype, which was diminished to the wildtype level upon sulfate supply. CLE2 and CLE3 transcript levels decreased under S deficiency and through CLV1-mediated feedback regulations. It is suggested that under S-deficient conditions CLV1 directs a signal to inhibit LR development, and the levels of CLE peptide signals are adjusted in the course of LR development. The study demonstrated a fine-tuned mechanism for LR development coordinately regulated by CLE-CLV1 signaling and in response to changes in S availability.

## 3. Role of S in Plants Grown under Drought Conditions

Abscisic acid (ABA) is the canonical trigger for stomatal closure upon drought. Soil-drying is known to facilitate root-to-shoot transport of sulfate, whereas sulfate and sulfide have been independently shown to promote stomatal closure. For induction of stomatal closure, sulfate must be incorporated into cysteine, which triggers ABA biosynthesis by transcriptional activation of NCED3. Rajab et al. (2019) [9] applied reverse genetics to unravel if the canonical ABA signal transduction machinery is required for sulfate-induced stomata closure, and if cysteine biosynthesis is also mandatory for the induction of stomatal closure by sulfide. The importance of reactive oxygen species (ROS) production by the plasma membrane-localized NADPH oxidases, RBOHD, and RBOHF is documented, during the sulfate-induced stomatal closure. In agreement with the established role of ROS as the second messenger of ABA-signaling, the SnRK2-type kinase OST1 and the protein phosphatase ABI1 are essential for sulfate-induced stomata closure, whilst sulfide failed to close stomata in a cysteine-biosynthesis depleted mutant. The presented data support the hypothesis that the two mobile signals, sulfate and sulfide, induce stomatal closure by stimulating cysteine synthesis to trigger ABA production.

## 4. Sulfur Nutrition under Field Conditions

Determination of the S status is very important to detect S deficiency and to prevent losses of yield and seed quality. Etienne et al. (2018) [10] studied the possibility of using the $([Cl^-]+[NO_3^-]+[PO_4^{3-}]):[SO_4^{2-}]$ ratio as an indicator of S nutrition in *Brassica napus* under field conditions, and whether this could be applied to other species. Different S and nitrogen (N) fertilization schemes were applied on a S-deficient field cultivated with oilseed rape. Mature leaves were harvested, and their anions and element contents were analyzed in order to evaluate the aforementioned S nutrition indicator, along with useful threshold values. Large sets of commercial varieties were used to test S deficiency scenarios. Under field conditions, the leaf $([Cl^-]+[NO^{3-}]+[PO_4^{3-}])/[SO_4^{2-}]$ ratio was increased by lowering S fertilization, indicating S deficiency. The usefulness of this ratio was also found for other species grown under controlled conditions. Furthermore, the ratio could be simplified by using the ratio based on the elemental concentrations ([Cl]+[P])/[S] determined by X-ray fluorescence (XRF). Based on the ratio quantified under field conditions, threshold values were determined and used for the clustering of commercial varieties within three groups: S-deficient, at risk of S deficiency, and S sufficiency. It is suggested that the elemental ([Cl]+[P])/[S] ratio can be used as an early and accurate diagnostic tool to manage S fertilization.

The demand to develop fertilizers with higher S use efficiency has been intensified over the last decade, since S deficiency in crops has become more widespread. Bouranis et al. (2018) [11] investigated whether fertilizers enriched with 2% elemental sulfur (ES) via a binding material of organic nature improve yield when compared to the corresponding conventional ones. Several fertilization schemes with or without incorporated ES were tested in various durum wheat varieties, cultivated in commercial fields. The Olsen-P content of each commercial field was found to be correlated with the corresponding relative change in the yields with a strong positive relationship and the content of 8 ppm of available soil phosphorus (P) was a turning point. At higher values the incorporation of ES in the fertilization scheme resulted in higher yield, while at lower values it resulted in lower yield, compared with the conventional one. The application of fertilizer mixtures containing the urease inhibitor N-(n-butyl) thiophosphoric triamide (NBPT), ES, and ammonium sulfate resulted in the highest relative yields. The yield followed a positive linear relationship with the number of heads per square meter. In this correlation, the Olsen-P content separated the results of the two groups of blocks, where the applied linear trend line in each group presented the same slope.

In order to assess and evaluate the effect of fertilizer granules enriched with ES on the dynamics of the rhizospheric bacteria in relation to the corresponding conventional ones, Bouranis et al. (2019) [12] performed comparative experiments. The enriched fertilization scheme was applied to a soil with Olsen-P at 7.8 mg $kg^{-1}$, and to a control treatment where Olsen-P accounted for 16.8 mg $kg^{-1}$ in the soil. Rhizospheric soil at various developmental stages of the crops was analyzed and the agronomic profile of the rhizospheric cultivable bacteria was characterized and monitored. Additionally, the dynamics of P, Fe, organic S, and organic N were studied in both the rhizosoil and in the aerial part of the plant during crop development. Both crops were characterized by a comparable dry mass accumulation per plant throughout development, while the yield of the ES enriched crop was 3.4% less compared to the control crop. The ES enriched crop's aerial parts showed a transient higher P and Fe concentration, while its organic N and S concentrations followed the pattern of the control crop. The incorporation of ES into the conventional fertilizer increased the percentage of arylsulfatase (ARS)-producing bacteria in the total bacterial population, suggesting an enhanced release of sulfate from the soil's organic S pool, which the plant could readily utilize. A large fraction of the population also possessed phosphate solubilization, and/or siderophore production, and/or ureolytic traits, thus improving the crop's P, Fe, S, and N balance. It is proposed that the used ES substantially improved the quality of the rhizosoil at the available P limiting level by modulating the abundance of the bacterial communities in the rhizosphere and effectively enhancing the microbially mediated nutrient mobilization towards improved plant nutritional dynamics.

The aforementioned original research articles clearly illustrate the dynamic nature of the topic, presenting a good example of the current progress in both the basic and applied aspects in the field of plant S research.

**Author Contributions:** All authors have contributed equally in the preparation of the manuscript. All authors have read and agreed to the published version of the manuscript.

**Funding:** This work received no external funding.

**Acknowledgments:** We would like to thank the editorial office of Plants MDPI for their constant support in the compilation of this special issue and all the colleagues that contributed to this Special Issue, giving their perspectives and ideas to increase translation of fundamental knowledge from model plants to crops.

**Conflicts of Interest:** The authors declare no conflict of interest.

## References

1. Kopriva, S.; Malagoli, M.; Takahashi, H. Sulfur nutrition: Impacts on plant development, metabolism, and stress responses. *J. Exp. Bot.* **2019**, *70*, 4069–4073. [CrossRef] [PubMed]
2. D'Hooghe, P.; Picot, D.; Brunel-Muguet, S.; Kopriva, S.; Avice, J.-C.; Trouverie, J. Germinative and Post-Germinative Behaviours of Brassica napus Seeds Are Impacted by the Severity of S Limitation Applied to the Parent Plants. *Plants* **2019**, *8*, 12. [CrossRef] [PubMed]

3. Armbruster, L.; Uslu, V.V.; Wirtz, M.; Hell, R. The Recovery from Sulfur Starvation Is Independent from the mRNA Degradation Initiation Enzyme PARN in Arabidopsis. *Plants* **2019**, *8*, 380. [CrossRef] [PubMed]
4. Morikawa-Ichinose, T.; Kim, S.-J.; Allahham, A.; Kawaguchi, R.; Maruyama-Nakashita, A. Glucosinolate Distribution in the Aerial Parts of sel1-10, a Disruption Mutant of the Sulfate Transporter SULTR1;2, in Mature Arabidopsis thaliana Plants. *Plants* **2019**, *8*, 95. [CrossRef] [PubMed]
5. Saboori-Robat, E.; Joshi, J.; Pajak, A.; Solouki, M.; Mohsenpour, M.; Renaud, J.; Marsolais, F. Common Bean (Phaseolus vulgaris L.) Accumulates Most S-Methylcysteine as Its γ-Glutamyl Dipeptide. *Plants* **2019**, *8*, 126. [CrossRef] [PubMed]
6. Saridis, G.; Chorianopoulou, S.N.; Ventouris, Y.E.; Sigalas, P.P.; Bouranis, D.L. An Exploration of the Roles of Ferric Iron Chelation-Strategy Components in the Leaves and Roots of Maize Plant. *Plants* **2019**, *8*, 133. [CrossRef] [PubMed]
7. Kimura, Y.; Ushiwatari, T.; Suyama, A.; Tominaga-Wada, R.; Wada, T.; Maruyama-Nakashita, A. Contribution of Root Hair Development to Sulfate Uptake in Arabidopsis. *Plants* **2019**, *8*, 106. [CrossRef] [PubMed]
8. Dong, W.; Wang, Y.; Takahashi, H. CLE-CLAVATA1 Signaling Pathway Modulates Lateral Root Development under Sulfur Deficiency. *Plants* **2019**, *8*, 103. [CrossRef] [PubMed]
9. Rajab, H.; Khan, M.S.; Malagoli, M.; Hell, R.; Wirtz, M. Sulfate-Induced Stomata Closure Requires the Canonical ABA Signal Transduction Machinery. *Plants* **2019**, *8*, 21. [CrossRef] [PubMed]
10. Etienne, P.; Sorin, E.; Maillard, A.; Gallardo, K.; Arkoun, M.; Guerrand, J.; Cruz, F.; Yvin, J.-C.; Ourry, A. Assessment of Sulfur Deficiency under Field Conditions by Single Measurements of Sulfur, Chloride and Phosphorus in Mature Leaves. *Plants* **2018**, *7*, 37. [CrossRef] [PubMed]
11. Bouranis, D.L.; Gasparatos, D.; Zechmann, B.; Bouranis, L.D.; Chorianopoulou, S.N. The Effect of Granular Commercial Fertilizers Containing Elemental Sulfur on Wheat Yield under Mediterranean Conditions. *Plants* **2019**, *8*, 2. [CrossRef] [PubMed]
12. Bouranis, D.L.; Venieraki, A.; Chorianopoulou, S.N.; Katinakis, P. Impact of Elemental Sulfur on the Rhizospheric Bacteria of Durum Wheat Crop Cultivated on a Calcareous Soil. *Plants* **2019**, *8*, 379. [CrossRef] [PubMed]

Article

# Germinative and Post-Germinative Behaviours of *Brassica napus* Seeds Are Impacted by the Severity of S Limitation Applied to the Parent Plants


**Philippe D'Hooghe [1], Dimitri Picot [1], Sophie Brunel-Muguet [1], Stanislav Kopriva [2], Jean-Christophe Avice [1] and Jacques Trouverie [1,*]**

1 Normandie Univ, UNICAEN, INRA, UMR EVA, SFR Normandie Végétal FED4277, Esplanade de la Paix, F-14032 Caen, France; philippe.dhooghe@live.fr (P.D.); dimi.picot@gmail.com (D.P.); sophie.brunel-muguet@unicaen.fr (S.B.-M.); jean-christophe.avice@unicaen.fr (J.-C.A.)

2 Botanical Institute and Cluster of Excellence on Plant Sciences (CEPLAS), University of Cologne, 50674 Cologne, Germany; skopriva@uni-koeln.de

* Correspondence: jacques.trouverie@unicaen.fr; Tel.: +33-231-565-166



Received: 13 November 2018; Accepted: 27 December 2018; Published: 5 January 2019



**Abstract:** In oilseed rape (*Brassica napus* L.), sulphur (S) limitation leads to a reduction of seed yield and nutritional quality, but also to a reduction of seed viability and vigour. S metabolism is known to be involved in the control of germination sensu stricto and seedling establishment. Nevertheless, how the germination and the first steps of plant growth are impacted in seeds produced by plants subjected to various sulphate limitations remains largely unknown. Therefore, this study aimed at determining the impact of various S-limited conditions applied to the mother plants on the germination indexes and the rate of viable seedlings in a spring oilseed rape cultivar (cv. Yudal). Using a $^{34}$S-sulphate pulse method, the sulphate uptake capacity during the seedling development was also investigated. The rate of viable seedlings was significantly reduced for seeds produced under the strongest S-limited conditions. This is related to a reduction of germination vigour and to perturbations of post-germinative events. Compared to green seedlings obtained from seeds produced by well-S-supplied plants, the viable seedlings coming from seeds harvested on plants subjected to severe S-limitation treatment showed nonetheless a higher dry biomass and were able to enhance the sulphate uptake by roots and the S translocation to shoots.



**Keywords:** S metabolism; $^{34}$S labelling; sulphate uptake; sulphate translocation; seed viability; oilseed rape


---

## 1. Introduction

Oilseed rape (*Brassica napus* L.) is particularly sensitive to sulphur (S) limitation that results in reduced yield and seed quality. This lack of seed quality may have unexpected impacts on seed germination and/or on the viability of subsequent seedlings. Indeed, mature seeds produced by oilseed rape subjected to a sulphate restriction applied early in the cycle of development (i.e., at the bolting stage), were characterised by a low ability to produce normal seedlings [1] and by reduced germination vigor and viability [2]. Seed germination sensu stricto is a two-phase process that starts with the uptake of water by the quiescent dry seed and ends with the elongation of the embryonic axis. The completion of germination is visible by the penetration of the structures surrounding the embryo by the radicle (radicle protrusion), and subsequent events are more specifically associated with seedling establishment [3–5]. Germination sensu stricto and seedling establishment are complex processes that possess multiple and variable levels of regulation that depend on the studied species [1,3,5–7]. In oilseed rape, mature seeds are non-dormant, but can enter into secondary dormancy if environmental conditions are not favourable for germination [8]. From this late growth stage, the sulphate previously stored by the plant and actively remobilised to seeds can meet the needs in S for a good seed development and quality [1,2]. However, if the sulphate shortage occurs earlier in plant development, S reserves are not sufficient to satisfy the needs for plant growth and seed development, and the mature seeds produced can display reduced accumulation of S and lipid reserves and profound metabolic changes that impact seed viability and do not favour seed germination of the still-viable seeds [2].

Sulphur metabolism is involved at various levels in the control of germination sensu stricto and seedling establishment [7]. For example, S-adenosylmethionine (SAM), synthesised by SAM synthetase from methionine, is a methyl-donor group but also a precursor of biotin and ethylene and thus could interplay at multiple levels in the regulation of these key steps [7,9–11]. Sulphur limitation is well-known for modulating the S management in Brassicacea species by increasing the sulphate uptake and remobilisation through the induction of the corresponding transporters (e.g., the *BnSultr1* and *BnSultr4* genes for root uptake and remobilisation of sulphate stored in a vacuole, respectively) and by increasing assimilation capacities [12,13]. S restriction could also cause a profound disturbance in seed metabolism, reduce the accumulation of seed reserves and reduce seed viability and germination vigour [2]. According to our knowledge, no study has sufficiently considered the development and the modulations of sulphate management of seedlings generated from seeds harvested of S-limited plants. In order to investigate these processes, we performed experiments using seeds harvested from plants subjected to various S-limitations during their growth. Germination indexes (such as speed of germination, time to reach 50% of germinated seeds, germination capacity) and the rate of normal seedlings produced were determined. The seeds were used (i) to verify if the germination and the subsequent seedlings' establishment were affected and (ii) to evaluate the sulphate uptake capacity of the green seedlings developed from these seeds by using a specific $^{34}$S-sulphate pulse method.

## 2. Results

### 2.1. Impacts of Suboptimal Levels of Sulphate Applied from the Beginning of the Plant Growth Cycle on S Status of Seeds and Its Consequence for Germination

The S400%, S70%, S20% and S5% seeds correspond to seeds harvested from plants supplied with 400%, 70%, 20% and 5%, respectively, of the optimal level of sulphate they need for growth. The mature seed dry weights and the amounts of C and N were not significantly different between the four seed lots. Compared to S400% seeds, the S70% seeds had similar S, S-sulphate and S-glutathione amounts but a higher S-protein amount (Table 1). The S20% seeds showed lower S and S-sulphate amounts than S400% seeds. Similarly, S5% seeds showed reduced S and S-sulphate amounts, and a reduced S-glutathione amount, compared to S400% seeds (Table 1).

The germination of S400% seeds began around 20 h after sowing and reached its maximum around 50 h after sowing (Figure 1). Compared to S400% seeds, germination was not affected for S70% and S20% seeds. However, S5% seeds showed a lower germination capacity (by 24.7%) and speed of germination (by 32.6%) than S400% seeds (Table 2). While the time to reach 50% of the germination capacity (T'50) was not significantly different from S400% seeds, the time to reach 50% of germination (T50) calculated for S5% seeds was significantly higher.

**Table 1.** Seed dry weight and amounts of C, N, S, S-sulphate, S-protein and S-glutathione in mature *Brassica napus* seeds obtained from plants well-supplied with S and grown under S-limited conditions. Values correspond to means ± standard error (SE) (n = 3). Seeds (without fragments and aborted seeds) were obtained from plants subjected to plethoric S nutrition (400% of plant S needs provided, S400% seeds) or subjected to a S limitation from the beginning of plant growth with 70%, 20% and 5% of plant S needs provided (S70%, S20% and S5% seeds, respectively).

| Seed Lot | Seed Dry Weight (mg·seed$^{-1}$) | C Amount (mg·seed$^{-1}$) | N Amount (µg·seed$^{-1}$) | S Amount (µg·seed$^{-1}$) | S-Sulphate Amount (µg·seed$^{-1}$) | S-Protein Amount (µg·seed$^{-1}$) | S-Glutathione Amount (ng·seed$^{-1}$) |
|---|---|---|---|---|---|---|---|
| S400% | 4.02 ± 0.2 | 2.62 ± 0.1 | 220.35 ± 13.61 | 58.95 ± 5.43 | 11.31 ± 1.76 | 7.21 ± 0.55 | 284.1 ± 28.2 |
| S70% | 4.27 ± 0.12 | 2.63 ± 0.07 | 235.24 ± 4.3 | 52.18 ± 3.06 | 9.19 ± 2.6 | 9.42 ± 0.49 * | 231.5 ± 51.8 |
| S20% | 4.1 ± 0.25 | 2.55 ± 0.18 | 207.17 ± 16.89 | 18.2 ± 2.53 * | 2.5 ± 0.41 * | 6.26 ± 0.39 | 333.8 ± 46.6 |
| S5% | 3.99 ± 0.43 | 2.54 ± 0.3 | 209.4 ± 28.53 | 10.56 ± 2.25 * | 0.91 ± 0.28 * | 6.6 ± 0.64 | 161.9 ± 9.6 * |

* Significant difference from the S400% seeds ($p < 0.05$).

**Table 2.** Germination indexes and normal seedling rate of *Brassica napus* seeds obtained from plants well S supplied and grown under S-limited conditions. Values correspond to means ± SE (n = 3). Seeds were obtained from plants subjected to plethoric S nutrition (400% of plant S needs provided, S400% seeds) or subjected to a S limitation from the beginning of plant growth with 70%, 20% and 5% of plant S needs provided (S70%, S20% and S5% seeds, respectively).

| SEED Lot | Germination Capacity (%) [a] | T50 (hours) [b] | T'50 (hours) [b] | Speed of Germination (hour$^{-1}$) [c] | CRG [d] | Normal Seedling Establishment Rate (%) [e] |
|---|---|---|---|---|---|---|
| S400% | 100 ± 0.0 | 28.7 ± 1.0 | 28.7 ± 1.0 | 3.37 ± 0.1 | 2.03 ± 0.03 | 86.7 ± 1.8 ° |
| S70% | 100 ± 0.0 | 26.6 ± 0.9 | 26.6 ± 0.9 | 3.62 ± 0.11 | 2.1 ± 0.03 | 74.7 ± 2.3 *° |
| S20% | 98 ± 1.2 | 24.6 ± 1.1 | 24.3 ± 1.2 | 3.91 ± 0.26 | 2.19 ± 0.05 | 66 ± 8.1 *° |
| S5% | 75.3 ± 4.7 * | 35.8 ± 1.2 * | 31.6 ± 1.1 | 2.27 ± 0.13 * | 1.96 ± 0.01 | 69 ± 2.8 * |

[a] Germination capacity determined at 71 h; [b] calculated from the Gompertz equation [14]; [c] as calculated by Bradbeer [15]; [d] Coefficient of the Rate of Germination (CRG) as calculated by Bewley and Black [3]; [e] determined at 14 days after sowing (DAS). * Significant difference from the value obtained for the S400% seeds ($p < 0.05$). ° Significant difference between germination capacity and normal seedling rate ($p < 0.05$).

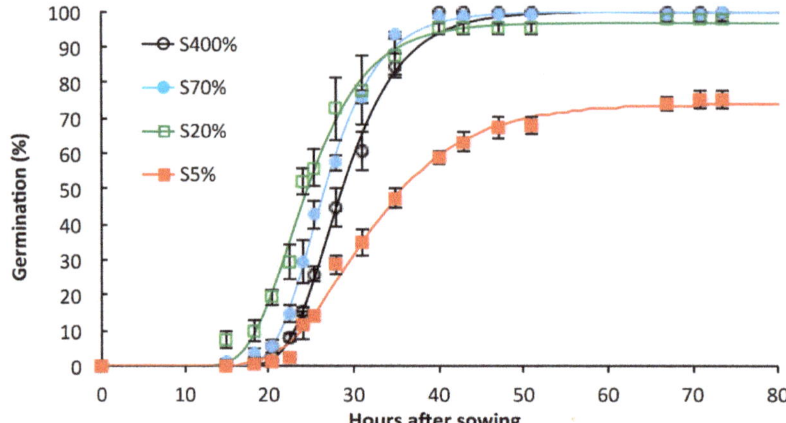


**Figure 1.** Evolution of germination percentage of *Brassica napus* seeds obtained from plants well S supplied and grown under S-limited conditions. Values shown by symbols correspond to means ± SE (n = 3). Seeds were obtained from plants subjected to plethoric S nutrition (400% of plant S needs provided, S400% seeds) or subjected to a S limitation from the beginning of plant growth with 70%, 20% and 5% of plant S needs provided (S70%, S20% and S5% seeds, respectively). The curves are calculated from the non-linear Gompertz regression model [14]. The germination indexes determined from these observations are given in Table 2.

### 2.2. Development and S Management of Green Seedlings

The normal seedling establishment rate was significantly reduced for S70%, S20% and S5% seeds compared to S400% seeds (Table 2). Fourteen days after sowing on perlite (14 DAS), the green seedlings from S400%, S70% and S20% seeds weighted about 5 mg dry weight, with a shoot:root ratio around 4. Interestingly, the S5% green seedlings had a significantly higher dry weight in both shoots and roots (Figure 2a). The shoot:root ratio was not significantly different between S400% and S5% green seedlings. The S amounts in seedlings produced from all of the seed lots (Supplemental Figure S1a) were not different to the S amounts measured in mature seeds (Table 1), showing that no S was absorbed during the 14 days of seedling establishment.

In order to minimise the differences linked to variations in biomass between S400%, S70%, S20% and S5% green seedlings, but also between 14 and 15 DAS, the results were expressed as contents. However, the expression of results in terms of quantity (see Supplemental Figures S1 and S2) reflects the same observations and does not change the interpretations. The S contents were similar between shoot and root for S400% and S70% green seedlings. These contents were significantly lower in S20% and S5% green seedlings, particularly for the shoot (Figure 2b,c). At 14 DAS, the S was principally allocated to the shoot, with 83% of the total S in S400% green seedlings; however, this preferential allocation to the aerial parts tended to be reduced in S70%, S20% and S5% green seedlings (with up to 32% of the total S in the root for S5% green seedlings; Supplemental Figure S1b).

The feeding of 14 DAS green seedlings from S400%, S70%, S20% and S5% seeds for 24 h with the nutrient solution containing $^{34}$S-sulphate (5 atom% excess; 15 DAS + $^{34}SO_4^{2-}$) or containing no S (15 DAS-S) did not change significantly the roots and shoot dry weights (Figure 2a). The S contents in the shoot and roots were not significantly affected by the -S treatment for all of the lots (Figure 2b,c). The supply with sulphate for 24 h ($^{+34}SO_4^{2-}$) increased the S content in roots and shoots of S5% green seedlings, but did not change the S contents in roots and shoots for S400%, S70% and S20% green seedlings (Figure 2b,c). As expected, no significant difference in $^{34}$S contents in root and shoot was observed between 14 and 15 DAS in response to the -S treatment, and the $^{+34}SO_4^{2-}$ treatment increased markedly the $^{34}$S content in both shoot and root of all the green seedlings lots (Figure 3a,b). In terms of quantity, the $^{34}S$ absorbed between 14 and 15 DAS was principally allocated to shoots (Supplemental Figure S2). Interestingly, the difference in $^{34}S$ content between 15 DAS + $^{34}SO_4^{2-}$ and 14 DAS was higher in shoots of S5% green seedlings than in S400% green seedlings (Figure 3b).

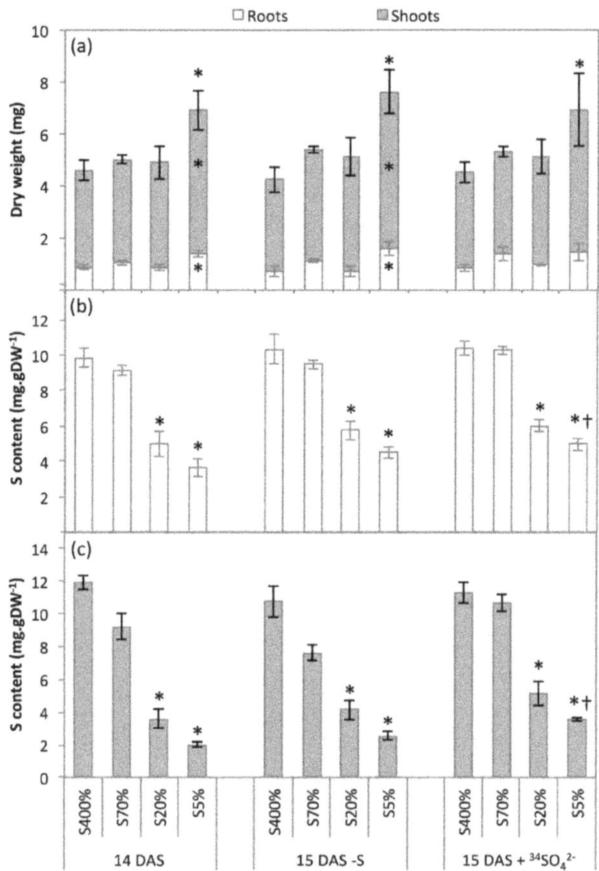


**Figure 2.** Dry weight and S content of green seedlings shoots and roots. (**a**) Dry weight of green seedlings shoots and roots. (**b**) S content of green seedlings roots. (**c**) S content of green seedlings shoots. Data were acquired 14 days after sowing on water (14 DAS) and after 24-h treatment of 14 DAS seedlings with 508.7 μM $^{34}SO_4^{2-}$ at 5 atom% excess (15 DAS + $^{34}SO_4^{2-}$) and with 8.7 μM $SO_4^{2-}$ (15 DAS-S). Seedlings were grown from seeds harvested on plants subjected to plethoric S nutrition (400% of plant S needs provided, S400% seeds) or subjected to a S limitation from the beginning of plant growth with 70%, 20% and 5% of plant S needs provided (S70%, S20% and S5% seeds, respectively). Values correspond to means ± SE (n = 3). * Significant difference from the value of S400% green seedlings ($p < 0.05$). † Significant difference from the corresponding value at 14 DAS ($p < 0.05$). In panel (**a**), signs (* and/or †) on the top of the bars are for the whole plant (shoot and root), while signs inside the bar are for root or shoot specifically.

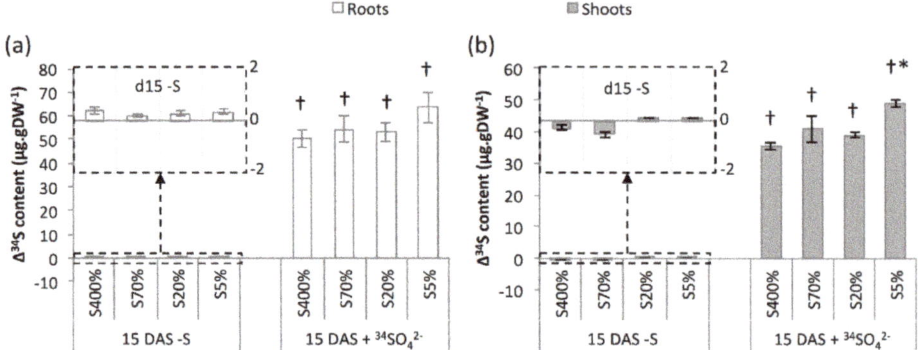


**Figure 3.** Differences in $^{34}S$ ($\Delta^{34}S$) content in excess in roots (**a**) and shoots (**b**) of oilseed rape green seedlings between 14 DAS and 15 DAS-S and between 14 DAS and 15 DAS + $^{34}SO_4{}^{2-}$. Values correspond to means ± SE (n = 3). Seedlings were grown from seeds harvested on plants subjected to plethoric S nutrition (400% of plant S needs provided, S400% seeds) or subjected to a S limitation from the beginning of plant growth with 70%, 20% and 5% of plant S needs provided (S70%, S20% and S5% seeds, respectively). † Significant difference from the 15 DAS-S value ($p < 0.05$). * Significant difference from the corresponding value of S400% seedlings ($p < 0.05$).

To test whether the difference in $^{34}S$ content is caused by different regulation of sulphate transporters, the transcript levels of genes encoding major sulphate transporters were compared by qPCR. As compared to S400% green seedlings at 14 DAS, no significant difference was observed for *BnSultr1;2* expression in roots of S70%, S20% and S5% green seedlings (Figure 4a) and for *BnSultr4;2* expression in shoots of S20% and S5% green seedlings (Figure 4b). *BnSultr1;1* gene expression was higher in roots of S20% and S5% green seedlings, and *BnSultr4;1* gene expression was higher in shoots of S5% green seedlings only. The expression of both *BnSultr4* genes was lower in shoots of S70% green seedlings compared to S400% seedlings (Figure 4).

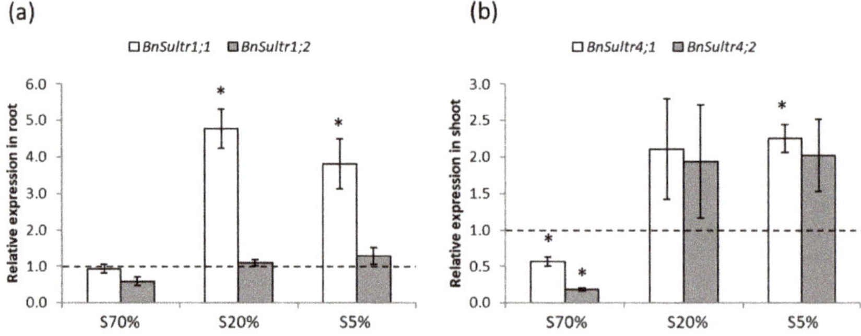


**Figure 4.** Relative expression of *BnSultr1;1* and *BnSultr1;2* in roots (**a**) and *BnSultr4;1* and *BnSultr4;2* in shoots (**b**) of S70%, S20% and S5% oilseed rape green seedlings 14 days after sowing on water (14 DAS). The expression level of *BnSultr1;1*, *BnSultr1;2*, *BnSultr4;1* and *BnSultr4;2* is relative to the result obtained in S400% green seedlings at 14 DAS, shown by the dotted line. Seedlings were grown from seeds harvested on plants subjected to plethoric S nutrition (400% of plant S needs provided, S400% seeds) or subjected to a S limitation from the beginning of plant growth with 70%, 20% and 5% of plant S needs provided (S70%, S20% and S5% seeds, respectively). Values correspond to means ± SE (n = 3) ($p < 0.05$). * Significant difference from the value of S400% green seedlings ($p < 0.05$).

## 3. Discussion

### 3.1. S-Limitation Impacts on Germination Sensu Stricto and on Post-Germinative Events

Under our experimental conditions, the germination kinetics (Figure 1) and the germination indexes (Table 2) of the seeds harvested on the spring genotype Yudal well-supplied with sulphate (i.e., S400% seeds) are not significantly different to the winter cv. Capitol used in our previous work [2]. The seeds produced by plants grown under severe S-limited conditions (i.e., S5% seeds) are characterised by a low viable seedlings establishment rate that is related to a reduction in germination vigour (Figure 1 and Table 2). The different germination indexes used in this study highlight that both speed of germination and germination capacity are reduced for the genotype studied in response to S5% treatment (Table 2). This observation is reinforced by similar findings in another oilseed rape genotype (cv. Capitol) and under a different severe S-limited condition applied to the mother plant [2] (i.e., a S restriction applied from the bolting stage [2] versus a severe S limitation from the beginning of growth in the present study). Interestingly, the low rate of viable seedlings for the S70% and S20% seeds was also associated with impacts on post-germinative events, since the normal seedling rates were significantly lower than the germination capacities.

The decline of germination capacity and vigor that is observed in S5% seeds could be related to various factors. First, the disturbances of C metabolism in developing and mature seeds in response to severe S limitation could reduce seed germination vigor [2]. Second, the modifications of synthesis, accumulation and perception of phytohormones in seeds with a low S status could be involved in the decrease of germination capacity and vigor. Among the candidates, ABA has an inhibitory role on testa rupture during seed germination of *Brassica napus* [16,17] and its accumulation is affected by sulphate starvation in *Arabidopsis thaliana* [18]. While a change in the ABA:GA balance to the benefit of ABA promotes dormancy, a modification of this balance in favour of GA promotes germination [6]. A sulphate restriction applied from the beginning of pollination did not impact ABA or GA contents significantly, but slightly increased the ABA:GA ratio without negative effect on germination vigour [19]. In addition, ethylene may also interact with ABA upon germination, counteracting its inhibitory effect in *Arabidospis thaliana* [20,21], and is involved in post-germinative events [22]. Under a S-limited condition, a lower availability of methionine could lead to a reduced production of SAM in seeds and subsequently to a decrease of the ethylene production that normally negatively impacts seed germination and seedling establishment. Therefore, the potential involvement of phytohormones on the seed germination capacity and vigor may deserve to be explored in a specific study in relationship with the level of S fertilization.

### 3.2. Seedling Sulphate Uptake Capacity Is Impacted by S Limitation Applied to the Parent Plant

The roots and shoots dry weights observed at 14 DAS for the S5% green seedlings were higher than for the S400% green seedlings (Figure 2a). These unexpected data highlight that seedling establishment could be strongly stimulated in seeds produced from plants severely limited in sulphate. Even though this enhanced biomass in offspring of plants submitted to S limitation is surprising, similar observations were made for *Brassica napus* in response to drought stress [23] and for *Arabidopsis thaliana* in response to salt stress [24]. Although it was not the matter of the present study, several hypotheses could be proposed to explain this positive transgenerational effect on seedling growth after germination completion in response to sulphate restriction, such as changes induced by stress in the epigenome of the seeds or maternal imprinting. Indeed, in addition to lipid and protein alterations [2], profound metabolic modifications may occur during seed development due to the lower S abundance, such as alteration of SAM, glutathione, acetyl-CoA or glucosinolate content [2,22,25,26] or other perturbations of homeostasis in seeds.

As seedlings were grown on water and under light for 14 DAS, an enhanced photosynthetic capacity of S5% green seedlings compared to S400% green seedlings could be a physiological process susceptible to explain the apparent discrepancy between similar seed and higher green seedling dry weights. Interestingly, this explanation is in agreement with previous hypothesis on maternal imprinting, since a lower SAM availability would impair ethylene production, which has been shown to be essential for achieving normal photosynthetic capacity in Arabidopsis leaves [27] and which is known to inhibit *Arabidopsis thaliana* root and shoot growth [28,29]. Moreover, due to its significant role in the modification of the epigenome through DNA methylation, SAM may also be potentially involved in the peculiar behaviour observed in S5% green seedlings.

Interestingly, as indicated by the higher S and $^{34}$S amounts and contents between 14 and 15 DAS (Figure 2b,c and Figure 3 and Supplemental Figures S1 and S2), the S5% green seedlings also showed a significantly higher sulphate uptake compared to S400% green seedlings. However, the role of the *BnSultr1;1* and *BnSultr1;2* genes' expression in this higher sulphate uptake capacity cannot be clarified, since *BnSultr1;1* gene expression in the root was higher for S5% green seedlings, but also for S20% green seedlings (Figure 4), that showed no difference in $^{34}$S content compared to S400% green seedlings (Figure 3). Although this result might suggest a post-transcriptional regulation of this transporter, the involvement of other sulphate transporters in the higher sulphate uptake capacity of S5% green seedlings deserves to be studied. In addition to increasing the quantity and content of $^{34}$S more significantly in S5% green seedlings than in S400% green seedlings, the supply with 500 μM sulphate labelled with $^{34}$S (5 atom% excess) appears to change the allocation of S in S5% green seedlings in favour of the shoot. Consistent with this observation, the shoot of *Brassica oleracea* functioned as the primary sink for sulphate taken up after sulphate resupply to sulphate-deprived plants [30]. According to all of these results, it could be hypothesised that *BnSultr2;1*, a sulphate transporter responsible for the translocation of sulphate between roots and shoots and upregulated in roots under S-limiting conditions [31], is highly expressed in root of S5% green seedlings, explaining both their better sulphate uptake capacity and the translocation of the S taken up to the shoot.

This study highlights an induced-intergenerational effect of sulfur deficiency applied on mother plants on the uptake and translocation sulfur capacities of the offspring green seedlings. Seeds harvested on the most severe S-limited plants showed lower germination capacity; however, surprisingly, those that were able to produce normal seedlings displayed a higher growth. These behaviors raised questions about the S effect that could contribute to the stress memory of the offspring.

## 4. Materials and Methods

To address the objectives of the present work, the experiments were conducted on a spring oilseed rape genotype (cv. Yudal) with the aim to verify if the germination and the subsequent development of seedlings were affected in seeds that were obtained from plants supplied with different levels of sulphate.

### *4.1. Production of Seeds from Plants Supplied with Different Levels of Sulphate*

The mature seeds from cv. Yudal were harvested on plants grown into individual pots (a mix of 2:3 perlite and 1:3 vermiculite) under greenhouse conditions (natural day/light conditions from January until July) and supplied with four different levels of S nutrition. Nutrient solutions were provided with an increasing amount of $MgSO_4$ according to the relative-addition rate nutrient-dosing system to reach a constant relative growth rate during exponential plant growth as described by Brunel-Muguet et al. [32]. The four treatments were as follows: plants were supplied with 400% (i.e., plethoric S nutrition), 70%, 20% and 5% of the optimal S needs as sulphate from the beginning of the plant growth cycle until mature seed stage. Optimal S requirement was calculated according to the S needs of Capitol under a non-limiting condition from the study reported by Dubousset et al. [1]. Table S1 summarises the effects of the sulfur treatments on the mother plants. The seeds of each treatment, i.e., S400%, S70%, S20% and S5% seeds, were finally harvested at maturity and stored at 12 °C under vacuum in a desiccator until their use. For germination tests and investigations on seedling development, the mature seeds obtained for each treatment were sorted based on a diameter higher than 1.6 mm in order to exclude the few aborted seeds and the seeds fragments.

### 4.2. Germination Tests and Indexes

Fifty mature seeds per biological repetition (n = 3) were sown on Whatman filter paper soaked with 10 mL sterile water within Petri dishes (12 × 12 cm). The Petri dishes were then closed and placed in darkness inside a growth chamber at 20 °C and 70% relative humidity. Seeds that showed a completed radicle protrusion through the seed coat were counted as germinated. The germination percentage of the different seed lots was represented using Gompertz functions [14].

Five germination indexes were selected for this study. The germination capacity (%) was determined at 71 h. The time to reach 50% germination of the total number of seeds (T50) and 50% of the germination capacity (T'50) were calculated from the Gompertz functions. The speed of germination was calculated according to Bradbeer [15] using the following equation:

$$\text{Speed of germination} = \sum[(N_t - N_{(t-1)})/t]$$

where Nt is the percentage of germinated seeds observed at t hours.

The Coefficient of the Rate of Germination (CRG) was calculated according to Bewley and Black [3] with the following equation:

$$\text{CRG} = 100 \times \sum N_t / \sum (t \times N_t).$$

### 4.3. Determination of Normal Seedling Rate

One hundred and fifty (150) mature seeds per biological repetition (n = 3) were germinated on perlite medium soaked with water for 14 days after sowing (14 DAS) with a cycle of 8 h dark (18 °C)/16 h light (25 °C) and a photon flux density of 400 $\mu mol \cdot m^{-2} \cdot s^{-1}$. Seedlings displaying at 14 DAS a radicle but no hypocotyl or a thick glassy radicle and/or hypocotyl were considered as abnormal. The percentage of seedlings with normal development is indicated for each S treatment in Table 2. The viable seedlings at 14 DAS were used for testing the sulphate uptake capacity of seedlings using a $^{34}SO_4{}^{2-}$ pulse method.

### 4.4. Protocol for Studying the Development and Sulphate Uptake Capacities of Seedlings Coming from Seeds Produced by Plants Supplied with Suboptimal Levels of Sulphate from the Beginning of the Plant Growth Cycle

Fourteen days after sowing on perlite soaked with water (14 DAS), normally developed seedlings obtained from each treatment (i.e., S400%, S70%, S20% and S5% seedlings) were fed with $^{34}SO_4{}^{2-}$ (pulse of 24 h) for testing their S uptake capacity. The $^{34}SO_4{}^{2-}$ was obtained by oxidation of 100 mg $^{34}$S-sulphur (Sigma Aldrich, St. Quentin Fallavier, France) with 10 mL of nitric acid ($HNO_3$ at 65%) in a tube of mineralisation at 200 °C for 2 h as described in Salon et al. [33]. Seedlings were transferred for 24 h on perlite soaked with 25% Hoagland nutrient solution with 508.7 µM $^{34}SO_4{}^{2-}$ (5 atom% excess; 15 DAS + $^{34}SO_4{}^{2-}$) or under a S-restricted condition (8.7 µM $SO_4{}^{2-}$; 15 DAS-S). After this period, seedlings were collected and profusely rinsed with water for 1 min. Shoot and roots were separated, and, for 14 DAS seedlings, an aliquot of each was stored at −80 °C for RNA extraction and q-PCR analysis. For 14 and 15 DAS seedlings, the remaining root and shoot material was freeze-dried for the determination of dry weight, total S content, $^{34}$S labelling, sulphate and $^{34}$S-sulphate before (14 DAS) and after application of $^{34}SO_4{}^{2-}$ feeding (15 DAS + $^{34}SO_4{}^{2-}$) or under a S-restricted condition (15 DAS-S).

### 4.5. Biochemical and Molecular Analyses

#### 4.5.1. C, N, S and $^{34}$S Analysis

Freeze-dried samples were ground to a fine powder, weighed and prepared into tin capsules for determination of total C, N and S contents using an elemental analyser (EA3000, EuroVector, Milan, Italy) connected to a continuous flow isotope ratio mass spectrometer (IRMS; Isoprime, GV

Instruments, Manchester, UK). The IRMS analysis provided the changes of the relative amount of $^{34}S$ in excess in each sample derived from the tracer fed to the test plant.

#### 4.5.2. Determination of S-Sulphate Amount

Sulphate was extracted from 45 mg dry weight (DW) of seed samples ground to a fine powder, incubated twice with 2 mL of 50% ethanol at 40 °C for 1 h, centrifuged (10,000× *g*, 20 min), incubated twice with water at 95 °C for 1 h and centrifuged (10,000× *g*, 20 min). The supernatants were pooled and evaporated under vacuum (Concentrator Evaporator RC 10.22, Jouan, Saint-Herblain, France). The dry residue was resuspended in 2 mL of ultra-pure water. The sulphate concentration was directly determined by high performance liquid chromatography (HPLC, ICS3000, Dionex Corp., Sunnyvale, CA, USA).

#### 4.5.3. Determination of S-Protein Amount

For determination of S-protein amounts, soluble proteins were extracted by grinding 30 mg fresh weight of mature seeds in 0.5 mL of citrate Na-phosphate buffer (20 mM citrate and 160 mM $Na_2HPO_4$, pH 6.8). The homogenate was centrifuged at 12,000× *g*, 4 °C for 1 h and the resulting supernatant was used to determine the concentration in soluble proteins by protein-dye staining [34] using bovine serum albumin as standard. Proteins were then precipitated by adding four volumes of 10% TCA in acetone to one volume of extract. After storage at −20 °C overnight, the extract was centrifuged (12,000× *g*, 4 °C, 20 min) and the resulting pellet was washed twice with 1 mL 80% acetone and centrifuged (16,000× *g*, 4 °C, 3 min). Residual acetone was evaporated under vacuum at 45 °C, and the resulting pellet, resuspended in 0.1 mL of ultra-pure water, was placed into a tin capsule. The water was then evaporated under vacuum at 45 °C (Speedvac Concentrator 5301, Eppendorf, France), and dry protein extract was analysed for S content using an elemental analyser combined with an IRMS as described above for total S.

#### 4.5.4. Determination of Glutathione (GSH) Amounts by HPLC

Thiols were extracted by grinding 10 mg dry weight of seed in 0.2 mL of 0.1 M HCl. After centrifugation (20,000× *g*, 10 min), the supernatant was used to measure the content of total GSH after reduction with DTT by HPLC using the monobromobimane derivatisation method as described by Koprivova et al. [35].

### *4.6. Relative Expression of BnSultr1 and BnSultr4 Genes Using Q-PCR*

Total RNA was extracted from 80 mg fresh weight of root or shoot samples at 14 DAS using the Mini Kit RNeasy from Qiagen (Qiagen, Courtaboeuf, France) according to the manufacturer's protocol. Quantification of total RNA was performed by spectrophotometer at 260 nm (BioPhotometer, Eppendorf, Le Pecq, France) before Reverse Transcription (RT) and Quantitative PCR (Q-PCR) analyses. For RT, 1 µg of total RNA was converted to cDNA with an 'iScript cDNA synthesis kit' according to the manufacturer's protocol (Bio-Rad, Marne-la-Coquette, France).

Q-PCR amplifications were performed using primers of *BnSultr1;1* (Accession No: AJ41640) and *BnSultr1;2* (Accession no: AJ311388) encoding transporters involved in sulphate uptake by roots, and primers of *BnSultr4;1* (Accession No: AJ416461) and *BnSultr4;2* (Accession No: FJ688133) encoding vacuolar sulphate transporters in *Brassica napus*. Q-PCR amplifications were performed by using *BnSultr4;1* forward primer: 5′-GACCAGACCCGTTAAGGTCA-3′ and reverse primer: 5′-TTGGAATCCATGTGAAGCAA-3′, *BnSultr4;2* forward primer: 5′-AGCAAGATCAGGGATTGTGG-3′ and reverse primer: 5′-TGCAACATTTGTGGGTGTCT-3′, *BnSultr1;1* forward primer: 5′-AGATATTGCGATCGGACCAG-3′ and reverse primer: 5′-GAAAACGCCAGCAAAGAAAG-3′, and *BnSultr1;2* forward primer: 5′-GGTGTAGTCGCTGGAATGGT-3′ and reverse primer: 5′-AACGGAGTGAGGAAGAGCAA-3′. *EF 1α* (Accession No: DQ312264) and 18S RNA (Accession No: X16077) were used as internal control genes and were amplified using EF1.1 forward primer: 5′-GCCTGGTATGGTTGTGACCT-3′ and reverse primer 5′-GAAGTTAGCAGCACCCTTGG-3′, and 18S forward primer: 5′-CGGATAACCGTAGTAATTCTAG-3′ and reverse primer 5′-GTACTCATTCCAATTACCAGAC-3′. Q-PCR reactions were performed with 4 µL of 200× diluted cDNA, 500 nM of primers and 1× SYBR Green PCR Master Mix (Bio-Rad, Marne-la-Coquette, France) in a ChromoFour System (Bio-Rad, Marne-la-Coquette, France). The specificity of PCR amplification was examined by monitoring the presence of the single peak in the melting curves after Q-PCR reactions. For each sample, the subsequent Q-PCR reactions were performed in duplicate, and the relative expression of the BnSultr transporters in each sample was compared to the control sample (corresponding to S400% seeds) and was determined with the ΔΔCt method using the following equation:

$$\text{Relative expression} = 2^{-(\Delta\text{Ct sample} - \Delta\text{Ct control})}$$

with

$$\Delta\text{Ct} = \text{CtBnSultr} - \text{Ctreference genes}$$

where Ct refers to the threshold cycle determined for each gene in the exponential phase of PCR amplification. Using this analysis method, relative expression of the *BnSultr* genes in the S400% seedlings samples was equal to one, by definition [36].

### 4.7. Statistics

The variability of the results is expressed by the average values for all biological replicates (n = 3 or 4) ± standard error (SE). The data were compared between treatments with the Mann–Whitney's test, used for unpaired data, with a statistical significance fixed at $p < 0.05$. The data were also compared between dates with the Wilcoxon test, used therefore for paired data, with a statistical significance fixed at $p < 0.05$. These statistical tests were performed with Microsoft® Excel 2011/XLSTAT©-Pro (Version 2013.3.01, Addinsoft, Inc., Brooklyn, NY, USA).

**Supplementary Materials:** The following are available online at http://www.mdpi.com/2223-7747/8/1/12/s1, Figure S1: S amount (a) and S repartition (b) of *Brassica napus* cv. Yudal green seedlings, 14 days after sowing on water (14 DAS) and after 24 h treatment of 14 DAS green seedlings with 508.7 µM $^{34}SO_4^{2-}$ at 5 atom% excess (15 DAS + $^{34}SO_4^{2-}$) and with 8.7 µM $SO_4^{2-}$ (15 DAS-S), Figure S2: Differences in $^{34}$S amount in excess in roots and shoots of oilseed rape green seedlings (cv. Yudal) between 14 DAS and 15 DAS-S and between 14 DAS and 15 DAS + $^{34}SO_4^{2-}$, Table S1: Plant dry weight, seed yield, number of seeds per plant and seed dry weight at harvest for plants well-supplied with S and grown under S-limited conditions. Values correspond to means ± SE (n = 3). Plants were either grown under plethoric S nutrition (400% of plant S needs provided, S400% seeds) or subjected to a S limitation from the beginning of plant growth with 70%, 20% and 5% of plant S needs provided (S70%, S20% and S5% seeds, respectively). Seed yield was determined without any sorting. The slight discrepancy observed between the calculation of seed yield from 'seed dry weight' × 'number of seeds per plant' and the reported values of seed yield is due to the manual removal of some seed fragments when seeds were counted and weighted. * Significant difference from the value of S400% ($p < 0.05$).

**Author Contributions:** Conceptualisation, P.D., J.-C.A. and J.T.; Formal analysis, P.D., D.P., J.-C.A. and J.T.; Investigation, P.D. and D.P.; Methodology, P.D., D.P. and J.T.; Resources, S.B.-M. and S.K.; Supervision, J.-C.A. and J.T.; Visualisation, P.D. and D.P.; Writing (Original Draft), P.D.; Writing (Review & Editing), P.D., S.B.-M., S.K., J.-C.A. and J.T.

**Funding:** This work was supported by the national SERAPIS program (Optimization of the Sulphur Fertilization: Development of Innovative Fertilizers and Indicators of Sulphur Nutrition in Plants). The SERAPIS program is coordinated by the private company TIMAC AGRO INTERNATIONAL–ROULLIER Group and benefits from financial support from the Région Basse-Normandie, Région Bretagne, Région Pays-de-la-Loire, FEDER Basse-Normandie, Fonds Unique Interministériel, Roullier, Angers Loire Métropole (grant number 12P03057). This work was also supported by a grant to Philippe D'Hooghe from the French Ministry of Research. Research in S.K.'s lab is funded by the Deutsche Forschungsgemeinschaft (EXC 1028).

**Acknowledgments:** The authors acknowledge the PLATIN' (Plateau d'Isotopie de Normandie) core facility for the element analysis used in this study.

**Conflicts of Interest:** The authors declare no conflict of interest. The founding sponsors had no role in the design of the study; in the collection, analyses, or interpretation of data; in the writing of the manuscript, and in the decision to publish the results.

## References

1. Dubousset, L.; Etienne, P.; Avice, J.-C. Is the remobilization of S and N reserves for seed filling of winter oilseed rape modulated by sulphate restrictions occurring at different growth stages? *J. Exp. Bot.* **2010**, *61*, 4313–4324. [CrossRef]
2. D'Hooghe, P.; Dubousset, L.; Gallardo, K.; Kopriva, S.; Avice, J.-C.; Trouverie, J. Evidence for proteomic and metabolic adaptations associated with alterations of seed yield and quality in sulfur-limited *Brassica napus* L. *Mol. Cell. Proteomics* **2014**, *13*, 1165–1183. [CrossRef] [PubMed]
3. Bewley, J.D.; Black, M. *Seeds: Physiology of Development and Germination*, 2nd ed.; Plenum Press: New York, NY, USA, 1994.
4. Bewley, J.D. Seed germination and dormancy. *Plant Cell* **1997**, *9*, 1055–1066. [CrossRef]
5. Nonogaki, H.; Bassel, G.W.; Bewley, J.D. Germination—Still a mystery. *Plant Sci.* **2010**, *179*, 574–581. [CrossRef]
6. Finch-Savage, W.E.; Leubner-Metzger, G. Seed dormancy and the control of germination. *New Phytol.* **2006**, *171*, 501–523. [CrossRef] [PubMed]
7. Rajjou, L.; Duval, M.; Gallardo, K.; Catusse, J.; Bally, J.; Job, C.; Job, D. Seed germination and vigor. *Annu. Rev. Plant. Biol.* **2012**, *63*, 507–533. [CrossRef]
8. López-Granados, F.; Lutman, P. Effect of environmental conditions on the dormancy and germination of volunteer oilseed rape seed (*Brassica napus*). *Weed Sci.* **1998**, *46*, 419–423.
9. Nakabayashi, K.; Okamoto, M.; Koshiba, T.; Kamiya, Y.; Nambara, E. Genome-wide profiling of stored mRNA in *Arabidopsis thaliana* seed germination: Epigenetic and genetic regulation of transcription in seed. *Plant J.* **2005**, *41*, 697–709. [CrossRef]
10. Alban, C.; Job, D.; Douce, R. Biotin metabolism in plants. *Annu. Rev. Plant. Biol.* **2000**, *51*, 17–47. [CrossRef]
11. Subbiah, V.; Reddy, K.J. Interactions between ethylene, abscisic acid and cytokinin during germination and seedling establishment in *Arabidopsis*. *J. Biosci.* **2010**, *35*, 451–458. [CrossRef]
12. Davidian, J.-C.; Kopriva, S. Regulation of sulfate uptake and assimilation—The same or not the same? *Mol. Plant* **2010**, *3*, 314–325. [CrossRef] [PubMed]
13. Takahashi, H.; Kopriva, S.; Giordano, M.; Saito, K.; Hell, R. Sulfur assimilation in photosynthetic organisms: Molecular functions and regulations of transporters and assimilatory enzymes. *Annu. Rev. Plant. Biol.* **2011**, *62*, 157–184. [CrossRef]
14. Brown, R.F.; Mayer, D.G. Representing cumulative germination. 2. The use of the Weibull function and other empirically derived curves. *Ann. Bot.* **1988**, *61*, 127–138. [CrossRef]
15. Bradbeer, J.W. *Seed Dormancy and Germination*; Blackie and Son Ltd.: Glasgow, Scotland; London, UK, 1988.
16. Schopfer, P.; Plachy, C. Control of Seed Germination by Abscisic Acid: II. Effect on Embryo Water Uptake in *Brassica napus* L. *Plant Physiol.* **1984**, *76*, 155–160. [CrossRef] [PubMed]
17. Schopfer, P.; Plachy, C. Control of Seed Germination by Abscisic Acid: III. Effect on Embryo Growth Potential (Minimum Turgor Pressure) and Growth Coefficient (Cell Wall Extensibility) in *Brassica napus* L. *Plant Physiol.* **1985**, *77*, 676–686. [CrossRef] [PubMed]
18. Cao, M.-J.; Wang, Z.; Zhao, Q.; Mao, J.-L.; Speiser, A.; Wirtz, M.; Hell, R.; Zhu, J.-K.; Xiang, C.-B. Sulfate availability affects ABA levels and germination response to ABA and salt stress in *Arabidopsis thaliana*. *Plant J.* **2014**, *77*, 604–615. [CrossRef] [PubMed]
19. Brunel-Muguet, S.; D'Hooghe, P.; Bataillé, M.-P.; Larré, C.; Kim, T.-H.; Trouverie, J.; Avice, J.-C.; Etienne, P.; Dürr, C. Heat stress during seed filling interferes with sulfur restriction on grain composition and seed germination in oilseed rape (*Brassica napus* L.). *Front. Plant Sci.* **2015**, *6*, 1236. [CrossRef]
20. Kucera, B.; Cohn, M.A.; Leubner-Metzger, G. Plant hormone interactions during seed dormancy release and germination. *Seed Sci. Res.* **2005**, *15*, 281–307. [CrossRef]

21. Linkies, A.; Muller, K.; Morris, K.; Tureckova, V.; Wenk, M.; Cadman, C.S.C.; Corbineau, F.; Strnad, M.; Lynn, J.R.; Finch-Savage, W.E.; et al. Ethylene interacts with abscisic acid to regulate endosperm rupture during germination: A comparative approach using *Lepidium sativum* and *Arabidopsis thaliana*. *Plant Cell* **2009**, *21*, 3803–3822. [CrossRef]
22. Gallardo, K.; Job, C.; Groot, S.P.C.; Puype, M.; Demol, H.; Vandekerckhove, J.; Job, D. Importance of methionine biosynthesis for *Arabidopsis* seed germination and seedling growth. *Physiol. Plant.* **2002**, *116*, 238–247. [CrossRef]
23. Hatzig, S.V.; Nuppenau, J.-N.; Snowdon, R.J.; Schießl, S.V. Drought stress has transgenerational effects on seeds and seedlings in winter oilseed rape (*Brassica napus* L.). *BMC Plant Biol.* **2018**, *18*, 297. [CrossRef] [PubMed]
24. Groot, M.P.; Kooke, R.; Knoben, N.; Vergeer, P.; Keurentjes, J.J.B.; Ouborg, N.J.; Verhoeven, K.J.F. Effects of Multi-Generational Stress Exposure and Offspring Environment on the Expression and Persistence of Transgenerational Effects in *Arabidopsis thaliana*. *PLoS ONE* **2016**, *11*, e0151566. [CrossRef] [PubMed]
25. Ahmad, A.; Khan, I.; Abdin, M.Z. Effect of sulfur fertilisation on oil accumulation, acetyl-CoA concentration, and acetyl-CoA carboxylase activity in the developing seeds of rapeseed (*Brassica campestris* L.). *Aust. J. Agric. Res.* **2000**, *51*, 1023. [CrossRef]
26. Zhao, F.; Evans, E.J.; Bilsborrow, P.E.; Syers, J.K. Influence of nitrogen and sulphur on the glucosinolate profile of rapeseed (*Brassica napus* L). *J. Sci. Food Agric.* **1994**, *64*, 295–304. [CrossRef]
27. Ceusters, J.; Van de Poel, B. Ethylene Exerts Species-Specific and Age-Dependent Control of Photosynthesis. *Plant Physiol.* **2018**, *176*, 2601–2612. [CrossRef] [PubMed]
28. Guzmán, P.; Ecker, J.R. Exploiting the triple response of *Arabidopsis* to identify ethylene-related mutants. *Plant Cell* **1990**, *2*, 513–523. [CrossRef] [PubMed]
29. Achard, P.; Vriezen, W.H.; Van Der Straeten, D.; Harberd, N.P. Ethylene regulates *Arabidopsis* development via the modulation of DELLA protein growth repressor function. *Plant Cell* **2003**, *15*, 2816–2825. [CrossRef]
30. Koralewska, A.; Buchner, P.; Stuiver, C.E.E.; Posthumus, F.S.; Kopriva, S.; Hawkesford, M.J.; De Kok, L.J. Expression and activity of sulfate transporters and APS reductase in curly kale in response to sulfate deprivation and re-supply. *J. Plant Physiol.* **2009**, *166*, 168–179. [CrossRef]
31. Takahashi, H.; Watanabe-Takahashi, A.; Smith, F.W.; Blake-Kalff, M.M.A.; Hawkesford, M.J.; Saito, K. The roles of three functional sulphate transporters involved in uptake and translocation of sulphate in *Arabidopsis thaliana*. *Plant J.* **2000**, *23*, 171–182. [CrossRef]
32. Brunel-Muguet, S.; Mollier, A.; Kauffmann, F.; Avice, J.-C.; Goudier, D.; Sénécal, E.; Etienne, P. SuMoToRI, an Ecophysiological Model to Predict Growth and Sulfur Allocation and Partitioning in Oilseed Rape (*Brassica napus* L.) Until the Onset of Pod Formation. *Front. Plant Sci.* **2015**, *6*, 993. [CrossRef]
33. Salon, C.; Bataillé, M.-P.; Gallardo, K.; Jeudy, C.; Santoni, A.-L.; Trouverie, J.; Voisin, A.-S.; Avice, J.-C. $^{34}$S and $^{15}$N labelling to model S and N flux in plants and determine the different components of N and S use efficiency. *Methods Mol. Biol.* **2014**, *1090*, 335–346. [PubMed]
34. Bradford, M.M. A rapid and sensitive method for the quantitation of microgram quantities of protein utilizing the principle of protein-dye binding. *Anal. Biochem.* **1976**, *72*, 248–254. [CrossRef]
35. Koprivova, A.; North, K.A.; Kopriva, S. Complex signaling network in regulation of adenosine 5′-phosphosulfate reductase by salt stress in Arabidopsis roots. *Plant Physiol.* **2008**, *146*, 1408–1420. [CrossRef] [PubMed]
36. Livak, K.J.; Schmittgen, T.D. Analysis of relative gene expression data using real-time quantitative PCR and the $2^{-\Delta\Delta Ct}$ method. *Methods* **2001**, *25*, 402–408. [CrossRef] [PubMed]

*Article*

# The Recovery from Sulfur Starvation Is Independent from the mRNA Degradation Initiation Enzyme PARN in Arabidopsis

**Laura Armbruster, Veli Vural Uslu, Markus Wirtz and Rüdiger Hell ***

Centre for Organismal Studies, Heidelberg University, 69120 Heidelberg, Baden-Württemberg, Germany; laura.armbruster@cos.uni-heidelberg.de (L.A.); veli.uslu@cos.uni-heidelberg.de (V.V.U.); markus.wirtz@cos.uni-heidelberg.de (M.W.)

* Correspondence: ruediger.hell@cos.uni-heidelberg.de; Tel.: +49-6221-54-6284

Received: 26 August 2019; Accepted: 26 September 2019; Published: 27 September 2019

**Abstract:** When plants are exposed to sulfur limitation, they upregulate the sulfate assimilation pathway at the expense of growth-promoting measures. Upon cessation of the stress, however, protective measures are deactivated, and growth is restored. In accordance with these findings, transcripts of sulfur-deficiency marker genes are rapidly degraded when starved plants are resupplied with sulfur. Yet it remains unclear which enzymes are responsible for the degradation of transcripts during the recovery from starvation. In eukaryotes, mRNA decay is often initiated by the cleavage of poly(A) tails via deadenylases. As mutations in the poly(A) ribonuclease PARN have been linked to altered abiotic stress responses in *Arabidopsis thaliana*, we investigated the role of PARN in the recovery from sulfur starvation. Despite the presence of putative PARN-recruiting AU-rich elements in sulfur-responsive transcripts, sulfur-depleted PARN hypomorphic mutants were able to reset their transcriptome to pre-starvation conditions just as readily as wildtype plants. Currently, the subcellular localization of PARN is disputed, with studies reporting both nuclear and cytosolic localization. We detected PARN in cytoplasmic speckles and reconciled the diverging views in literature by identifying two PARN splice variants whose predicted localization is in agreement with those observations.

**Keywords:** AGS1; AHG2; sulfur starvation; PARN; recovery; sulfate transporters; sulfate resupply; mRNA degradation; rapid recovery downregulation

---

## 1. Introduction

Sulfur is one of six essential macronutrients plants absorb from the soil in large quantities to sustain growth and survival [1]. In the last decade, insufficient sulfate nutrition has been reported with increasing frequency in widely cultivated crops such as wheat, soybean and rapeseed [2–4]. Since prolonged sulfur depletion results in severe stunting and impaired resistance to biotic stress, this translates into significant losses in crop yield [5,6]. Understanding the mechanisms by which plants respond to and recover from sulfur deficiency is an essential step towards improving agricultural productivity.

Plants adapt to sulfur depletion by upregulating the expression of genes involved in sulfate uptake and reduction [6]. Additionally, the expression of negative regulators of glucosinolate biosynthesis is induced to prioritize sulfur usage for primary metabolism [7]. The reverse processes by which sulfur-deficient plants reshape their transcriptome upon sulfur resupply are, however, only poorly understood.

In resupply studies, Bielecka and coworkers identified so-called genuine sulfur-responsive transcripts that directly reflect the sulfur status of *Arabidopsis thaliana*. Most (30 out of 35) of those transcripts accumulate upon sulfur starvation and display rapid decay rates in the first hours after the resupply of the macronutrient [8]. Adopting an exponential decay model, the average half-life of those starvation-induced transcripts can be determined to amount to 2.3 hours during the recovery phase [8]. This is considerably shorter than the mean mRNA half-life of 5.9 hours measured in global studies of Arabidopsis mRNA stability under standard growth conditions [9,10]. Taken together, these findings suggest that during the recovery from sulfur limitation, the transcriptome is cleared of starvation-responsive mRNAs by active degradation rather than the regular turnover of transcripts. The clearance of stress-induced transcripts is also required for the recovery from high-light stress. In this context, the term "rapid recovery downregulation" has been coined. It describes the quick return of transcript abundance to pre-stress levels upon recovery [11]. Yet for both sulfur starvation and excess-light stress, it remains unclear which enzymes are mediating the targeted degradation of mRNAs upon recovery.

In plants, a set of endonucleases, decapping enzymes and deadenylases govern the degradation of mRNA. While endonucleases cleave phosphodiester bonds within transcripts, decapping enzymes and deadenylases remove the methylguanine cap and the stabilizing poly(A) tail from the 5′ and 3′ ends of mRNAs [12–15]. The aforementioned enzymes are candidates to initiate the degradation of transcripts during the recovery from sulfur starvation. In this study, we focus on the role of the poly(A)-specific ribonuclease (*At*PARN) for the following reasons:

Unlike in *Schizosaccharomyces pombe* and *Caenorhabditis elegans*, PARN is essential for embryogenesis in *Arabidopsis thaliana*, indicating a unique role of PARN-mediated mRNA decay in higher plants [16]. *At*PARN hypomorphic mutants (*parn*) display diminished growth accompanied by an increased resistance to bacterial pathogens and a decreased tolerance towards osmotic stress [17–20]. These altered stress responses can be attributed to the accumulation of the phytohormone abscisic acid (ABA) in *parn* mutants, resulting from the disturbed equilibrium between polyadenylation and the deadenylation of specific stress-related transcripts [17,18,20,21]. Remarkably, sulfate and cysteine have recently been shown to trigger ABA biosynthesis in plants. Hence, the elevated ABA levels observed in *parn* mutants could be the result of increased sulfate assimilation [22,23].

Mammalian PARNs have been observed to preferentially degrade the poly(A) tails of transcripts harboring AU-rich signal elements (AREs) in their 3′ untranslated regions (UTRs). AREs range from 40 to 150 nucleotides in length and typically contain one or more AUUUA motifs within AU-rich sequence stretches [24–26]. The close evolutionary relationship between mammalian PARNs and *At*PARN suggests that *At*PARN might be recruited by AREs as well. In line with this assumption, *At*PARN was shown to target a specific subset of mRNAs rather than the entire mRNA population [16].

Here, we report that the AUUUA motif is present in the 3′UTRs of many transcripts induced by sulfur depletion, including the *O*-acetylserine cluster genes *SDI1*, *GGCT* and *SHM7* [27] as well as the high-affinity sulfur transporter genes *SULTR1;1* and *SULTR1;2*. This finding makes *At*PARN a potential candidate for the regulation of active mRNA degradation during the recovery from sulfur depletion.

However, analysis of transcript stability by qRT-PCR in PARN hypomorphic mutants demonstrates that PARN is not required for the targeted degradation of sulfur deficiency-induced transcripts in Arabidopsis. To understand the biological role of PARN, we determine the subcellular localization of PARN and its antagonist, the poly(A) polymerase AGS1. To this end we image stable Arabidopsis lines expressing PARN-GFP or AGS1-GFP fusions under the control of the respective endogenous promoters. Unlike the predominantly nuclear localized AGS1, PARN accumulates in cytoplasmic speckles. So far, PARN has been observed to localize to processing bodies (p-bodies) when transiently expressed under the control of the 35S promoter in tobacco leaves [28]. This, however, contradicts earlier PARN localization studies in onion epidermal peels reporting a nuclear–cytoplasmic localization [16,29]. By detecting two PARN splice variants, which were bioinformatically predicted to localize to different cellular compartments, we offer an explanation for the diverging accounts of PARN localization in the literature.

## 2. Results

Upon recovery from sulfur starvation, transcripts of many stress-induced genes are rapidly degraded. We reasoned that recognition signals embedded in those transcripts might provide specificity to the process and link them to the active degradation machinery of plants. AREs in the 3′UTRs of transcripts have been known to target mammalian mRNAs for rapid degradation by recruiting the deadenylase PARN [30]. This observation prompted us to search for AREs in the 3′UTRs of transcripts upregulated upon sulfur starvation.

### 2.1. The Sulfur-Responsive Transcripts SULTR1;1, SULTR1;2, SDI1, SHM7 and GGCT Contain ARE Sites

We could identify AREs in many transcripts that are involved in sulfur metabolism and are upregulated upon sulfur starvation, including the *O*-acetylserine dependent cluster genes *SDI1*, *GGCT* and *SHM7* [27] as well as the high-affinity sulfate transporter genes *SULTR1;1* and *SULTR1;2* (Figure 1). In contrast, transcripts that were not induced upon sulfur starvation but are involved in sulfur metabolism did in many cases lack AREs (e.g., *SHM1-4*, *SERAT 2;1* and *SERAT2;2*, *OAS-TL A* and *SIR*). There was, however, no significant difference in ARE frequency between sulfur metabolism-related and general transcripts. Given the broad presence in mRNAs encoding for the sulfur metabolism pathway, we hypothesized that the degradation of sulfur-responsive transcripts upon the recovery from starvation might depend on PARN.

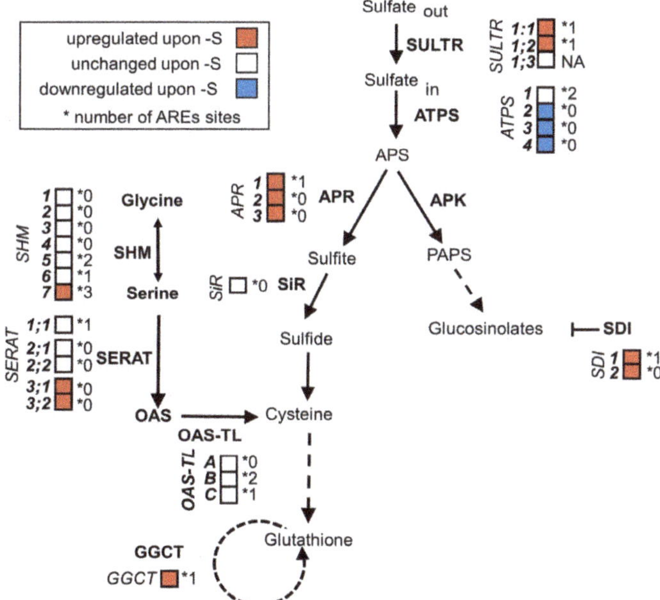


**Figure 1.** Plants regulate the transcription of genes involved in sulfate assimilation and glucosinolate biosynthesis in response to sulfur supply. When plants are exposed to sulfur limitation, they upregulate the expression of genes involved in sulfate uptake and assimilation. Simultaneously, the expression of genes implicated in the synthesis of sulfur-containing secondary metabolites is downregulated to prioritize sulfate usage for primary metabolites. In this scheme of plant sulfate metabolism, the transcript levels of genes (italics) encoding for enzymes mediating sulfate assimilation (bold) are indicated by a color code. Red and blue represent significant ($p < 0.05$; ≥2-fold) up- and downregulations under sulfur limitation. White represents no significant change in comparison to full nutrient supply. Many sulfur responsive transcripts harbor AU-rich signal elements (AREs) in their 3′ untranslated regions (UTRs). The number of AREs found in each 3′UTR is indicated next to the asterisk. (Transcript data from [31]).

## 2.2. The Degradation of Sulfur Metabolism-Related Transcripts Is Independent of AtPARN

To determine whether PARN degrades starvation-induced transcripts after sulfur resupply in *Arabidopsis thaliana*, we depleted six-week-old hydroponically grown *parn*, *parn-ags1* and wildtype plants of sulfur. After two weeks of starvation (0 µM sulfur), the plants were transferred back to $\frac{1}{2}$ Hoagland medium (500 µM sulfur) for three hours to allow recovery. Subsequently, the transcript levels of the sulfur-starvation marker genes *SULTR1;1*, *SULTR1;2*, *SDI1*, *SHM7* and *GGCT* were assessed via qRT-PCR in roots (Figure 2).

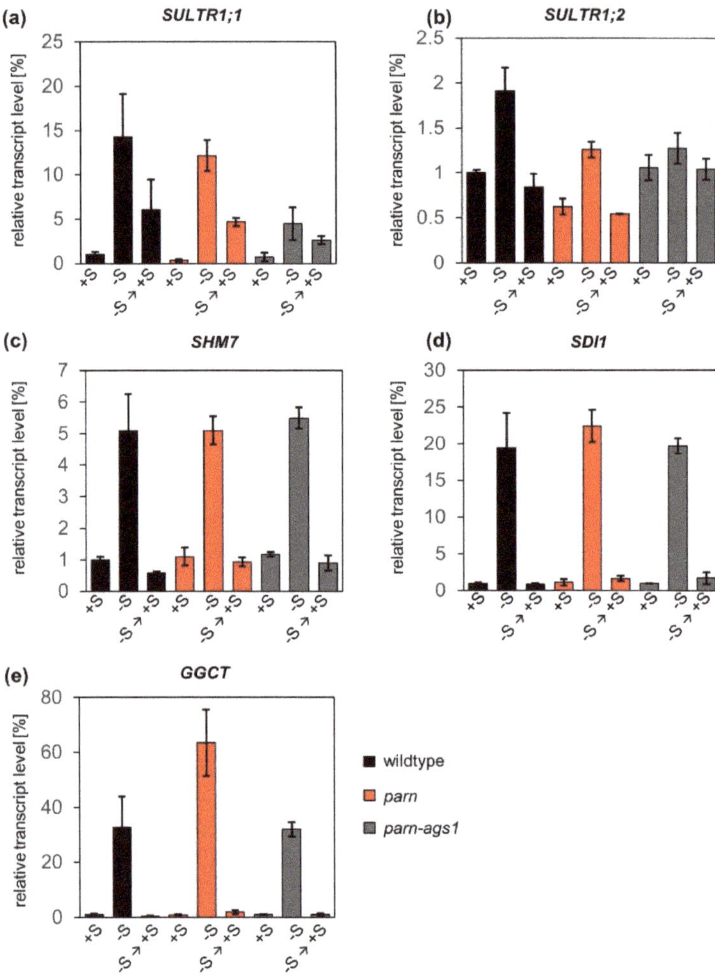


**Figure 2.** *At*PARN is not required for the degradation of sulfur starvation-induced transcripts upon the resupply of the macronutrient. Relative transcript levels of *SULTR1;1* (**a**), *SULTR1;2* (**b**), *SHM7* (**c**), *SDI1* (**d**) and *GGCT* (**e**) upon regular sulfur supply (+S, 500 µM), starvation (-S, two weeks at 0 µM) and recovery (-S → +S, 3 h at 500 µM) in roots of *parn* (red), *parn-ags1* (grey) and wildtype plants (black). Results were normalized to the expression values measured for wildtype plants under full nutrient supply. Bars represent standard errors (n = 3).

The *AGS1* gene encodes for a poly(A) polymerase, which acts as an antagonist of PARN. Since all of the known *parn* phenotypes are suppressed by loss-of-function mutations in *AGS1* [20,21], we expected the *parn-ags1* double mutants to reset their transcriptome to pre-starvation conditions just as readily as wildtype plants.

As suggested by publicly available microarray data [17,18], the transcript levels of *SULTR1;1*, *SULTR1;2*, *SHM7*, *SDI1* and *GGCT* did not differ considerably between wildtype plants and *parn* hypomorphic or *parn-ags1* double mutants under full nutrient supply. Upon starvation, a clear upregulation of the previously mentioned transcripts was observed in all genotypes. The strongest induction was measured for *GGCT* (33-fold for wildtype, 63-fold for *parn* and 32-fold for *parn-ags1*), whereas the transcript levels of *SULTR1;2* increased to a lesser extent (2-fold for wildtype, 2-fold for *parn* and 1.3-fold for *parn-ags1*). This is well in line with published transcript data from sulfur-starved wildtype plants [31,32] and supports the validity of the nutrient starvation conditions used. After sulfate resupply, the abundance of the five sulfur starvation marker transcripts decreased rapidly, not only in wildtype plants, but also in *parn* and *parn-ags1* mutants. With the exception of *SULTR1;1*, three hours of recovery were sufficient for the transcript levels to return to pre-starvation conditions. When this single time point was used as a basis for a rough estimation of transcript half-life, the measurements taken for wildtype plants indicated transcript half-lives of 151 minutes for *SULTR1;2*, 144 min for *SULTR1;1*, 58 min for *SHM7*, 40 min for *SDI1* and 29 min for *GGCT*.

Taken together, these results indicate that *parn* and *parn-ags1* mutants clear their transcriptomes of surplus sulfur-responsive transcripts just as readily as wildtype plants. This finding excludes a significant function of PARN and its antagonist AGS1 in the clearance of sulfur starvation-induced transcripts. Furthermore, it puts a note of caution on the identification of functional ARE sites in plants based on the currently available prediction tools (or data on mammalian ARE sites).

### 2.3. PARN Accumulates in Cytoplasmic Speckles

The subcellular localization of enzymes provides the physiological context for their activity and determines their access to substrates and interaction partners [33]. Therefore, the identification of the subcellular localization of PARN and its antagonist—the poly(A) polymerase AGS1—is critical to understanding the biological role of the PARN-AGS1 mRNA degradation system.

In order to elucidate the subcellular localization of PARN and AGS1, we used stable transgenic lines expressing either PARN-GFP or AGS1-GFP under the control of their endogenous promoters [21]. These lines were germinated on agar-medium in the presence and absence of sulfur. Root sections from the tip to the elongation zone of ten-day-old seedlings were imaged for GFP signals after incubation, with dyes staining the mitochondria (MitoTracker, 100 nM for 15 min) and the nucleus (DAPI, 2 $\mu$g mL$^{-1}$ for 15 min). Under full nutrient supply, PARN-GFP localized exclusively to cytoplasmic speckles that were partly, but not entirely, overlapping with the mitochondrial signal. AGS1-GFP was found in cytosolic speckles, as well as in the nucleus, where it was evenly distributed (Figure 3). In order to demonstrate that the observed GFP signals were not bleed-through signals from the DAPI or the MitoTracker channel, wildtype plants were stained with both dyes (Figure S1). Since AGS1 is an antagonist of PARN, we hypothesized that PARN might also localize to the nucleus under certain conditions.

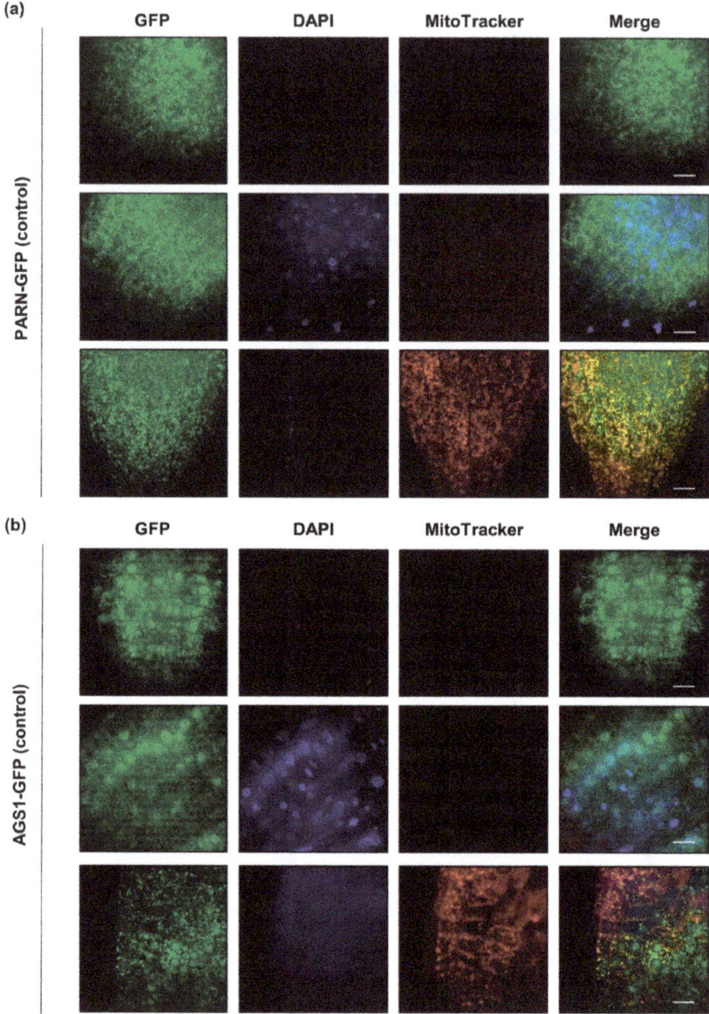


**Figure 3.** Under full nutrient supply, PARN-GFP (**a**) localizes to cytoplasmic speckles, whereas AGS1-GFP (**b**) is confined to the nucleus. Roots (tip and elongation zones) of ten-day-old PARN-GFP and AGS1-GFP seedlings grown under control conditions were left untreated (first row) or incubated with DAPI (second row) and MitoTracker Orange (third row). Each column represents a different channel (GFP, DAPI, MitoTracker). The last column shows a merge of all channels. Pictures are the result of maximum intensity z-projections of slices in a z-stack. Scale bar 10 μm.

Indeed, when the growth media were prepared without sulfur, in rare cases the subcellular localization of PARN-GFP shifted from cytoplasmic speckles to the nucleus (Figure S2a). AGS1-GFP, on the other hand, did not display any changes in localization upon nutrient starvation. To determine whether the starvation-induced relocalization of PARN to the nucleus was sulfur-specific or a general adaptation to nutrient starvation, the experiment was repeated with seedlings depleted of nitrogen (Figure S2b). Both nitrogen and sulfur are important macronutrients required for the synthesis of essential amino acids. As observed for sulfur depletion, nitrogen starvation induced a relocalization of PARN to the nucleus at a comparably low frequency. Similarly, reductive stress induced by 30 min of

10 mM dithiothreitol (DTT) caused PARN to shift its subcellular localization from cytosolic speckles to the nucleus (Figure S2c). Similar to nutrient starvation, the reductive stress treatment did not induce a comprehensive relocalization. However, the occasional relocalization of PARN-GFP under the applied stress conditions was never observed for the AGS1-GFP fusion under identical conditions (Figure S3).

### *2.4. PARN is Encoded for by Two Alternative Splice Variants Predicted to Localize to Different Cellular Compartments*

One possible explanation for the dual localization of PARN observed in our experiments is the existence of alternative PARN splice isoforms encoding for different protein variants confined to distinct cellular compartments. The Arabidopsis Information Resource (TAIR) provides sequences of four *At*PARN splice variants, A–D (Figure 4a). Since the splice variants A and D give rise to the same protein, the four splice variants encode for three distinct protein species. These species differ only in the length of their N-terminus (Figure 4b). Since the N-terminus is an important determinant of subcellular protein localization, the localization of the isoforms A–D was predicted by seven algorithms. While the splice variants A and B were predicted to localize to the nucleus or the cytoplasm (denoted as "N.A."), splice variant C was predicted to be confined to the mitochondria or the chloroplasts (Figure 4g).

To determine which of the four *PARN* transcripts were present in planta under full nutrient supply and sulfur starvation, the primers I–V detecting the splice variants A–D (Figure 4a) were used for a qRT-PCR analysis of cDNA extracted from starved and non-starved wildtype plants. By comparing the expected and observed fragments produced by the isoform-specific primers, the presence of the splice isoforms B and C could be verified in vivo. Isoforms A and D, however, were not detected. These results were further corroborated by the immunodetection of the splice variants B and C, but not A and D, in protein extracts from 10-day-old seedlings (Figure 4f). Remarkably, no difference in patterns between the seedlings grown under sulfur deficiency and full nutrient supply could be observed. Under both conditions, isoform C was expressed at an approximately 4-fold higher level than isoform B. No free GFP was detected, indicating that the nuclear GFP signal observed in PARN-GFP roots cannot be attributed to cleaved fluorophores, but is indeed an authentic GFP-PARN signal. Furthermore, the qRT-PCR analysis revealed that neither the total PARN transcript levels nor the ratio of isoform B to isoform C changed significantly when comparing full and limited nutrient supply (Figure 4e). The relative abundance of isoform B was calculated using the primer pair IV, whereas the primer pair V was used to determine the relative abundance of isoform C. Primer pair II was used to quantify the total PARN transcript level.

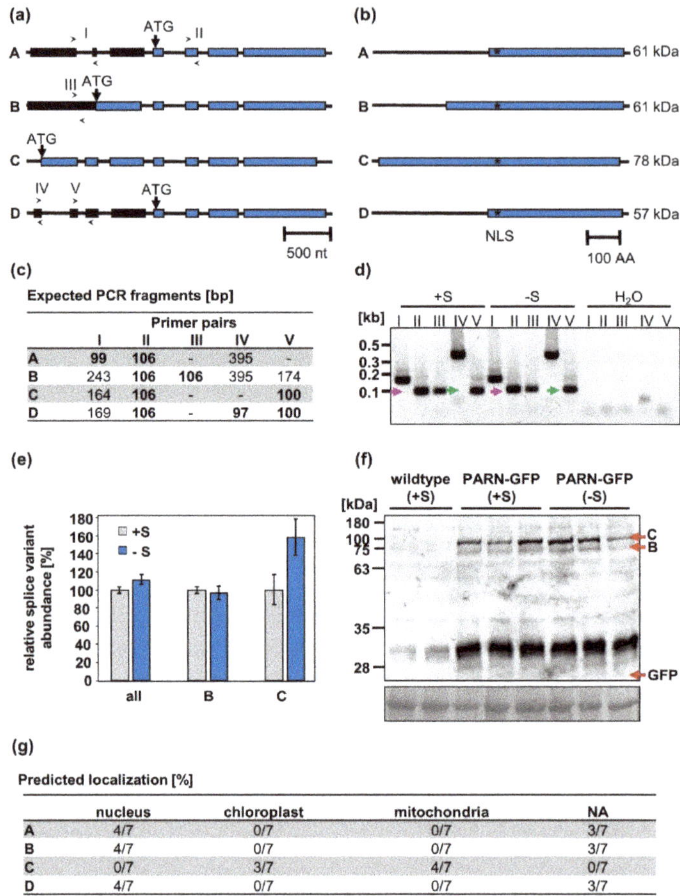


| | Primer pairs | | | | |
|---|---|---|---|---|---|
| | I | II | III | IV | V |
| A | **99** | **106** | - | 395 | - |
| B | 243 | **106** | **106** | 395 | 174 |
| C | **164** | **106** | - | - | **100** |
| D | 169 | **106** | - | **97** | **100** |

| | nucleus | chloroplast | mitochondria | NA |
|---|---|---|---|---|
| A | 4/7 | 0/7 | 0/7 | 3/7 |
| B | 4/7 | 0/7 | 0/7 | 3/7 |
| C | 0/7 | 3/7 | 4/7 | 0/7 |
| D | 4/7 | 0/7 | 0/7 | 3/7 |

**Figure 4.** *At*PARN is encoded for by four splice variants. (**a**) Schematic structure of splice variants A–D. Their presence was verified with primer pairs I–V (indicated by arrows). While black boxes represent untranslated exons, blue boxes indicate translated exons. Introns are represented by black lines. (**b**) Proteins encoded for by transcripts A–D. The predicted nuclear localization signal (NLS) is marked with an asterisk. (**c**) Expected PCR products for primer pairs I–V in the presence of splice variants A–D. Most primer pairs may generate several amplicons of differing lengths, depending on which isoforms are actually present. Due to the extremely short annealing/extension step however, only the shortest amplicons (bold) are expected to be produced. (**d**) qRT-PCR products generated using primers I–V on cDNA from roots of starved (-S, 0 µM sulfur) and non-starved (+S, 500 µM sulfur) wildtype plants. As a negative control, the cDNA was substituted with $H_2O$. (**e**) Relative abundance of splice variants B and C as well as total PARN transcripts under full nutrient supply and sulfur starvation in wildtype roots. (**f**) Immunodetection of the PARN-GFP isoforms B and C in protein extracts from sulfur-starved and non-starved seedlings with a polyclonal rabbit α-GFP antibody (# A-6455, Thermo Fisher Scientific). Amido black-stained protein served as the loading control. (**g**) Localization of isoforms A–D as predicted by seven independent algorithms (see 4.7). Since the cytosol is the default localization of a protein, cytosolic proteins will not yield any prediction by the aforementioned algorithms (denoted by "N.A.").

## 3. Discussion

Plants as sessile organisms rely on transcriptional reprogramming to adapt to a constantly changing macronutrient supply. When exposed to sulfur limitation, they upregulate sulfate assimilation pathways, resulting in an accumulation of transcripts encoding for sulfur deficiency marker genes [1]. Upon cessation of the stress, however, those transcripts are subjected to rapid recovery downregulation [8,11]. In eukaryotes, mRNA decay is generally initiated by the deadenylation of transcripts [25]. As mutations in the poly(A) ribonuclease *At*PARN have been linked to altered abiotic stress responses in *Arabidopsis thaliana* [17–19], this manuscript investigated a few aspects of the role of PARN in the recovery from sulfur starvation.

### *3.1. The Degradation of Sulfur Starvation-Induced Genes during the Recovery from Starvation is Independent of AtPARN*

Although we identified putative PARN-recruiting AREs [25] in several sulfur starvation-induced transcripts, PARN is not required for their degradation upon recovery from starvation. When mutants of *parn* and its antagonist *ags1* were subjected to sulfur depletion followed by resupply, they degraded surplus sulfur-induced transcripts just as effectively as wildtype plants (Figure 2). The mRNA half-life estimations inferred from the observed transcript degradation rates of *GGCT, SDI1* and *SHM7* amounted to less than an hour. Since those calculations were based on only one time point, they can only provide a rough estimate of the upper limit of the mRNA half-life. In Arabidopsis, the average mRNA half-life is estimated to be in the order of several hours [10]. This indicates that even though PARN is not mediating the degradation of *GGCT, SDI1* and *SHM7*, other active mRNA degradation enzymes might act on those transcripts upon sulfur resupply. The sulfur-responsive transcripts might, for instance, be degraded by the cytoplasmic exoribonuclease XRN4. XRN4 has been shown to degrade the mRNA of heat shock factor HSFA2 and thereby represses heat stress responses after the return to normal temperature [34]. The fact that XRN4 targets specific transcripts involved in the response to abiotic and biotic stimuli [35] supports the notion that XRN4, rather than PARN, might be involved in the recovery from sulfur starvation in *Arabidopsis thaliana*.

### *3.2. The Presence of Two Alternative PARN Splice Variants Reconciles the Diverging Views on PARN Localization in the Literature*

We selected PARN as a candidate for rapid transcript degradation upon sulfur resupply because we identified putative PARN-recruiting AREs in mRNAs upregulated upon sulfur starvation. Transcripts harboring AREs are known to localize to the cytosolic sites of mRNA degradation, termed processing-bodies (p-bodies), where they are rapidly degraded [36].

According to translational fusions with GFP, PARN localizes to p-bodies when it is transiently expressed under the control of the 35S promoter in tobacco leaves [28]. This finding is discussed controversially in the literature, since it contradicts earlier PARN localization studies in onion epidermal peels, reporting a nuclear–cytoplasmic localization [16,29]. Both observations do however agree with the predominantly nuclear–cytosolic localization of PARN reported for metazoans. When *At*PARN-GFP expression is driven by the native PARN promoter, the fusion protein localizes to the mitochondria, indicating a unique function of PARN in higher plants [20,21].

Here we made use of stable transgenic PARN-GFP lines to show that under optimal nutrient supply, PARN localizes to cytoplasmic speckles. Unlike previously reported by Hirayama and co-workers [21], we did not observe full colocalization of those speckles with the mitochondria. Our findings suggest that PARN might localize to the mitochondria, but a considerable portion of the observed PARN-GFP signal is localized to p-bodies [28]. Since p-bodies are involved in the degradation and translational arrest of transcripts during development and the adaptation to stress, this subcellular localization agrees with the biological function of PARN observed in mammals.

Furthermore, we found that mineral nutrient deficiency and reductive stress in rare cases induces a dynamic delocalization of PARN from cytoplasmic speckles to the nucleus. The fact that this delocalization was observed under several stress conditions points to a general cellular mechanism. Although the function of PARN in the nucleus remains unknown, the stress-induced delocalization system we describe here opens new avenues to study the function of nuclear-localized PARN.

A potential mechanism for the stress-induced delocalization of the PARN protein is the existence of different PARN splice variants as evidenced by The Arabidopsis Information Resource (TAIR). These variants encode for proteins that differ only in the length of their N-terminus. Whereas the full-length *At*PARN splice variant carries an N-terminal extension that distinguishes *At*PARN from putative animal homologs [20], the shorter *At*PARN variants lack this non-conserved N-terminus. When subjected to intracellular targeting algorithms, the different *At*PARN splice variants are bioinformatically predicted to localize to distinct subcellular compartments. It is worth noting that three of the seven applied prediction algorithms (iPSORT, TargetP and Predotar) focus on N-terminal sorting signals to determine the localization of proteins [37–39]. Thus, their predictions are biased towards mitochondrial and chloroplastic localizations. While the full-length protein is predicted to localize to the mitochondria, the shorter variants are thought to localize to the nucleus and the cytoplasm. We detected two out of the four *At*PARN splice variants in roots of wildtype Arabidopsis plants. These splice variants are predicted to localize to the nucleus or the cytoplasm and the mitochondria.

Profiling of *PARN* transcripts from roots revealed that both splice forms were present at similar levels under optimal or sulfur-depleted conditions. Hence, changes in the transcription or processing of both variants cannot directly explain the sulfur limitation-induced delocalization of PARN in roots. However, both splice variants were present in roots under full nutrient supply as well as starvation conditions, which potentially enables stress-induced differential loading of the transcripts to ribosomes. Indeed, in yeast and mammalian somatic cells, thousands of untranslated mRNAs were shown to be targeted to p-bodies, where they are translationally repressed, suggesting that p-bodies provide a reservoir for quick adaptation of gene expression [40]. Similarly, plant p-bodies might act as reservoirs for the two *PARN* transcript variants, enabling plants to react to stresses without the delay caused by a transcriptional regulation of *PARN*.

Immunodetection of *At*PARN under standard growth conditions and sulfur depletion, however, revealed no shift in the ratio between isoform B and C on the protein level. Most likely, the expression of isoform B is not sufficient to induce nuclear localization but requires stress-induced post-translational modifications of PARN or the binding of PARN to interaction partners. Both mechanisms have previously been described for other proteins. The Arabidopsis leucine zipper transcription factor VIP1, for instance, is dephosphorylated upon mechanical and hypo-osmotic stress and subsequently changes its localization from the cytosol to the nucleus [41]. In mammalian cells, RNA-binding proteins move from the cytoplasm to the nucleus in response to accelerated mRNA decay associated with cellular stress. In some cases, this delocalization is mediated via direct interaction of those proteins with the nuclear transport machinery. In other cases, interactions with proteins containing NLS are responsible for the import into the nucleus. On the other hand, PARN might be sequestered in the cytosol via interaction with cytosolic proteins. In accordance with this hypothesis, the BioAnalyticResource Tool Arabidopsis Interaction Viewer predicts PARN to interact with two cytosolic proteins (AT5G47010 and AT2G39260) required for nonsense-mediated mRNA decay. The mechanism of PARN delocalization will be the subject of further studies.

## 4. Materials and Methods

### *4.1. Plant Material and Growth Conditions*

All work was performed with *Arabidopsis thaliana* ecotype Columbia-0 (Col-0). The transgenic *parn* knockdown, *parn-ags1* double mutant, PARN-GFP and AGS1-GFP lines are characterized in [19,21]. All experiments except the subcellular localization study and the sulfur resupply assay (described below) were conducted with plants grown under short-day conditions (8.5 h light, 100 µE light photon flux density, 24 °C by day, 18 °C by night and 50% humidity) on a medium containing one half soil and one half substrate 2 (Klasmann-Deilmann, Germany).

### 4.2. Sulfur Resupply Assay

In order to determine the role of *At*PARN in the recovery from sulfur starvation, *parn*, *parn-ags1* and wildtype seeds were surface-sterilized with 70% (v/v) ethanol for 5 min followed by 6% sodium hypochlorite for 3 min and a second wash with 70% (v/v) ethanol for 5 min. Afterwards, the seeds were washed trice with $ddH_2O$. Individual seeds were placed in microcentrifuge tubes containing $\frac{1}{2}$ Hoagland medium (0.5 mM $KH_2PO_4$, 0.05 µM $(NH_4)_6Mo_7O_{24} \cdot 4\ H_2O$, 0.5 mM $MgSO_4/MgCl_2$, 0.15 µM $CuSO_4 \cdot 5\ H_2O$, 2.5 mM $Ca(NO_3)_2 \cdot 4\ H_2O$, 1.9 µM $ZnSO_4 \cdot 7\ H_2O$, 2.5 mM $KNO_3$, 10 µM NaCl, 2.25 µM $MnCl_2 \cdot 4\ H_2O$, 25 µM $H_3BO_3$, 40 µM Fe-EDTA; pH 5.8) supplemented with 0.6% (w/v) agar. Subsequently, the tubes were inserted in standard 1 mL pipette tip racks. Plants were stratified at 4 °C for three days before being germinated in a short-day growth cabinet. After two weeks, individual plants were transferred to 6 liter boxes containing $\frac{1}{2}$ Hoagland medium. After an additional two weeks of growth on full medium, a subset of the plants was starved for sulfur by replacing $MgSO_4$ in the Hoagland medium with $MgCl_2$. The control group continued to receive full medium. Starvation lasted for two weeks, with media being exchanged on a weekly basis. Subsequently, root and shoot material were collected and snap-frozen in liquid nitrogen.

### 4.3. Genotyping by PCR

In order to identify the transgenic plants, gDNA was extracted from 50–100 mg Arabidopsis leaf material. The fresh tissue of four-week-old plants was ground with a plastic pestle for 10–15 sec. Subsequently, 400 µL Edwards buffer (200 mM Tris/HCl, 25 mM EDTA, 250 mM NaCl, 0.5% SDS) were added and the mixture was vortexed for 5 sec. After centrifugation at 13.000 rpm for 5 min at room temperature, 300 µL of supernatant were transferred to a fresh microfuge tube. 300 µL 100% isopropanol were added and the mixture was left to incubate for two minutes at room temperature. After centrifugation at 13,000 rpm for 10 min at room temperature, the supernatant was discarded and the pellet was washed with 700 µL 70% ethanol. After a final centrifugation step at 13,000 rpm for 10 min, the ethanol was discarded and the DNA pellet was resuspended in 40 µL $ddH_2O$. Subsequently, 20 ng of the harvested gDNA were used for PCR reactions performed with the GoTaq® Green Master Mix (Promega) and specific primers (see Appendix A Table A1).

### 4.4. Quantifying Gene Expression by qRT-PCR

To analyze the transcript levels of sulfur starvation-induced genes, total RNA was extracted from frozen leaf and root material using the peqGOLD total RNA kit (PeqLab). Subsequently, total RNA was transcribed into complementary DNA (cDNA) with the RevertAid H Minus First Strand cDNA Synthesis Kit using oligo(dT) primers (Thermo Fisher Scientific). All reactions were conducted according to the supplier's protocol. The cDNA was analyzed by qRT-PCR with the SqPCRBIO SyGreen Mix Lo-ROX (Nippon Genetics Europe GmbH) and *TIP41* (AT4G34270, [42]) and *PP2A* (AT1G69960, [43]) as reference genes. The corresponding primer sequences are listed in Table A3 of the Appendix A. Data was analyzed via the Rotor-Gene Q Series Software.

In order to quantify the amount of transcripts that encode for alternatively spliced forms of *At*PARN, primers that discriminated between the mRNA models A–D were designed and used for qRT-PCR (see Table A2 of the Appendix A for sequences). The four models encode for the cDNA clones AK227465 and AB223028 (A), AB223029 (B), AB19466 (C) and AB223027 (D). To ensure that each primer pair produced only the shortest possible fragment, the annealing time was reduced to 20 sec. As a quality control measure, a melting curve (ramp from 63 °C–95 °C rising by 1 °C per step) was recorded. Additionally, the qRT-PCR products were visualized on agarose gels (0.8% agarose in 40 mM Tris, 20 mM acetic acid, 1 mM EDTA, pH 8.0).

The shares of the individual isoforms A–D of the total transcript amount of *At*PARN were calculated based on the ΔΔCT method [44].

### 4.5. Immunodetection of PARN-GFP

Total proteins were extracted from 10-day-old PARN-GFP and wildtype seedlings sown on $\frac{1}{2}$ Hoagland medium supplemented with 0.8% agarose and either 0 µM (-S) or 500 µM (+S) sulfate. After extraction in 80 mM Tris-HCl (pH 7.5), 300 mM NaCl, 1 mM EDTA, 1 mM PMSF, 10 mM DTT, 1% TritonX and 1 tablet EDTA-free protease Inhibitor cocktail per 50 mL (Roche), the samples were subjected to discontinuous SDS–PAGE in Mini-ProteanTM II cells (BioRad), followed by immunoblotting with a polyclonal rabbit α-GFP antibody (# A-6455, Thermo Fisher Scientific) diluted 1:5000 in TBS-T (50 mM Tris pH 7.6, 150 mM NaCl, 0.05% Tween-20). After blocking for 1 h with 5% BSA in TBS-T, the blot was washed trice with TBS-T before the addition of the primary antibody, which was left to incubate overnight at 4 °C. Subsequently, the blot was washed trice with TBS-T and the secondary horseradish peroxidase-linked anti-rabbit antibody (#AS10 852, Agrisera) diluted 25,000-fold in TBS-T was left to incubate for 1 h. Membranes were developed using SuperSignal West Dura Extended Duration Substrate (Thermo Fisher Scientific) according to the manufacturer's protocol. The resulting signals were recorded using the ImageQuant LAS 4000 (GE Healthcare).

### 4.6. Subcellular Localization

To assess the subcellular localization of *At*PARN and its antagonist AGS1, *At*PARN-GFP and AGS1-GFP seeds were sterilized as described previously and sown on $\frac{1}{2}$ Hoagland medium supplemented with 0.8% agarose. The plants were stratified at 4 °C in the dark for three days before they were transferred to long-day conditions (16 h light, 100 µE light photon flux density, 24 °C by day, 18 °C by night and 50% humidity). In order to visualize the mitochondria, 10-day-old seedlings were incubated with 100 nM MitoTracker™ Orange CMTMRos (Thermo Fisher Scientific) in $\frac{1}{2}$ Hoagland medium for 15 min as described in [21]. For nuclear staining, samples were incubated for 15 min with 2 µg $mL^{-1}$ DAPI (Sigma-Aldrich) in $\frac{1}{2}$ Hoagland medium supplemented with 1:20,000 Triton-X. Samples grown on -S plates were incubated in staining solutions prepared with -S $\frac{1}{2}$ Hoagland medium. For the DTT treatment, seedlings were floated for 30 min in $\frac{1}{2}$ Hoagland medium supplemented with 10 mM DTT before staining with MitoTracker or DAPI. Subsequently, the roots were separated from the seedlings and imaged with a Nikon automated Ti inverted microscope equipped with a Yokagawa CSU-X1 confocal scanning unit, a Hamamatsu C9100-02 EMCCD camera and a Nikon Plan Apo VC 100x NA 1.4 oil immersion objective (Nikon). Images were taken as z-stacks with an approximate thickness of 1 µm. Images were taken in three different channels (DAPI: 405 nm/445 nm; GFP 488 nm/527 nm; MitoTracker 561 nm/615 nm). Additionally, a brightfield image was recorded. The resulting z-stacks were processed with the open-source image analysis software Fiji [45]. For each channel, a maximum intensity z-projection image was calculated. Subsequently, the background fluorescence intensity was measured and subtracted for each channel.

### 4.7. Identification and Functional Annotation of Genes with mRNA Destabilizing Motifs

Sequence data for all known 3′UTRs of cytosolic mRNAs was downloaded from The Arabidopsis Information Resource (TAIR) (TAIR10 blastsets, TAIR10_3_utr_20101028, as of 10.11.2010). Subsequently, pattern match algorithms were devised to search the sequence strings for the occurrence of the core sequence "AUUUA", characteristic for AU-rich elements.

### 4.8. Prediction of Subcellular Protein Localization

The subcellular localization of *At*PARN was predicted based on its amino acid sequence. For that purpose, several bioinformatical tools, including Predotar [37], iPSORT [38], Target [39], SherLock2 [46], BaCelLo [47], WoLF PSORT [48] and YLoc [49] were applied.

The presence of nuclear localization signals was predicted with the public NLS db server (to be found at www.rostlab.org/services/nlsdb/, as of 13.04.2018).

**Supplementary Materials:** The following are available online at http://www.mdpi.com/2223-7747/8/10/380/s1, Figure S1: Wildtype seedlings stained with DAPI and MitoTracker display no signal in the GFP channel, Figure S2: AGS1-GFP does not change its subcellular localization upon nutrient starvation, Figure S3: AGS1-GFP does not change its subcellular localization upon nutrient starvation.

**Author Contributions:** Conceptualization, R.H., V.V.U. and L.A.; methodology, L.A.; formal analysis, L.A.; investigation, L.A.; data curation, L.A.; writing—original draft preparation, L.A.; writing—review and editing, M.W. and L.A.; visualization, L.A.; supervision, R.H., M.W. and V.V.U.; project administration, M.W. and R.H.; funding acquisition, M.W. and R.H.

**Funding:** This research benefitted from funding grants held by M.W. (WI3560/1-1, -/2-1) and R.H. (HE1848/115-1,).

**Acknowledgments:** The authors want to thank Takashi Hirayama (Okayama University, Japan) for the kind gift of *parn* hypomorphic and *parn-ags1* mutants, as well as the AGH2-GFP and AGS1-GFP lines.

**Conflicts of Interest:** The authors declare no conflict of interest.

## Appendix A

**Table A1.** Primers used to genotype transgenic plant lines.

| Allele | TAIR Identifier | Sequence |
|---|---|---|
| *ACTIN7* | AT5G09810 | CAACCGGTATTGTGCTCGATTC |
| | | GAGTGAGTCTGTGAGATCCCG |
| *AGS1* | AT2G17580 | CTAGCAAATTCGACAGCTTTGC |
| | | ATTTTAGAGGTTATTCTCCAATATGG |
| *AGS1* mutated | AT2G17580 | ATTTTAGAGGTTATTCTCCAATATGA |
| *AtPARN* | AT1G55870 | CTGATTCAGATTCCGACAAGGA |
| | | CTTTGCCTCCTTCTGTGAAAAG |
| *AtPARN* mutated | AT1G55870 | GTATACTGATTCAGATTCCGACAAA |
| *GFP* | N.A. | GCGGATCTTGAAGTTGGCC |
| *GFP* | N.A. | CGACGGCAACTACAAGACC |

**Table A2.** Primers used to identify *At*PARN splice variants.

| Primer | Sequence |
|---|---|
| I | CCCTTTCGTTGGGATTCTCG |
| | GTATGCAGGTGTTGAAATCAAAC |
| II | CACGAATTTCTCAGCTGTTGAAG |
| | TGCTTCAGAGAAAAAGCTGATCG |
| III | CTACCCGTTAGTCTCTCTTTC |
| | GACGAGGAAATACAAAGAAATTGTG |
| IV | GCAAAACCTAAAAATGGTCGTTTG |
| | CGAGAATCCCAACGAAAGGG |
| V | CCCTTTCGTTGGGATTCTCG |
| | CATGAGCTGGTGGATCAAATG |

**Table A3.** Primers used to assess abundance of genuine sulfur-responsive transcripts.

| Allele | TAIR Identifier | Sequence |
|---|---|---|
| *GGCT* | AT1G44790 | CCGGAGCTATTTGCTGGGGTG<br>GTCGTATTCACACTCTCTTCGTTCC |
| *PP2A* | AT1G69960 | CTTCTCGCTCCAGTAATGGGATCC<br>GCTTGGTCGACTATCGGAATGAGAG |
| *SDI1* | AT5G48850 | CCCTTGACAATGTCCTCATCG<br>GCTTCTCCTTGATAGATCTGCC |
| *SHM7* | AT1G36370 | CTATACAGCCTCGGGTTGTCATTG<br>AACTAACGTCATTACATACACATCTTG |
| *SULTR1;1* | AT5G04590 | GTCCGGGACTATTAATCCC<br>CGTACCCCATGCTCAGCG |
| *SULTR1;2* | AT1G78000 | GGATCCAGAGATGGCTACATGA<br>TCGATGTCCGTAACAGGTGAC |
| *TIP41* | AT3G54000 | GATGAGGCACCAACTGTTCTTCGTG<br>CTGACTGATGGAGCTCGGGTCG |

## References


1. Kopriva, S.; Mugford, S.G.; Baraniecka, P.; Lee, B.R.; Matthewman, C.A.; Koprivova, A. Control of sulfur partitioning between primary and secondary metabolism in Arabidopsis. *Front. Plant Sci.* **2012**, *3*, 163. [CrossRef] [PubMed]
2. Jackson, T.L.; Baker, G.W.; Wilks, F.R.; Popov, V.A.; Mathur, J.; Benfey, P.N. Large cellular inclusions accumulate in Arabidopsis roots exposed to low-sulfur conditions. *Plant Physiol.* **2015**, *168*, 1573–1589. [CrossRef] [PubMed]
3. Zhao, F.J.; Hawkesford, M.J.; McGrath, S.P. Sulphur assimilation and effects on yield and quality of wheat. *J. Cereal. Sci* **1999**, *30*, 1–17. [CrossRef]
4. Ding, Y.; Zhou, X.; Zuo, L.; Wang, H.; Yu, D. Identification and functional characterization of the sulfate transporter gene *GmSULTR1;2b* in soybean. *BMC Genom.* **2016**, *17*, 373. [CrossRef] [PubMed]
5. Lewandowska, M.; Sirko, A. Recent advances in understanding plant response to sulfur-deficiency stress. *Acta Biochim. Pol.* **2008**, *55*, 457–471.
6. Nikiforova, V.; Freitag, J.; Kempa, S.; Adamik, M.; Hesse, H.; Hoefgen, R. Transcriptome analysis of sulfur depletion in *Arabidopsis thaliana*: Interlacing of biosynthetic pathways provides response specificity. *Plant J.* **2003**, *33*, 633–650. [CrossRef] [PubMed]
7. Maruyama-Nakashita, A. Metabolic changes sustain the plant life in low-sulfur environments. *Curr. Opin. Plant Biol.* **2017**, *39*, 144–151. [CrossRef] [PubMed]
8. Bielecka, M.; Watanabe, M.; Morcuende, R.; Scheible, W.-R.; Hawkesford, M.J.; Hesse, H.; Hoefgen, R. Transcriptome and metabolome analysis of plant sulfate starvation and resupply provides novel information on transcriptional regulation of metabolism associated with sulfur, nitrogen and phosphorus nutritional responses in Arabidopsis. *Front. Plant Sci.* **2015**, *5*, 805. [CrossRef] [PubMed]
9. Crisp, P.A.; Ganguly, D.; Eichten, S.R.; Borevitz, J.O.; Pogson, B.J. Reconsidering plant memory: Intersections between stress recovery, RNA turnover, and epigenetics. *Sci. Adv.* **2016**, *2*, e1501340. [CrossRef] [PubMed]
10. Narsai, R.; Howell, K.A.; Millar, A.H.; O'Toole, N.; Small, I.; Whelan, J. Genome-wide analysis of mRNA decay rates and their determinants in *Arabidopsis thaliana*. *Plant Cell* **2007**, *19*, 3418–3436. [CrossRef]
11. Crisp, P.A.; Ganguly, D.; Smith, A.B.; Murray, K.D.; Estavillo, G.M.; Searle, I.R.; Ford, E.; Bogdanović, O.; Lister, R.; Borevitz, J.O.; et al. Rapid recovery gene downregulation during excess-light stress and recovery in Arabidopsis. *Plant Cell* **2017**, *29*, 1836–1863. [CrossRef] [PubMed]
12. Nakaminami, K.; Matsui, A.; Shinozaki, K.; Seki, M. RNA regulation in plant abiotic stress responses. *Biochim. Biophys. Acta* **2012**, *1819*, 149–153. [CrossRef] [PubMed]
13. Christie, M.; Brosnan, C.A.; Rothnagel, J.A.; Carroll, B.J. RNA decay and RNA silencing in plants: Competition or collaboration? *Front. Plant Sci.* **2011**, *2*, 99. [CrossRef] [PubMed]
14. Belostotsky, D.A.; Sieburth, L.E. Kill the messenger: mRNA decay and plant development. *Curr. Opin. Plant Biol.* **2009**, *12*, 96–102. [CrossRef] [PubMed]
15. Chiba, Y.; Green, P.J. mRNA degradation machinery in plants. *J. Plant Biol.* **2009**, *52*, 114–124. [CrossRef]

16. Reverdatto, S.V.; Dutko, J.A.; Chekanova, J.A.; Hamilton, D.A.; Belostotsky, D.A. mRNA deadenylation by PARN is essential for embryogenesis in higher plants. *RNA* **2004**, *10*, 1200–1214. [CrossRef]
17. Nishimura, N.; Kitahata, N.; Seki, M.; Narusaka, Y.; Narusaka, M.; Kuromori, T.; Asami, T.; Shinozaki, K.; Hirayama, T. Analysis of *ABA hypersensitive germination 2* revealed the pivotal functions of PARN in stress response in Arabidopsis. *Plant J.* **2005**, *44*, 972–984. [CrossRef] [PubMed]
18. Nishimura, N.; Okamoto, M.; Narusaka, M.; Yasuda, M.; Nakashita, H.; Shinozaki, K.; Narusaka, Y.; Hirayama, T. *ABA hypersensitive germination 2-1* causes the activation of both abscisic acid and salicylic acid responses in Arabidopsis. *Plant Cell Physiol.* **2009**, *50*, 2112–2122. [CrossRef]
19. Nishimura, N.; Yoshida, T.; Murayama, M.; Asami, T.; Shinozaki, K.; Hirayama, T. Isolation and characterization of novel mutants affecting the abscisic acid sensitivity of Arabidopsis germination and seedling growth. *Plant Cell Physiol.* **2004**, *45*, 1485–1499. [CrossRef]
20. Hirayama, T. A unique system for regulating mitochondrial mRNA poly(A) status and stability in plants. *Plant Singal. Behav.* **2014**, *9*, e973809. [CrossRef]
21. Hirayama, T.; Matsuura, T.; Ushiyama, S.; Narusaka, M.; Kurihara, Y.; Yasuda, M.; Ohtani, M.; Seki, M.; Demura, T.; Nakashita, H.; et al. A poly(A)-specific ribonuclease directly regulates the poly(A) status of mitochondrial mRNA in Arabidopsis. *Nat. Commun.* **2013**, *4*, 2247. [CrossRef] [PubMed]
22. Malcheska, F.; Ahmad, A.; Batool, S.; Muller, H.M.; Ludwig-Muller, J.; Kreuzwieser, J.; Randewig, D.; Hansch, R.; Mendel, R.R.; Hell, R.; et al. Drought-enhanced xylem sap sulfate closes stomata by affecting ALMT12 and guard cell ABA synthesis. *Plant Physiol.* **2017**, *174*, 798–814. [CrossRef] [PubMed]
23. Batool, S.; Uslu, V.V.; Rajab, H.; Ahmad, N.; Waadt, R.; Geiger, D.; Malagoli, M.; Hedrich, R.; Rennenberg, H.; Herschbach, C.; et al. Sulfate is incorporated into cysteine to trigger ABA production and stomata closure. *Plant Cell* **2018**, *30*, 2973–2987. [CrossRef] [PubMed]
24. Lai, W.S.; Kennington, E.A.; Blackshear, P.J. Tristetraprolin and its family members can promote the cell-free deadenylation of AU-rich element-containing mRNAs by poly(A) ribonuclease. *Mol. Cell. Biol.* **2003**, *23*, 3798–3812. [CrossRef] [PubMed]
25. Chen, C.Y.; Shyu, A.B. Mechanisms of deadenylation-dependent decay. *Wiley Interdiscip. Rev. RNA* **2011**, *2*, 167–183. [CrossRef] [PubMed]
26. Chen, C.Y.; Gherzi, R.; Ong, S.E.; Chan, E.L.; Raijmakers, R.; Pruijn, G.J.; Stoecklin, G.; Moroni, C.; Mann, M.; Karin, M. AU binding proteins recruit the exosome to degrade ARE-containing mRNAs. *Cell* **2001**, *107*, 451–464. [CrossRef]
27. Aarabi, F.; Hubberten, H.-M.; Heyneke, E.; Watanabe, M.; Hoefgen, R. OAS Cluster Genes: A Tightly Co-regulated Network. In *Molecular Physiology and Ecophysiology of Sulfur*; De Kok, L.J., Hawkesford, M.J., Rennenberg, H., Saito, K., Schnug, E., Eds.; Springer: Cham, Germany, 2015; pp. 125–132. [CrossRef]
28. Moreno, A.B.; Martinez de Alba, A.E.; Bardou, F.; Crespi, M.D.; Vaucheret, H.; Maizel, A.; Mallory, A.C. Cytoplasmic and nuclear quality control and turnover of single-stranded RNA modulate post-transcriptional gene silencing in plants. *Nucleic Acids Res.* **2013**, *41*, 4699–4708. [CrossRef] [PubMed]
29. Chiba, Y.; Johnson, M.A.; Lidder, P.; Vogel, J.T.; van Erp, H.; Green, P.J. AtPARN is an essential poly(A) ribonuclease in Arabidopsis. *Gene* **2004**, *328*, 95–102. [CrossRef] [PubMed]
30. Barreau, C.; Paillard, L.; Osborne, H.B. AU-rich elements and associated factors: Are there unifying principles? *Nucleic Acids Res.* **2005**, *33*, 7138–7150. [CrossRef] [PubMed]
31. Forieri, I.; Sticht, C.; Reichelt, M.; Gretz, N.; Hawkesford, M.J.; Malagoli, M.; Wirtz, M.; Hell, R. System analysis of metabolism and the transcriptome in Arabidopsis thaliana roots reveals differential co-regulation upon iron, sulfur and potassium deficiency. *Plant Cell Environ.* **2017**, *40*, 95–107. [CrossRef]
32. Forieri, I.; Wirtz, M.; Hell, R. Toward new perspectives on the interaction of iron and sulfur metabolism in plants. *Front. Plant Sci.* **2013**, *4*, 357. [CrossRef] [PubMed]
33. Shen, J.; Song, Z.; Sun, Z. GO molecular function coding based protein subcellular localization prediction. *Chin. Sci. Bull.* **2007**, *52*, 2240–2245. [CrossRef]
34. Nguyen, A.H.; Matsui, A.; Tanaka, M.; Mizunashi, K.; Nakaminami, K.; Hayashi, M.; Iida, K.; Toyoda, T.; Nguyen, D.V.; Seki, M. Loss of Arabidopsis 5′-3′ exoribonuclease AtXRN4 function enhances heat stress tolerance of plants subjected to severe heat stress. *Plant Cell Physiol.* **2015**, *56*, 1762–1772. [CrossRef] [PubMed]

35. Rymarquis, L.A.; Souret, F.F.; Green, P.J. Evidence that XRN4, an Arabidopsis homolog of exoribonuclease XRN1, preferentially impacts transcripts with certain sequences or in particular functional categories. *RNA* **2011**, *17*, 501–511. [CrossRef] [PubMed]
36. Von Roretz, C.; Di Marco, S.; Mazroui, R.; Gallouzi, I.E. Turnover of AU-rich-containing mRNAs during stress: A matter of survival. *Wiley Interdisc. Rev. RNA* **2011**, *2*, 336–347. [CrossRef] [PubMed]
37. Small, I.; Peeters, N.; Legeai, F.; Lurin, C. Predotar: A tool for rapidly screening proteomes for N-terminal targeting sequences. *Proteomics* **2004**, *4*, 1581–1590. [CrossRef] [PubMed]
38. Bannai, H.; Tamada, Y.; Maruyama, O.; Nakai, K.; Miyano, S. Extensive feature detection of N-terminal protein sorting signals. *Bioinformatics* **2002**, *18*, 298–305. [CrossRef]
39. Emanuelsson, O.; Nielsen, H.; Brunak, S.; von Heijne, G. Predicting subcellular localization of proteins based on their N-terminal amino acid sequence. *J. Mol. Biol.* **2000**, *300*, 1005–1016. [CrossRef]
40. Hubstenberger, A.; Courel, M.; Benard, M.; Souquere, S.; Ernoult-Lange, M.; Chouaib, R.; Yi, Z.; Morlot, J.B.; Munier, A.; Fradet, M.; et al. P-body purification reveals the condensation of repressed mRNA regulons. *Mol. Cell* **2017**, *68*, 144–157.e145. [CrossRef]
41. Takeo, K.; Ito, T. Subcellular localization of VIP1 is regulated by phosphorylation and 14-3-3 proteins. *FEBS Lett.* **2017**, *591*, 1972–1981. [CrossRef]
42. Czechowski, T.; Stitt, M.; Altmann, T.; Udvardi, M.K.; Scheible, W.-R. Genome-wide identification and testing of superior reference genes for transcript normalization in Arabidopsis. *Plant Physiol.* **2005**, *139*, 5–17. [CrossRef] [PubMed]
43. Zhu, J.; Zhang, L.; Li, W.; Han, S.; Yang, W.; Qi, L. Reference gene selection for quantitative real-time PCR normalization in *Caragana intermedia* under different abiotic stress conditions. *PLoS ONE* **2013**, *8*, e53196. [CrossRef] [PubMed]
44. Livak, K.J.; Schmittgen, T.D. Analysis of relative gene expression data using real-time quantitative PCR and the 2(-ΔΔC(T)) Method. *Methods* **2001**, *25*, 402–408. [CrossRef] [PubMed]
45. Schindelin, J.; Arganda-Carreras, I.; Frise, E.; Kaynig, V.; Longair, M.; Pietzsch, T.; Preibisch, S.; Rueden, C.; Saalfeld, S.; Schmid, B.; et al. Fiji: An open-source platform for biological-image analysis. *Nat. Methods* **2012**, *9*, 676–682. [CrossRef]
46. Briesemeister, S.; Blum, T.; Brady, S.; Lam, Y.; Kohlbacher, O.; Shatkay, H. SherLoc2: A high-accuracy hybrid method for predicting subcellular localization of proteins. *J. Proteome Res.* **2009**, *8*, 5363–5366. [CrossRef] [PubMed]
47. Pierleoni, A.; Martelli, P.L.; Fariselli, P.; Casadio, R. BaCelLo: A balanced subcellular localization predictor. *Bioinformatics* **2006**, *22*, e408–e416. [CrossRef] [PubMed]
48. Horton, P.; Park, K.J.; Obayashi, T.; Fujita, N.; Harada, H.; Adams-Collier, C.J.; Nakai, K. WoLF PSORT: Protein localization predictor. *Nucleic Acids Res.* **2007**, *35*, W585–W587. [CrossRef]
49. Briesemeister, S.; Rahnenführer, J.; Kohlbacher, O. Yloc—An interpretable web server for predicting subcellular localization. *Nucleic Acids Res.* **2010**, *38*, W497–W502. [CrossRef]

*Article*

# Glucosinolate Distribution in the Aerial Parts of *sel1-10*, a Disruption Mutant of the Sulfate Transporter SULTR1;2, in Mature *Arabidopsis thaliana* Plants

**Tomomi Morikawa-Ichinose [1], Sun-Ju Kim [2], Alaa Allahham [1], Ryota Kawaguchi [1] and Akiko Maruyama-Nakashita [1,*]**

[1] Department of Bioscience and Biotechnology, Faculty of Agriculture, Kyushu University, 744 Motooka, Nishi-ku, Fukuoka 819-0395, Japan; ichinose226@agr.kyushu-u.ac.jp (T.M.-I.); 2BE17488K@s.kyushu-u.ac.jp (A.A.); 2BE19462N@s.kyushu-u.ac.jp (R.K.)

[2] Department of Bio-Environmental Chemistry, College of Agriculture and Life Sciences, Chungnam National University, Daejeon 34134, Korea; kimsunju@cnu.ac.kr

* Correspondence: amaru@agr.kyushu-u.ac.jp; Tel.: +81-92-802-4712

Received: 26 February 2019; Accepted: 4 April 2019; Published: 10 April 2019

**Abstract:** Plants take up sulfur (S), an essential element for all organisms, as sulfate, which is mainly attributed to the function of SULTR1;2 in *Arabidopsis*. A disruption mutant of *SULTR1;2*, *sel1-10*, has been characterized with phenotypes similar to plants grown under sulfur deficiency (−S). Although the effects of −S on S metabolism were well investigated in seedlings, no studies have been performed on mature *Arabidopsis* plants. To study further the effects of −S on S metabolism, we analyzed the accumulation and distribution of S-containing compounds in different parts of mature *sel1-10* and of the wild-type (WT) plants grown under long-day conditions. While the levels of sulfate, cysteine, and glutathione were almost similar between *sel1-10* and WT, levels of glucosinolates (GSLs) differed between them depending on the parts of the plant. GSLs levels in the leaves and stems were generally lower in *sel1-10* than those in WT. However, *sel1-10* seeds maintained similar levels of aliphatic GSLs to those in WT plants. GSL accumulation in reproductive tissues is likely to be prioritized even when sulfate supply is limited in *sel1-10* for its role in S storage and plant defense.

**Keywords:** mature *Arabidopsis thaliana* plants; sulfate transporter; SULTR1;2; *sel1-10* mutant; glucosinolates

---

## 1. Introduction

Sulfur (S) is an essential macronutrient for all organisms. Plants take up inorganic sulfate as the major S source and assimilate it into a variety of S-containing organic compounds [1,2]. As animals are unable to assimilate sulfate, the role of plants in the global S cycle on the earth is extremely important [2]. In addition, many of the S-containing compounds biosynthesized in plants are beneficial to health, such as methionine (an essential amino acid for animals), glutathione (a redox controller), and various secondary compounds with specific functions [2]. Glucosinolates (GSLs) are the major S-containing secondary compounds biosynthesized in *Brassicaceae*, that act as defense compounds against insects and pathogens [3–5]. Depending on their amino acid precursors, most GSLs accumulated in *Arabidopsis* are classified into aliphatic and indolic GSLs (iGSLs) synthesized from methionine and tryptophan, respectively [3–5]. Among them, some aliphatic GSLs (mGSLs) are known to be beneficial for humans as cancer-preventive chemicals [6,7]. Thus, understanding GSL accumulation in plant tissues would contribute to improved food quality in Brassica crops.

The composition and content of GSLs are different among plant parts in *Arabidopsis* [8–12]. Most GSLs accumulated in developing rosette leaves are mGSLs, and mainly consist of 4-methylsulfinylbutyl GSL (4MSOB, 34 to 60%), 3-methylsulfinylpropyl GSL (3MSOP, 4 to 9%), 4-methylthiobutyl GSL (4MTB, 1 to 23%), and 8-methylsulfinyloctyl GSL (8MSOO, 2 to 6%). The remaining GSLs are iGSLs, and mostly comprise indol-3-ylmethyl GSL (I3M, 11 to 23%) [8,9,13]. Cauline leaves and stems have a similar concentration and composition to that of rosette leaves [9]. GSL content in the seeds is 3.5- to 8.5-fold than that in the leaves, with the higher GSL variations characterized by a higher amount of 4MTB (37 to 41%); the long-chain mGSLs, such as 8MSOO (9.9 to 10%), 8-methylthiooctyl GSL (8MTO, 6.9 to 7.4%), 7-methylthioheptyl GSL (7MTH, 4.7 to 4.8%), and 7-methylsulfinylheptyl GSL (7MSOH, 1.8 to 2.4%); as well as with a relatively low amount of I3M (2.3 to 2.9%) [8,9]. mGSLs are structurally divided into methylsulfinylalkyl (MSOX) GSLs (3MSOP, 4MSOB, 7MSOH, and 8MSOO) and methylthioalkyl (MTX) GSLs (4MTB, 7MTH, and 8MTO) [3–5,7]. Seeds accumulated more MTX GSLs than MSOX GSLs compared to the other tissues [8,9]. The GSL concentration in the siliques is lower than that in the seeds, and the composition is intermediate of that in the rosette leaves and the seeds [8,9]. This plant part-specific variation in GSL concentration and composition suggests that GSL accumulation is controlled by different mechanisms in each part [8–12].

GSL content in plants is also influenced by environmental factors [5,14]. For example, it is stimulated by glucose and jasmonic acid [15,16], and is increased upon pathogen infection [17,18] and insect bite [19]. Among the environmental factors, nutritional conditions, particularly S status, greatly influence GSL accumulation in plants [13,20–22]. GSL synthesis and accumulation are stimulated under S sufficiency (+S) but suppressed under S deficiency (−S), which is regulated by specific transcriptional networks induced by −S in *Arabidopsis* [5,13,20,22–24]. However, these experiments were mostly undertaken on seedlings and the effects of –S on GSL accumulation in mature plants have not been reported.

Previous studies have shown a close correlation between the effects of −S and the disruption of SULTR1;2, a major sulfate transporter that facilitates sulfate uptake from roots [25–28]. In this study, we examined the accumulation of S-containing compounds in aerial tissues of mature SULTR1;2 mutants, known as *sel1-10*, and wild-type (WT) plants to clarify the distribution of sulfate as well as cysteine and glutathione in relation to the distribution of GSL in the mature plants.

## 2. Results

### 2.1. Growth Phenotypes of WT and sel1-10 Plants

To investigate the metabolic changes occurring in mature *sel1-10* plants, we initially observed the growth phenotypes of *sel1-10* plants (Figure 1). WT and *sel1-10* plants were grown for six weeks in vermiculite. Although visible differences in shoot phenotype were not observed between WT and *sel1-10* plants (Figure 1a,b), a significant decrease was observed in the primary stem diameters of *sel1-10* plants compared to those of the WT, while the plant heights were similar between WT and *sel1-10* plants (Figure 1c). Correlated with the decrease in primary stem diameter in *sel1-10*, dry weight of primary stems (PS) was decreased in *sel1-10* to 70% of that in WT plants (Figure 1d). Dry weights of rosette leaves (RL), cauline leaves (CL), lateral stems (LS), and siliques (Si) were not significantly lower but tended to be lower in *sel1-10* plants relative to those in WT plants (Figure 1d).

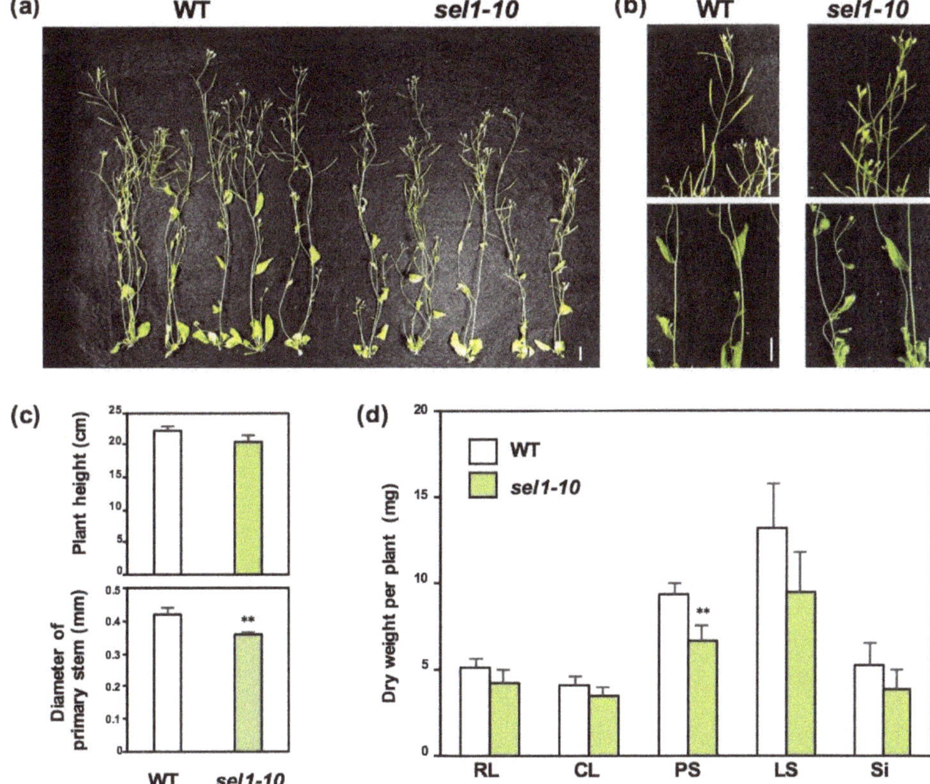


**Figure 1.** Growth phenotypes of wild-type (WT) and *sel1-10* plants. (**a**) WT and *sel1-10* plants grown for six weeks on vermiculite. (**b**) Siliques (upper panels) and primary stems (lower panels) of WT and *sel1-10* plants. Scale bar = 1 cm. (**c**) Plant heights and diameters of primary stems of WT and *sel1-10* plants. (**d**) Dry weight of rosette leaves (RL), cauline leaves (CL), primary stems (PS), lateral stems (LS), and siliques (Si) in WT and *sel1-10* plants. White and green bars represent WT and *sel1-10*, respectively, in (**c**) and (**d**). Data are shown as the averages with error bars denoting SEM (n = 5). Asterisks indicate significant differences (Student's *t*-test; ** $p < 0.05$) between WT and *sel1-10*.

### 2.2. Concentrations of Sulfate and Selected Sulfur-Containing Metabolites in Different Parts of WT and sel1-10 Plants

We harvested RL, CL, PS, LS, and Si separately and analyzed sulfate, cysteine, glutathione (GSH), and GSL in different parts of the *sel1-10* and WT plants (Figures 2–4).

Sulfate content in the RL of *sel1-10* plants was 26% higher than that in the WT plants. Both WT and *sel1-10* plants accumulated a similar level of sulfate in CL, PS, and LS. In Si, the sulfate content of *sel1-10* plants was 61% of that in the WT plants. These results indicated that the distribution of sulfate was modulated in *sel1-10* plants.

To examine the effects of modulated sulfate distribution in *sel1-10*, cysteine and GSH contents in WT and *sel1-10* plants were analyzed (Figure 3). Cysteine content was not significantly different between WT and *sel1-10* plants in all examined parts. The GSH content in Si of *sel1-10* plants was 29% lower than that in WT plants, suggesting that the dysfunction of SULTR1;2 affects GSH accumulation in reproductive tissues as observed in the seedlings [28]. GSH content in other parts of *sel1-10* plants was similar to that in the WT plants.

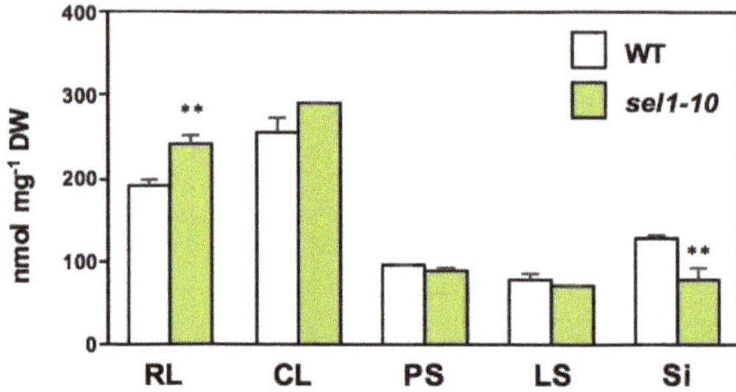


**Figure 2.** Sulfate concentrations in different parts of WT and *sel1-10* plants. Sulfate content in each part was determined by ion chromatography. WT and *sel1-10* seedlings were grown for six weeks in vermiculite, and each part was harvested. Rosette leaves (RL), cauline leaves (CL), primary stems (PS), lateral stems (LS), and siliques (Si). White and green bars represent the sulfate content in WT and *sel1-10*, respectively. Data are shown as averages with error bars denoting SEM (n = 3). Asterisks indicate significant differences (Student's *t*-test; * $0.05 < p < 0.1$, ** $p < 0.05$) between WT and *sel1-10* plants.

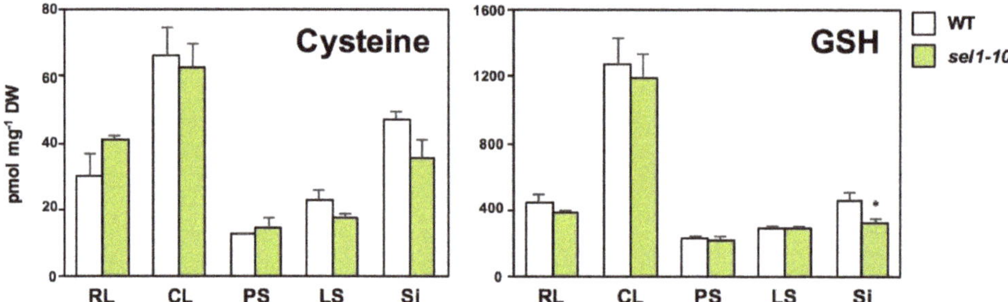


**Figure 3.** Cysteine and glutathione (GSH) concentrations in different parts of WT and *sel1-10* plants. The cysteine and GSH contents of different parts were measured using HPLC-fluorescence detection. WT and *sel1-10* seedlings were grown for 6 weeks in vermiculite, after which each part was harvested. Rosette leaves (RL), cauline leaves (CL), primary stems (PS), lateral stems (LS), and siliques (Si). White and green bars represent cysteine and GSH levels in WT and *sel1-10* plants, respectively. Data are shown as the averages with error bars denoting SEM (n = 3). Asterisks indicate significant differences (Student's *t*-test; * $0.05 < p < 0.1$, ** $p < 0.05$) between WT and *sel1-10* plants.

The following seven major GSLs were analyzed in both plants (Figure 4). These included six mGSLs: 3-methylsulfinylpropyl GSL (3MSOP), 4-methylsulfinylbutyl GSL (4MSOB), 8-methylsulfinyloctyl GSL (8MSOO), 4-methylthiobutyl GSL (4MTB), 7-methylthioheptyl GSL (7MTH), and 8-methylthiooctyl GSL (8MTO), and one iGSL, indol-3-ylmethyl GSL (I3M).

GSL levels were generally lower in RL, CL, PS, and LS of *sel1-10* relative to the same parts of WT plants (Figure 4). However, in Si, GSL levels did not significantly vary between *sel1-10* and WT plants, and some GSL levels were even higher in *sel1-10* plants relative to the WT plants, that is, the levels of MSOX GSLs and I3M were similar between *sel1-10* and WT plants, but the levels of MTX GSLs were higher in *sel1-10* plants relative to the WT plants (Figure 4).

Because GSL levels in Si were not affected in *sel1-10* plants except for 4MTB and 7MTH, GSL levels in mature dried seeds were analyzed to determine the effects of reduced sulfate uptake (Figure 5). Seeds contained much higher levels of MTX GSLs and 8MSOO and lower levels of I3M compared to other vegetative tissues in both plant lines, which is consistent with previous studies [8,9]. In seeds, GSL levels did not significantly vary between *sel1-10* and WT plants.

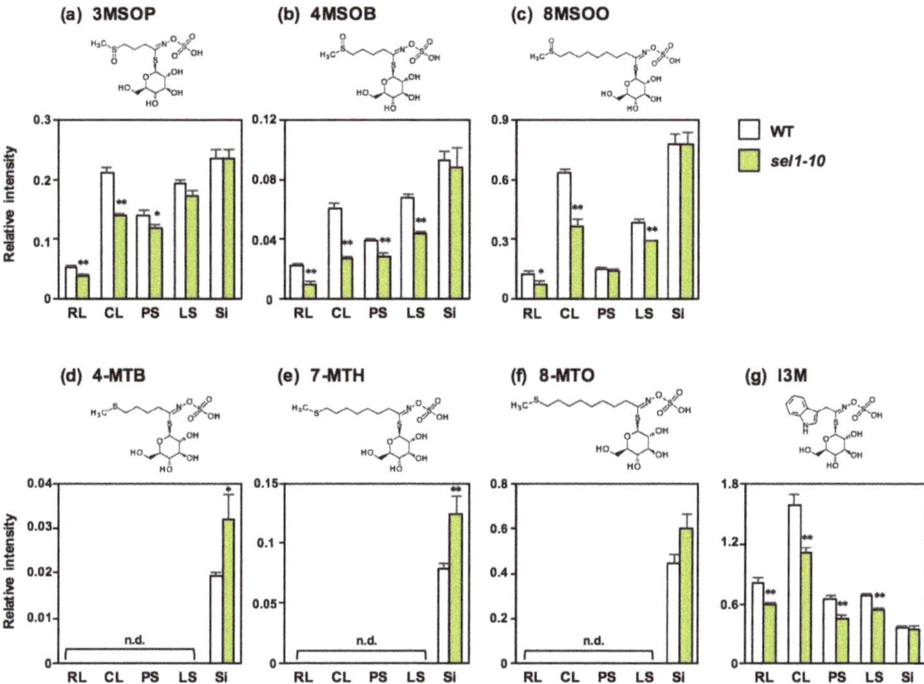


**Figure 4.** Glucosinolates (GSL) accumulation in different parts of WT and *sel1-10* plants. GSL levels in different parts were determined by LC-MS. The relative amount was calculated as the ratio of peak height of each GSL to that of the internal standard (L(+)-10-camphor sulfonic acid) and then divided by the dry weight of the sample. WT and *sel1-10* seedlings were grown for 6 weeks in vermiculite, after which, each part was harvested. (**a**) 3-methylsulfinylpropyl GSL (3MSOP), (**b**) 4-methylsulfinylbutyl GSL (4MSOB), (**c**) 8-methylsulfinyloctyl GSL (8MSOO), (**d**) 4-methylthiobutyl GSL (4MTB), (**e**) 7-methylthioheptyl GSL (7MTH), (**f**) 8-methylthiooctyl GSL (8MTO), (**g**) indol-3-ylmethyl GSL (I3M). Rosette leaves (RL), cauline leaves (CL), primary stems (PS), lateral stems (LS), and siliques (Si). White and green bars represent the relative GSL content in WT and *sel1-10* plants, respectively. Data are shown as averages with error bars denoting SEM (n = 3). Asterisks indicate significant differences (Student's *t*-test; * $0.05 < p < 0.1$, ** $p < 0.05$) between WT and *sel1-10* plants. n.d., not detected. The chemical structures of the GSLs were obtained from KEGG Databases in DBGET (https://www.genome.jp/dbget/).

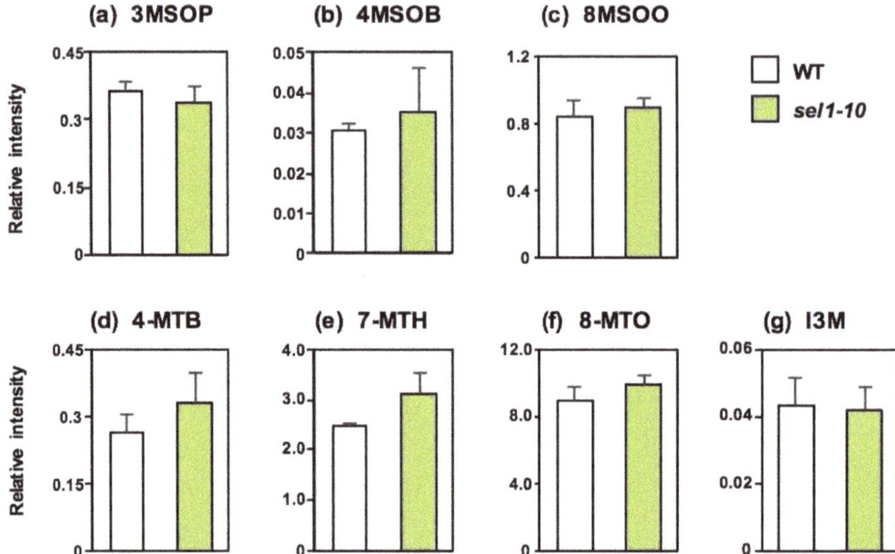


**Figure 5.** GSL accumulation in mature seeds of WT and *sel1-10* plants. GSL levels in seeds were determined by LC-MS. The relative amount was calculated as the ratio of peak height of each GSL to that of the internal standard (L(+)-10-camphor sulfonic acid) and then divided by the dry weight of the sample. (**a**) 3-methylsulfinylpropyl GSL (3MSOP), (**b**) 4-methylsulfinylbutyl GSL (4MSOB), (**c**) 8-methylsulfinyloctyl GSL (8MSOO), (**d**) 4-methylthiobutyl GSL (4MTB), (**e**) 7-methylthioheptyl GSL (7MTH), (**f**) 8-methylthiooctyl GSL (8MTO), (**g**) indol-3-ylmethyl GSL (I3M). White and green bars represent the relative GSL content in WT and *sel1-10* seeds, respectively. Data are shown as averages with error bars denoting SEM (n = 3). Statistical analysis was performed with Student's *t*-test between WT and *sel1-10* plants, but any significant differences were not detected.

## 3. Discussion

Growth phenotypes of mature *sel1-10* plants have not been well studied as regards their aerial part. The sulfate uptake rate in *sel1-10* plants was almost half of that in WT plants under both +S and –S conditions at the seedling stage [27,29]. In addition, the biomass and the levels of sulfate and GSH in *sel1-10* seedlings were significantly lowered relative to those in the WT plants under both +S and –S conditions [27–29]. Similar growth retardation in mature *sel1-10* plants observed in Figure 1 is assumed to be due to the reduction in sulfate uptake in *sel1-10* plants.

It is known that GSL accumulation is differentially regulated in plant parts and S status [8,9,13,21,22]. In our analysis, the levels of MSOX GSLs and I3M in the leaves and stems of *sel1-10* plants were significantly lower than those in the WT plants (Figure 4), in agreement with the –S-induced-like phenotypes observed in *sel1-10* seedlings [27,28]. In contrast, the GSL levels in Si and Se of *sel1-10* plants were similar or higher than those in WT plants (Figures 4 and 5). Considering the previous theory that GSLs accumulated in the seeds provide an S source for seedling growth [8–10,21], GSL accumulation should be prioritized in reproductive tissues even when the S supply is limited in *sel1-10* plants. Plants should have adapted to fluctuations in S availability by using GSLs as S storage substances in reproductive tissues. GSLs can be considered as a beneficial S storage compounds because of the relatively high molecular weight, that enable them to reduce osmotic pressure in the seeds. Additionally, GSLs can be a source of carbon and nitrogen, especially in the case of long-chain mGSLs highly accumulated in the seeds, and can also act as the defense compounds to protect the seeds from diseases or predators [11,21].

Unexpectedly, the levels of MTX GSLs were higher in Si of *sel1-10* plants compared to those in the WT plants, while MSOX GSL and iGSL levels in Si were similar between *sel1-10* and WT plants (Figure 4). Taking into account that GSL levels in Si were the sum of the levels in silique tissues, including the developing seeds, and all samples were collected on the same date, the increase of MTX GSLs could be because of the acceleration of seed maturation in *sel1-10* plants. In general, nutrient stress accelerates bud appearance and subsequent development of the siliques and seeds in *Arabidopsis* [30,31]. Considering that MTX GSLs in the seeds are continuously increased during the seed maturation period [8,9], the timing of flowering and seed development may occur earlier in *sel1-10* than in the WT plants. Lower levels of sulfate and GSH in Si of *sel1-10* plants relative to those in the WT plants also support this assumption (Figures 2 and 3).

Several maternal tissues have been suggested as source tissues for seed GSLs, including the leaves and siliques [10,11]. Although the GSL transport machinery in whole plants is not fully understood [10–12,32], GSL transporters, GTR1 and GTR2, that belong to the NRT/PTR family have been characterized for their roles [32,33]. In the double disruption lines of GTR1 and GTR2, most GSLs were not found in the seeds, whereas, mGSLs were highly accumulated in rosette leaves and siliques [32,33]. This suggested that seed GSLs are mostly transported from the source tissues. Decreased GSL levels in vegetative tissues and the maintenance of GSL levels in the seeds suggested that GSL transport to the seeds was not restricted or was even accelerated in *sel1-10* plants.

In conclusion, we found that GSL levels of the MSOX group were decreased in the leaves and stems, whilst all GSL were found to be maintained in the seeds in *sel1-10* plants. This shows that accumulation of mGSL characterizes the reproductive tissues, thus indicating that mGSL are destinated to store in the seeds in order to support the initial growth of the next generation.

## 4. Materials and Methods

### 4.1. Plant Materials and Growth Conditions

*Arabidopsis thaliana* were cultured in a growth chamber controlled at 23 ± 2 °C under constant illumination (40 μmol $m^{-2}$ $s^{-1}$). The *sel1-10* mutant, carrying a T-DNA insertion in the ninth exon of SULTR1;2 (At1g78000) [28], and the background Wassilewskija (Ws-0) wild-type plants (WT) were used as plant materials. Seeds of WT and *sel1-10* were sown on vermiculite as growth substrate supplemented with MGRL mineral nutrient media in 5 × 5 × 5 cm plastic pots [34,35]. After germination, the number of plants was adjusted to three plants per pot. Plants were grown for 6 weeks and the different parts of the plants, rosette leaves (RL), cauline leaves (CL), primary stems (PS), lateral stems (LS), and siliques (Si) were harvested separately from each pot and weighed for the fresh weights. Mature dried seeds collected from the former generation were used for the analysis. Right after harvest, plant tissues were frozen in liquid nitrogen, freeze-dried, ground into a fine powder using a Tissue Lyser (Retsch, Germany), and used for each metabolite analysis. Three independent samples for each part were used for metabolite analysis.

### 4.2. Measurement of Glucosinolates

Three milligrams of the plant powder was extracted with 300 μL of ice-cold 80% methanol containing 2 μM L(+)-10-camphor sulfonic acid (10CS, internal standard, Tokyo Kasei, Japan) using a Tissue Lyser. After homogenization, cell debris was separated by centrifugation (15,000 rpm, 10 min, 4 °C), and the supernatants were evaporated with a centrifugal evaporator (CVE-3110, EYELA, Japan) connected to a high vacuum pump (DAH-60, ULVAC, Japan) and a cold trap (UNI TRAP UT-1000, EYELA). Dried supernatants were dissolved into water, filtered with Millex-GV filter units (Millipore, USA), and analyzed by a high-performance liquid chromatograph connected to a triple quadrupole (LC-QqQ)-MS (LCMS8040, Shimadzu, Kyoto, Japan) using L-column 2 ODS (pore size 3 μm, length 2.1 × 150 mm, CERI, Japan). The mobile phase consisted of solvent A (0.1% formic acid, Wako Pure Chemicals, Osaka, Japan) and solvent B (0.1% formic acid in acetonitrile, Wako Pure Chemicals, Osaka,

Japan). The gradient elution program was as follows with a flow rate of 0.3 mL/min, 0–0.1 min, 1% B; 0.1–15.5 min, 99.5% B; 15.5–17 min, 99.5% B; 17–17.1 min, 1% B; and 17.1–20 min, 1% B as described previously [36]. For the MS, electrospray ionization mass spectrometry technique in negative ionization mode was used. The ionization parameters were as follows: the nebulizer gas flow was 1.5 L/min, the CDL temperature was 250 °C, heat block temperature was 400 °C. All GSLs were detected with optimized selective reaction monitoring transitions in negative ionization mode as follows (precursor ion [*m/z*]/product ion [*m/z*] scores are shown): 3MSOP GSL: 422.02/358.02, 422.02/96.9, 422.02/95.9; 4MSOB GSL: 436.05/96.9, 436.05/96.0, 436.05/177.9; 8MSOO GSL: 492.1/428.1, 492.1/96.9, 492.1/95.9; 4MTB GSL: 420.04/96.9, 420.04/95.9, 420.04/74.9; 7MTH GSL: 462.09/96.9, 462.09/95.9, 462.09/74.9; 8MTO GSL: 476.11/96.9, 476.11/95.9, 476.11/74.9; I3M GSL: 447.05/96.9, 447.05/95.9, 447.05/74.9. MRM transitions were determined by using standard compounds (Cfm Oskar Tropitzsch GmbH, Marktredwitz, Germany) or a database (MassBank, http://www.massbank.jp). The relative quantities of GSLs were calculated as the ratio of peak height to the height of 10CS.

### *4.3. Measurement of Sulfate, Cysteine and Glutathione*

One mg of the plant powder was extracted with 200 µL of 10 mM HCl. The cell debris was removed by centrifugation, and the supernatant was used for the analysis. The extracts were diluted 100 fold with extra pure water and analyzed by ion chromatography as described previously [29], using an eluent containing 1.9 mM $NaHCO_3$ and 3.2 mM $Na_2CO_3$.

Cysteine and GSH contents were determined by monobromobimane (Invitrogen) labeling of thiol bases after reduction of the extracts with dithiothreitol (Nacalai Tesque) as described [13,28,29]. The labeled products were then separated by HPLC (JASCO, Tokyo, Japan) using the TSKgel ODS-120T column (150 × 4.6 mm, TOSOH) and detected with a fluorescence detector FP-920 (JASCO), monitoring for fluorescence of thiol-bimane adducts at 478 nm under excitation at 390 nm. GSH and Cys standards were purchased from Nacalai Tesque (Kyoto, Japan).

### *4.4. Statistical Analysis*

The data were statistically analyzed using Student's *t*-test with Microsoft Excel. Significant differences between WT and *sel1-10* in biological replicates are shown in each Figure.

**Author Contributions:** A.M.-N. designed research. T.M.-I., A.A., R.K. and S.-J.K. performed experiments and analyzed data. T.M.-I. and A.M.-N. wrote the manuscript.

**Funding:** This work was supported by Grant-in-Aid for JSPS fellow 16J40073 (for T.M.-I.), JSPS KAKENHI Grant Number 15KT0028 and 17H03785 (for A.M.-N.) and Japan Foundation for Applied Enzymology (for A.M.-N.). This research was supported in part by the Science and Technology Incubation Program in Advanced Region from the funding program "Creation of Innovation Center for Advanced Interdisciplinary Research Areas" from the Japan Science and Technology Agency.

**Acknowledgments:** We thank Yukiko Okuo for technical support.

**Conflicts of Interest:** The authors declare no conflict of interest.

## References

1. Takahashi, H.; Kopriva, S.; Giordano, M.; Saito, K.; Hell, R. Sulfur assimilation in photosynthetic organisms: Molecular functions and regulations of transporters and assimilatory enzymes. *Annu. Rev. Plant Biol.* **2011**, *62*, 157–184. [CrossRef]
2. Long, S.R.; Kahn, M.; Seefeldt, L.; Tsay, Y.F.; Kopriva, S. Chapter 16 Nitrogen and Sulfur. In *Biochemistry & Molecular Biology of Plants*; Buchana, B.B., Gruissem, W., Jones, R.L., Eds.; WILEY Blackwell: Hoboken, NJ, USA, 2015; pp. 746–768.
3. Grubb, C.D.; Abel, S. Glucosinolate metabolism and its control. *Trends Plant Sci.* **2006**, *11*, 89–100. [CrossRef]
4. Halkier, B.A.; Gershenzon, J. Biology and biochemistry of glucosinolates. *Annu. Rev. Plant Biol.* **2006**, *57*, 303–333. [CrossRef]

5. Sonderby, I.E.; Geu-Flores, F.; Halkier, B.A. Biosynthesis of glucosinolates–gene discovery and beyond. *Trends Plant Sci.* **2010**, *15*, 283–290. [CrossRef] [PubMed]
6. Talalay, P.; Fahey, J.W. Phytochemicals from cruciferous plants protect against cancer by modulating carcinogen metabolism. *J. Nutr.* **2001**, *131*, 3027S–3033S. [CrossRef]
7. Ishida, M.; Hara, M.; Fukino, N.; Kakizaki, T.; Morimitsu, Y. Glucosinolate metabolism, functionality and breeding for the improvement of Brassicaceae vegetables. *Breed. Sci.* **2014**, *64*, 48–59. [CrossRef]
8. Petersen, B.L.; Chen, S.; Hansen, C.H.; Olsen, C.E.; Halkier, B.A. Composition and content of glucosinolates in developing *Arabidopsis thaliana*. *Planta* **2002**, *214*, 562–571. [CrossRef]
9. Brown, P.D.; Tokuhisa, J.G.; Reichelt, M.; Gershenzon, J. Variation of glucosinolate accumulation among different organs and developmental stages of *Arabidopsis thaliana*. *Phytochemistry* **2003**, *62*, 471–481. [CrossRef]
10. Nour-Eldin, H.H.; Halkier, B.A. Piecing together the transport pathway of aliphatic glucosinolates. *Phytochem. Rev.* **2009**, *8*, 53–67. [CrossRef]
11. Jørgensen, M.E.; Nour-Eldin, H.H.; Halkier, B.A. Transport of defence compounds from source to sink: Lessons learned from glucosinolates. *Trends Plant Sci.* **2015**, *20*, 508–514. [CrossRef]
12. Burow, M.; Halkier, B.A. How does a plant orchestrate defense in time and space? Using glucosinolates in Arabidopsis as case study. *Curr. Opin. Plant Biol.* **2017**, *38*, 142–147. [CrossRef]
13. Maruyama-Nakashita, A.; Nakamura, Y.; Tohge, T.; Saito, K.; Takahashi, H. *Arabidopsis* SLIM1 is a central transcriptional regulator of plant sulfur response and metabolism. *Plant Cell* **2006**, *18*, 3235–3251. [CrossRef] [PubMed]
14. Wittstock, U.; Burow, M. Glucosinolate breakdown in Arabidopsis: Mechanism, regulation and biological significance. *Arab. Book* **2010**, *8*, e0134. [CrossRef]
15. Guo, R.; Shen, W.; Qian, H.; Zhang, M.; Liu, L.; Wang, Q. Jasmonic acid and glucose synergistically modulate the accumulation of glucosinolates in *Arabidopsis thaliana*. *J. Exp. Bot.* **2013**, *64*, 5707–5719. [CrossRef]
16. Miao, H.; Cai, C.; Wei, J.; Huang, J.; Chang, J.; Qian, H.; Zhang, X.; Zhao, Y.; Sun, B.; Wang, B.; et al. Glucose enhances indolic glucosinolate biosynthesis without reducing primary sulfur assimilation. *Sci. Rep.* **2016**, *6*, 31854. [CrossRef] [PubMed]
17. Clay, N.K.; Adio, A.M.; Denoux, C.; Jander, G.; Ausubel, F.M. Glucosinolate metabolites required for an Arabidopsis innate immune response. *Science* **2009**, *323*, 95–101. [CrossRef]
18. Bednarek, P.; Pislewska-Bednarek, M.; Svatos, A.; Schneider, B.; Doubsky, J.; Mansurova, M.; Humphry, M.; Consonni, C.; Panstruga, R.; Sanchez-Vallet, A.; et al. A glucosinolate metabolism pathway in living plant cells mediates broad-spectrum antifungal defense. *Science* **2009**, *323*, 101–106. [CrossRef]
19. Mewis, I.; Appel, H.M.; Hom, A.; Raina, R.; Schultz, J.C. Major signaling pathways modulate Arabidopsis glucosinolate accumulation and response to both phloem-feeding and chewing insects. *Plant Physiol.* **2005**, *138*, 1149–1162. [CrossRef]
20. Maruyama-Nakashita, A. Metabolic changes sustain the plant life in low-sulfur environments. *Curr. Opin. Plant Biol.* **2017**, *39*, 144–151. [CrossRef]
21. Falk, K.L.; Tokuhisa, J.G.; Gershenzon, J. The effect of sulfur nutrition on plant glucosinolate content: Physiology and molecular mechanisms. *Plant Biol.* **2007**, *9*, 573–581. [CrossRef] [PubMed]
22. Aarabi, F.; Kusajima, M.; Tohge, T.; Konishi, T.; Gigolashvili, T.; Takamune, M.; Sasazaki, Y.; Watanabe, M.; Nakashita, H.; Fernie, A.R.; et al. Sulfur-deficiency-induced repressor proteins optimize glucosinolate biosynthesis in plants. *Sci. Adv.* **2016**, *2*, e1601087. [CrossRef]
23. Yokota-Hirai, M.; Sugiyama, K.; Sawada, Y.; Tohge, T.; Obayashi, T.; Suzuki, A.; Araki, R.; Sakurai, N.; Suzuki, H.; Aoki, K.; et al. Omics-based identification of *Arabidopsis* Myb transcription factors regulating aliphatic glucosinolate biosynthesis. *Proc. Nat. Acad. Sci. USA* **2007**, *104*, 6478–6483. [CrossRef]
24. Gigolashvili, T.; Yatusevich, R.; Berger, B.; Muller, C.; Flugge, U.I. The R2R3-MYB transcription factor HAG1/MYB28 is a regulator of methionine-derived glucosinolate biosynthesis in *Arabidopsis thaliana*. *Plant J.* **2007**, *51*, 247–261. [CrossRef]
25. Shibagaki, N.; Rose, A.; McDermott, J.P.; Fujiwara, T.; Hayashi, H.; Yoneyama, T.; Davies, J.P. Selenate-resistant mutants of *Arabidopsis thaliana* identify *Sultr1;2*, a sulfate transporter required for efficient transport of sulfate into roots. *Plant J.* **2002**, *29*, 475–486. [CrossRef]
26. Yoshimoto, N.; Takahashi, H.; Smith, F.W.; Yamaya, T.; Saito, K. Two distinct high-affinity sulfate transporters with different inducibilities mediate uptake of sulfate in *Arabidopsis* roots. *Plant J.* **2002**, *29*, 465–473. [CrossRef]

27. Yoshimoto, N.; Inoue, E.; Watanabe-Takahashi, A.; Saito, K.; Takahashi, H. Posttranscriptional regulation of high-affinity sulfate transporters in Arabidopsis by sulfur nutrition. *Plant Physiol.* **2007**, *145*, 378–388. [CrossRef]
28. Maruyama-Nakashita, A.; Inoue, E.; Watanabe-Takahashi, A.; Yamaya, T.; Takahashi, H. Transcriptome profiling of sulfur-responsive genes in Arabidopsis reveals global effect on sulfur nutrition on multiple metabolic pathways. *Plant Physiol.* **2003**, *132*, 597–605. [CrossRef]
29. Yamaguchi, C.; Takimoto, Y.; Ohkama-Ohtsu, N.; Hokura, A.; Shinano, T.; Nakamura, T.; Suyama, A.; Maruyama-Nakashita, A. Effects of cadmium treatment on the uptake and translocation of sulfate in *Arabidopsis thaliana*. *Plant Cell Physiol.* **2016**, *57*, 2353–2366. [CrossRef]
30. Kolár, J.; Senková, J. Reduction of mineral nutrient availability flowering of Arabidopsis thaliana. *J. Plant Physiol.* **2008**, *165*, 1601–1609. [CrossRef]
31. Kazan, K.; Lyons, R. The link between flowering time and stress tolerance. *J. Exp. Bot.* **2016**, *67*, 47–60. [CrossRef]
32. Andersen, T.G.; Nour-Eldin, H.H.; Fuller, V.L.; Olsen, C.E.; Burow, M.; Halkier, B.A. Integration of biosynthesis and long-distance transport establish organ-specific glucosinolate profiles in vegetative *Arabidopsis*. *Plant Cell* **2013**, *8*, 3133–3145. [CrossRef]
33. Nour-Eldin, H.H.; Andersen, T.G.; Burow, M.; Madsen, S.R.; Jørgensen, M.E.; Olsen, C.E.; Dreyer, I.; Hedrich, R.; Geiger, D.; Halkier, B.A. NRT/PTR transporters are essential for translocation of glucosinolates defence compounds to seeds. *Nature* **2012**, *488*, 531–534. [CrossRef]
34. Hirai, M.Y.; Fujiwara, T.; Chino, M.; Naito, S. Effects of sulfate concentrations on the expression of a soybean seed storage protein gene and its reversibility in transgenic *Arabidopsis thaliana*. *Plant Cell Physiol.* **1995**, *36*, 1331–1339.
35. Fujiwara, T.; Hirai, M.Y.; Chino, M.; Komeda, Y.; Naito, S. Effects of sulfur nutrition on expression of the soybean seed storage protein genes in transgenic petunia. *Plant Physiol.* **1992**, *99*, 263–268. [CrossRef]
36. Matsuda, F.; Yonekura-Sakakibara, K.; Niida, R.; Kuromori, T.; Shinozaki, K.; Saito, K. MS/MS spectral tag-based annotation of non-targeted profile of plant secondary metabolites. *Plant J.* **2009**, *57*, 555–577. [CrossRef]

*Article*

# Common Bean (*Phaseolus vulgaris* L.) Accumulates Most *S*-Methylcysteine as Its $\gamma$-Glutamyl Dipeptid

**Elham Saboori-Robat [1,2], Jaya Joshi [1,3,4], Aga Pajak [1], Mahmood Solouki [2], Motahhareh Mohsenpour [5], Justin Renaud [1] and Frédéric Marsolais [1,3,*]**

1 Genomics and Biotechnology, London Research and Development Centre, Agriculture and Agri-Food Canada, London, ON N5V 4T3, Canada; saboorielham@gmail.com (E.S.-R.); jayajoshi20@gmail.com (J.J.); Aga.Pajak@canada.ca (A.P.); Justin.Renaud@canada.ca (J.R.)

2 Department of Plant Breeding and Biotechnology, Faculty of Agriculture, University of Zabol, Zabol 538-98615, Iran; mahmood.solouki@uoz.ac.ir

3 Department of Biology, University of Western Ontario, London, ON N6A 3K7, Canada

4 Horticultural Sciences Department, University of Florida, Gainesville, FL 32611, USA

5 Agricultural Biotechnology Research Institute of Iran (ABRII), Agricultural Research Education and Extension Organization (AREEO), Karaj 31585-845, Iran; mthrhm@abrii.ac.ir

* Correspondence: Frederic.Marsolais@canada.ca; Tel.: +1-519-953-6718

Received: 26 March 2019; Accepted: 12 May 2019; Published: 14 May 2019

**Abstract:** The common bean (*Phaseolus vulgaris*) constitutes an excellent source of vegetable dietary protein. However, there are sub-optimal levels of the essential amino acids, methionine and cysteine. On the other hand, *P. vulgaris* accumulates large amounts of the $\gamma$-glutamyl dipeptide of *S*-methylcysteine, and lower levels of free *S*-methylcysteine and *S*-methylhomoglutathione. Past results suggest two distinct metabolite pools. Free *S*-methylcysteine levels are high at the beginning of seed development and decline at mid-maturation, while there is a biphasic accumulation of $\gamma$-glutamyl-*S*-methylcysteine, at early cotyledon and maturation stages. A possible model involves the formation of *S*-methylcysteine by cysteine synthase from *O*-acetylserine and methanethiol, whereas the majority of $\gamma$-glutamyl-*S*-methylcysteine may arise from *S*-methylhomoglutathione. Metabolite profiling during development and in genotypes differing in total *S*-methylcysteine accumulation showed that $\gamma$-glutamyl-*S*-methylcysteine accounts for most of the total *S*-methylcysteine in mature seed. Profiling of transcripts for candidate biosynthetic genes indicated that *BSAS4;1* expression is correlated with both the developmental timing and levels of free *S*-methylcysteine accumulated, while homoglutathione synthetase (*hGS*) expression was correlated with the levels of $\gamma$-glutamyl-*S*-methylcysteine. Analysis of *S*-methylated phytochelatins by liquid chromatography and high resolution tandem mass spectrometry revealed only small amounts of homophytochelatin-2 with a single *S*-methylcysteine. The mitochondrial localization of phytochelatin synthase 2—predominant in seed, determined by confocal microscopy of a fusion with the yellow fluorescent protein—and its spatial separation from *S*-methylhomoglutathione may explain the lack of significant accumulation of *S*-methylated phytochelatins.

**Keywords:** *Phaseolus vulgaris*; common bean; *S*-methylcysteine; homoglutathione; phytochelatin synthase; cysteine; methionine

## 1. Introduction

The common bean (*Phaseolus vulgaris* L.) is an important source of protein, dietary fiber, complex carbohydrates, vitamins, minerals and phenolic compounds [1–5]. However, its nutritional quality is restricted by a low concentration of the essential sulfur amino acids, methionine and cysteine. A large number of studies have focused on improving sulfur-containing amino acids in crops by transgenic development [6–10], synthetic protein synthesis [11] or traditional breeding [12]. Major seed proteins present in the common bean, including the 7S globulin phaseolin and lectin phytohaemagglutinin, have a low concentration of methionine and cysteine. On the other hand, the common bean accumulates non-protein sulfur amino acid derivatives such as γ–glutamyl-*S*-methylcysteine [13,14], *S*-methylcysteine [15–17] and *S*-methylhomoglutathione [18].

Two genetically related lines of the common bean, SARC1 and SMARC1N-PN1 differ in their composition of major storage proteins [19]. SMARC1N-PN1 is deficient in phaseolin and major lectins, through introgressions from *Phaseolus coccineus* and Great Northern 1140, respectively. SARC1 contains the lectin arcelin-1 introduced from a wild accession. SMARC1N-PN1 has an increased concentration of methionine and cysteine, by 10 and 70%, respectively. This increase occurs largely at the expense of total *S*-methylcysteine measured after acid hydrolysis, which is reduced by 70% [20]. This non-protein amino acid cannot replace methionine or cysteine in the diet [21]. Shifting sulfur from the non-protein amino acid pool to methionine and cysteine could be an effective strategy for protein quality improvement.

The biosynthesis of *S*-methylcysteine likely intersects with sulfur metabolism, particularly of cysteine. Sulfur is an essential macronutrient which plays a significant role in plant metabolism, increases yield and gives rise to the sulfur amino acids cysteine and methionine [22]. The most abundant environmental source of sulfur is sulfate ($SO_4^{2-}$). Utilizing this essential nutrient requires a series of steps for incorporation into metabolically active forms. This includes uptake of sulfate from the environment, reduction to sulfide and synthesis of cysteine [23–26]. Delivery of adequate sulfur to seed tissues is needed for maximizing yield and to improve protein quality [22]. During cysteine biosynthesis, several enzymes are active in developing seeds of the common bean (Figure 1) [18,27]. Serine acetyltransferase catalyzes the conversion of serine to *O*-acetylserine using acetyl-CoA [28,29]. Cysteine synthase is part of the β-substituted alanine synthase (BSAS) family, along with enzymes that use cysteine and cyanide to produce β-cyanoalanine [30]. One fate of cysteine is homoglutathione that is formed via a two-step process [31,32]. γ-Glutamylcysteine is synthesized from cysteine by glutamate–cysteine ligase, and homoglutathione from γ-glutamylcysteine by homoglutathione synthetase. Another fate of sulfur is to form homophytochelatins, through the reaction of phytochelatin synthase with homoglutathione as its substrate. Phytochelatins are cysteine-rich polypeptides that chelate toxic metals [33,34].

From past studies, a model emerges for possible pathways of *S*-methylcysteine biosynthesis. There appears to be two different metabolite pools, with a different pattern of temporal accumulation. Free *S*-methylcysteine concentration is relatively high at early developmental stages, and rapidly declines at mid-maturation, whereas γ–glutamyl-*S*-methylcysteine rapidly accumulates during the early cotyledon and maturation stages [35]. *BSAS4;1* represents the dominant cytosolic cysteine synthase expressed in developing seeds [36]. As can be inferred from data in Arabidopsis [37], this enzyme might be involved in the condensation of *O*-acetylserine and methanethiol, produced by methionine γ-lyase, to form *S*-methylcysteine (Figure 1). The presence of *S*-methylhomoglutathione suggests that a pathway similar to that of the alk(en)yl sulfoxides in *Allium* [38] might be involved in the formation of γ–glutamyl-*S*-methylcysteine in *P. vulgaris*. Since homoglutathione biosynthesis takes place in plastids, such a biosynthetic mechanism would ensure an absence of interference with the cytosolic pathway of cysteine biosynthesis, predominant in seed [39,40]. This hypothesis is consistent with the depletion of *O*-acetylserine in SMARC1N-PN1, associated with increased cysteine concentration [27].

In the present study, the different forms of *S*-methylcysteine and its possible precursors, including homoglutathione and *S*-methylhomoglutathione, were profiled during seed development and between genotypes that differ in total *S*-methylcysteine concentration. The relationship between metabolite levels and transcript levels of several candidate genes, including *BSAS4;1*, glutamate–cysteine ligase (*GCL1*), homoglutathione synthetase (*hGS*) and phytochelatin synthase (*PCS2*) was evaluated. The subcellular localization of PCS2 was determined and the possible presence of *S*-methylated phytochelatins analyzed by liquid chromatography and tandem mass spectrometry (MS/MS). The results of these experiments provide further insight into *S*-methylcysteine biosynthesis in *P. vulgaris*.

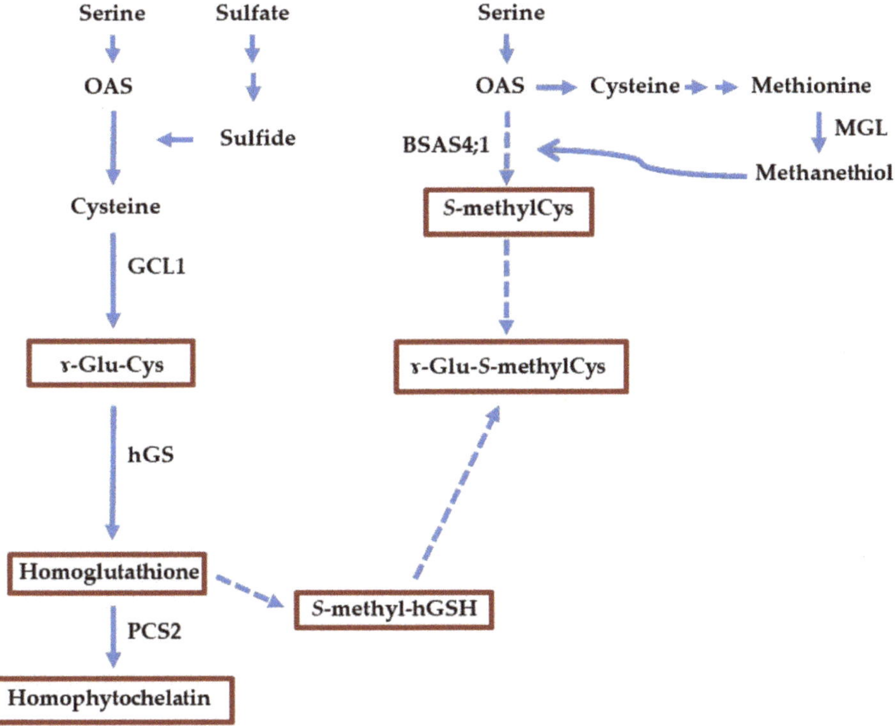


**Figure 1.** Possible biosynthetic pathway of sulfur amino acids in seed of the common bean. Solid arrows indicate steps taking place in seeds based on gene expression analysis. Multiple arrows refer to multiple steps. Broken arrows represent hypothetical steps. Metabolites in boxes are related to *S*-methylcysteine and were profiled in this study. The predominant pathway of cysteine biosynthesis in seed is cytosolic and *BSAS4;1* is the main cytosolic cysteine synthase expressed in seed. Formation of free *S*-methylcysteine by cysteine synthase is inferred from data in Arabidopsis. The presence of *S*-methylhomoglutathione suggests a pathway similar to that of the *S*-alk(enyl) sulfoxides in *Allium* leading to the formation of γ–glutamyl-*S*-methylcysteine (see text for details). OAS: *O*-acetylserine; γ–Glu-Cys, γ-glutamylcysteine; γ–Glu-*S*-methylCys, γ–glutamyl-*S*-methylcysteine; *S*-methyl-hGSH, *S*-methylhomoglutathione; BSAS4;1, β-substituted alanine synthase 4;1; PCS2, phytochelatin synthase 2; GCL1, glutamate–cysteine ligase 1; hGS, homoglutathione synthetase; MGL, methionine γ-lyase.

## 2. Results

### 2.1. Analysis of Sulfur Metabolite Profiles During Seed Development

To obtain more information on the accumulation of the different forms of *S*-methylcysteine and its precursors, free amino acids were profiled at seven developmental stages in BAT93 seed by high pressure liquid chromatography (HPLC) after derivatization with 3-mercaptopropionic acid and *O*-phthalaldehyde (Table 1). Developmental stages are designated after Walbot et al. [41]. Stages IV–early cotyledon to VI–early maturation are characterized by the presence of storage protein transcripts, while storage proteins accumulate from stages V–mid-cotyledon to VII–mid-maturation [42]. Stage VI–early maturation is marked by a decrease in photosynthetic capacity, with white cotyledons by stage VII–mid-maturation. Amino acids had been profiled previously using a different methodology involving derivatization with phenyl isothiocyanate [35]. This time, additional sulfur metabolites were quantified, namely homoglutathione and *S*-methylhomoglutathione, as possible precursors to γ-glutamyl-*S*-methylcysteine. Attempts to measure the homoglutathione precursor, γ-glutamyl-cysteine, were unsuccessful. Its levels were too low for accurate quantification using the present method. Accumulation of γ-glutamyl-*S*-methylcysteine followed a biphasic curve, as previously observed, with a steady rise from stage IV–early cotyledon until stage VI–early maturation, followed by a lag, and resumption of accumulation at stage VIII–late maturation. The levels of free *S*-methylcysteine were high at the first two stages of seed development, followed by a rapid decline during the late cotyledon and maturation stages. Concentrations of *S*-methylhomoglutathione were also higher at the beginning of seed development, when γ-glutamyl-*S*-methylcysteine accumulation is most rapid. Concentrations of homoglutathione were in the same range as those of *S*-methylhomoglutathione.

**Table 1.** Free amino acid profiles in developing seeds of BAT93. Metabolites related to *S*-methylcysteine are highlighted.

| Average Seed Weight (mg)/ Develop-Mental Stage | 25/IV-Cotyledon | 52/V-Cotyledon | 105/V-Cotyledon | 157/VI-Maturation | 200/VI-Maturation | 348/VIII-Maturation | 160/Mature |
|---|---|---|---|---|---|---|---|
| Asp | 1.6 ± 0.1 | 0.80 ± 0.10 | 0.79 ± 0.02 | 1.2 ± 0.1 | 1.3 ± 0.1 | 1.4 ± 0.2 | 3.4 ± 0.4 |
| Glu | 4.5 ± 0.2 | 4.1 ± 0.2 | 4.0 ± 0.1 | 3.3 ± 0.1 | 2.9 ± 0.1 | 2.8 ± 0.1 | 4.4 ± 0.3 |
| hGSH | 0.06 ± 0.01 | 0.06 ±0.01 | 0.06 ± 0.01 | 0.12 ± 0.01 | 0.12 ± 0.02 | 0.11 ± 0.01 | 0.35 ± 0.06 |
| Ser | 1.3 ± 0.1 | 0.77 ± 0.09 | 0.56 ± 0.01 | 0.50 ± 0.01 | 0.39 ± 0.01 | 0.52 ± 0.12 | 0.24 ± 0.02 |
| His | 1.8 ± 0.1 | 1.3 ± 0.2 | 0.84 ± 0.02 | 0.98 ± 0.07 | 0.11 ± 0.01 | 0. 45 ± 0.29 | 2.5 ± 0.2 |
| γ-Glu-*S*-methylCys | 2.4 ± 0.1 | 7.0 ± 0.1 | 9.3 ± 0.2 | 10.2 ± 0.2 | 8.6 ± 0.3 | 11.4 ± 2.7 | 38.0 ± 2.0 |
| Gly | 0.33 ± 0.01 | 0.30 ± 0.01 | 0.25 ± 0.01 | 0.25 ± 0.01 | 0.24 ± 0.01 | 0.27 ± 0.03 | 0.58 ± 0.36 |
| Thr | 3.2 ± 0.1 | 3.3 ± 0.2 | 1.8 ± 0.1 | 1.6 ± 0.1 | 0.66 ± 0.02 | 1.3 ± 0.6 | 0.66 ± 0.06 |
| *S*-methylhGSH | 0.35 ± 0.02 | 0.28 ± 0.03 | 0.18 ± 0.01 | 0.17 ± 0.01 | 0.12 ± 0.01 | 0.13 ± 0.01 | 0.44 ± 0.04 |
| Arg | 5.2 ± 0.4 | 5.6 ± 0.4 | 2.9 ± 0.1 | 4.2 ± 3.2 | 0.41 ± 0.02 | 1.7 ± 1. 1 | 2.4 ± 0.2 |
| Ala | 2.6 ± 0.3 | 1.4 ± 0.1 | 1.3 ± 0. 1 | 0.82 ± 0.04 | 0.50 ± 0.02 | 0.44 ± 0.04 | 1.7± 0.1 |
| γ-Glu-Leu | 0.26 ± 0.05 | 0.39 ± 0.01 | 2.0 ± 0.1 | 1.3 ± 0.1 | 1.9 ± 0.1 | 1.1 ± 0.6 | 1.9 ± 1.4 |
| Tyr | 0.18 ± 0.03 | 0.19 ± 0.02 | 0.17 ± 0.01 | 0.13 ± 0.01 | 0.10 ± 0.01 | 0.08 ± 0.02 | 0.22 ± 0.02 |
| *S*-methylCys | 1.3 ± 0.1 | 1.3 ± 0. 1 | 0.43 ± 0.01 | 0.28 ± 0.01 | 0.13 ± 0.01 | 0.18 ± 0.05 | 0.32 ± 0.02 |
| Val | 1.8 ± 0.1 | 3.0 ± 0.3 | 0.94 ± 0.02 | 0.90 ± 0.05 | 0.41 ± 0.01 | 0.64 ± 0.21 | 0.93 ± 0.08 |
| Met | 2.9 ± 0.5 | 2.4 ± 0.1 | 0.75 ± 0.02 | 0.55 ± 0.03 | 0.24 ± 0.01 | 0.31 ± 0.07 | 0.38 ± 0.01 |
| Trp | 0.70 ± 0.09 | 0.38 ± 0.11 | 0.95 ± 0.03 | 0.66 ± 0.01 | 0.66 ± 0.01 | 0.51 ± 0.12 | 0.49 ± 0.01 |
| Phe | 0.47 ± 0.01 | 0.39 ± 0.05 | 0.20 ± 0.03 | 0.14 ± 0.01 | 0.09 ± 0.01 | 0.11 ± 0.02 | 0.37 ± 0.03 |
| Ile | 1.9 ± 0.1 | 3.7 ± 0.3 | 1.6 ± 0.1 | 0.90 ± 0.02 | 0.90 ± 0.01 | 0.77 ± 0.11 | 0.28 ± 0.04 |
| Leu | 4.0 ± 2.3 | 1.4 ± 0.5 | 2.2 ± 0.1 | 0.39 ± 0.38 | 0.36 ± 0.01 | 0.42 ± 0.06 | 0.26 ± 0.02 |
| Lys | 0.30 ± 0.02 | 0.59 ± 0.07 | 0.19 ± 0.01 | 0.27 ± 0.02 | 0.14 ± 0.07 | 0.19 ± 0.02 | 0.35 ± 0.03 |

Values presented are in nmol per mg seed weight; average ± standard deviation; $n = 3$. hGSH: homoglutathione; γ-Glu-*S*-methylCys: γ-glutamyl-*S*-methylcysteine; *S*-methylhGSH: *S*-methylhomoglutathione; *S*-methylCys: *S*-methylcysteine.

### 2.2. *Differences in Sulfur Amino Acid Concentrations between SARC1 and SMARC1N-PN1 under Sulfate Sufficient Conditions*

The same sulfur metabolites were quantified at five developmental stages between SARC1 and SMARC1N-PN1 under defined, sulfate sufficient conditions. Free amino acids had been profiled before in SARC1 and SMARC1N-PN1, however, the levels of homoglutathione and *S*-methylhomoglutathione had not been determined [27]. The work by Pandurangan et al. [43] indicated that 2 mM is a suitable sulfate-sufficient concentration, at which the two genotypes accumulate different levels of total *S*-methylcysteine after hydrolysis. As expected, levels of γ-glutamyl-*S*-methylcysteine were higher in SARC1 than SMARC1N-PN1, at four out of five developmental stages (Figure 2). Similar to what was previously observed, the levels of free *S*-methylcysteine were higher in SARC1 than SMARC1N-PN1 at early developmental stages, when concentrations are most elevated. The concentration of *S*-methylhomoglutathione was higher in SMARC1N-PN1 at maturity, consistent with a reduced use of the putative γ-glutamyl-*S*-methylcysteine precursor in this genotype. Homoglutathione concentration was higher in SARC1 at stages IV and VIII, and *S*-methylhomoglutathione concentration at stage VIII, suggesting a possible enhanced flux through these putative precursors at these developmental stages.

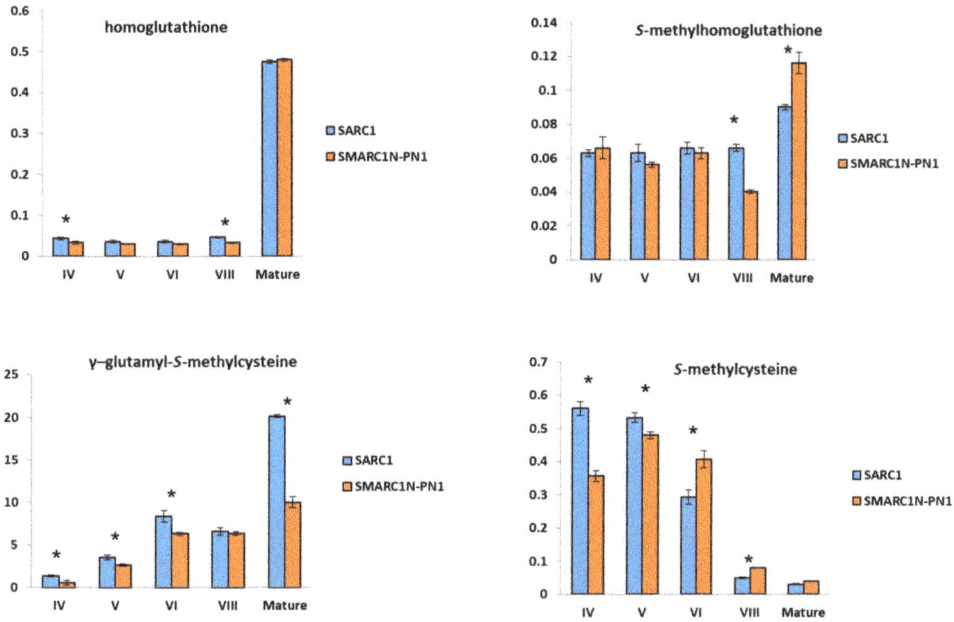


**Figure 2.** Quantification of sulfur metabolites at different developmental stages in SARC1 and SMARC1N-PN1. Concentration is expressed in nmol per mg seed weight; average ± standard deviation; $n = 3$. Asterisks indicate significant differences at *t*-test *p* value $\leq 0.01$. $n = 3$.

### 2.3. Expression Analysis of Genes Related to S-Methylcysteine and γ-Glutamyl-S-methylcysteine Biosynthesis

The expression patterns of candidate genes related with the biosynthesis or metabolism of *S*-methylcysteine or γ-glutamyl-*S*-methylcysteine, *BSAS4;1*, *GCL1*, *hGS* and *PCS2*, were examined (Figure 3). *BSAS4;1* is the predominantly expressed cytosolic cysteine synthase gene in developing seeds [36]. There are two *GCL* genes in *P. vulgaris*. *GCL2* is expressed at very low levels, whereas *GCL1* and *hGS* expression is developmentally correlated (Pearson correlation coefficient = 0.85) (visualized at https://phytozome.jgi.doe.gov) [44]. Expression of the two glutathione synthetase (*GS*) genes is marginal, which explains the low level of accumulation of glutathione in *P. vulgaris*. Among two *PCS* genes, expression of *PCS1* was very low. Attempts to detect its expression by reverse transcription quantitative polymerase chain reaction (RT-qPCR) were unsuccessful. Expression of *BSAS4;1* was higher in SARC1 at the first developmental stage, which correlates with the higher levels of free *S*-methylcysteine (Figure 3). There was also a correlation between the levels of *BSAS4;1* transcripts and free *S*-methylcysteine during development (Pearson correlation coefficient = 0.83 in SARC1 and 0.96 in SMARC1N-PN1). Expression of *hGS* was significantly higher in SARC1 at three out of four developmental stages and *GCL1* at two developmental stages. This correlates with the higher levels of γ-glutamyl-*S*-methylcysteine in this genotype. *PCS2* transcript levels were higher in SMARC1N-PN1 at stage VI, opposite to what was observed for *GCL1* and *hGS*.

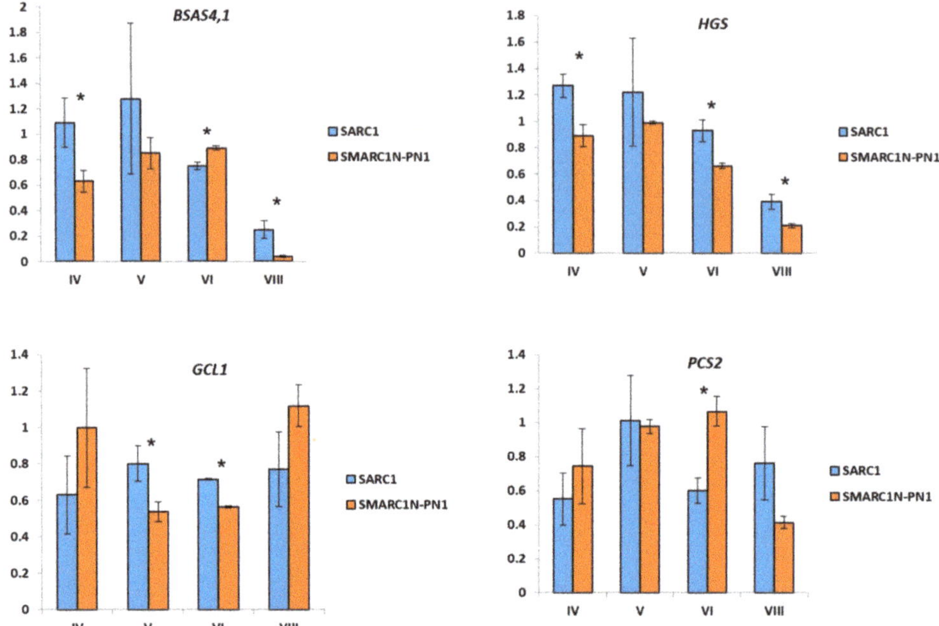


**Figure 3.** Relative expression of transcripts at different developmental stages in SARC1 and SMARC1N-PN1 determined by RT-qPCR. Average ± standard deviation. Asterisks indicate significant differences at t-test $p$ value ≤ 0.02. $n$ = 3.

*2.4. Subcellular Localization of PCS2*

*PCS2* is the major phytochelatin synthase gene expressed in developing seed. In vitro, *S*-methylglutathione can be used as substrate by phytochelatin synthase, in place of the metal thiolate complex with glutathione [45]. The subcellular localization of *PCS2* may therefore determine whether *S*-methylated phytochelatins can accumulate in seed. While analyzing the *PCS2* sequence from BAT93, a polymorphism specific to this genotype was uncovered which affects the length of the predicted PCS2 protein (Figure 4) [46]. This polymorphism was confirmed by RT-PCR and DNA sequencing. There is an insertion of a cytosine after position 109 downstream from the first start codon, as compared with the sequence from SARC1, SMARC1N-PN1 and the reference genome (accession G19833) [47]. The polymorphism results in a frameshift, such that PCS2 may only be translated from a downstream, alternative start codon, resulting in a shorter protein of 464 amino acid residues as compared with the longer PCS2 of 501 residues.

cDNAs encoding the long and short versions of *PCS2* were cloned from SARC1 and constructs were made to express C-terminal yellow fluorescent protein (YFP) fusions transiently in *Nicotiana benthamiana* epidermal cells. Figure 5 shows representative results obtained with the longer version of the protein. When PCS2-YFP was co-expressed with a CFP-tagged mitochondrial marker protein, PCS2-YFP was co-localized with the marker to the mitochondria. Similar results were obtained with the shorter version of the protein.

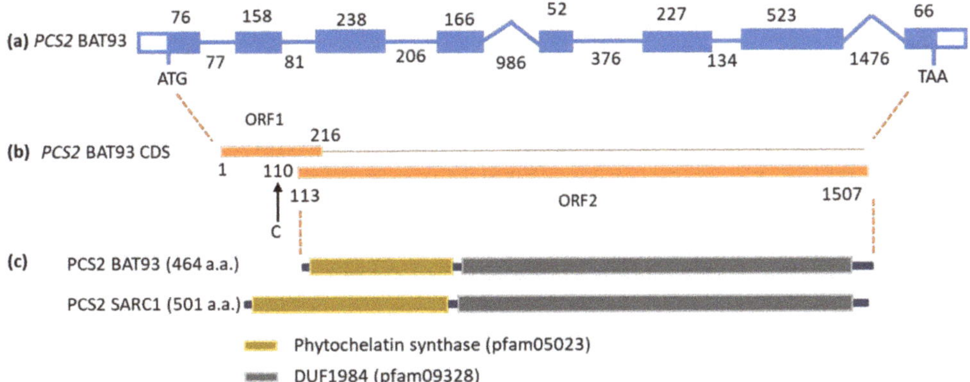


**Figure 4.** Naturally occurring variant of *PCS2* in BAT93. (**a**) Intron–exon structure of the BAT93 *PCS2* gene. The length of introns and exons is indicated starting from the first translation initiation codon and ending at the stop codon. (**b**) The gene gives rise to two open reading frames, due to the insertion of a cytosine at position 110 of the coding sequence (CDS), which results in a premature stop codon. Corresponding positions in the cDNA with respect to the first initiation codon is indicated. ORF1 encodes a predicted polypeptide of 71 amino acids. (**c**) Due to the insertion, *PCS2* encoded in BAT93 represents a shorter form translated from an alternative, downstream start codon as compared with *PCS2* encoded by SARC1, SMARC1N-PN1 and the reference bean genome, G19833. Pfam domains present in the phytochelatin synthase sequence are indicated.

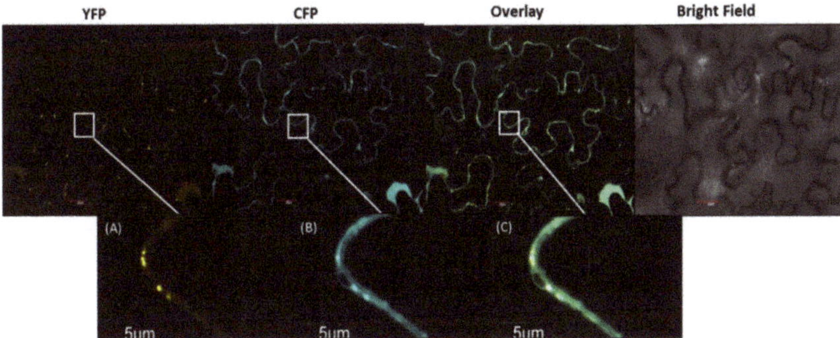


**Figure 5.** Subcellular localization of PCS2. (**a**) YFP-tagged PCS2; (**b**) CFP-tagged mitochondrial marker (Mt-CFP); (**c**) Co-localization of PCS2 with Mt-CFP. YFP: Yellow fluorescent protein; CFP: Cyan fluorescent protein.

### *2.5. Analysis of S-Methylated Phytochelatins*

To determine whether *S*-methylated phytochelatins could constitute a significant sink for *S*-methylcysteine, a systematic analysis of mature seed extracts from BAT93, SARC1 and SMARC1N-PN1 was performed by liquid chromatography and MS/MS. Similar results were obtained for the three genotypes. There are 12 distinct phytochelatins and homophytochelatins with 2–7 repeat units. However, allowing for the possibility that phytochelatins and homophytochelatins may incorporate variable numbers of *S*-methylcysteines instead of cysteines increases this number to 28 possible phytochelatins, homophytochelatins and *S*-methylated analogues (Table S1). The full MS data was screened for the theoretical *m/z* of these compounds (< 3 ppm). When putatively detected, the profiles were scrutinized for the presence of the $^{34}$S isotope and MS/MS was performed.

In these extracts, the most abundant compound based on relative peak areas was γ-glutamyl-S-methylcysteine, followed by homoglutathione and S-methylhomoglutathione. γ-Glutamylcysteine, S-methylglutathione and glutathione were present at lower concentrations. Homophytochelatin-2 and S-methylhomophytochelatin-2 with a single S-methylcysteine were detected (Figure 5). Their concentration was similar and in the same range as that of glutathione. Phytochelatin-2 was not detected (Figure 6a). A phytochelatin-2 that contains a single S-methylcysteine residue is isobaric with homophytochelatin-2 which contains an alanine residue in place of glycine and would not be distinguishable by high resolution MS. However, homophytochelatin and phytochelatin can be distinguished by their MS/MS product ions. Upon MS/MS, phytochelatins and homophytochelatins yield *m/z* 179.0486 and 193.0648 product ions, respectively. Within the analyzed samples, homophytochelatin-2 was detected and distinguished from S-methylphytochelatin-2 by observing this product ion (Figure 6b). Additionally, S-methylhomophytochelatin-2 with a single S-methyl group was detected with slightly more retention than homophytochelatin-2, as is expected by the additional methyl group (Figure 6c). Although a standard of S-methylhomophytochelatin-2 with two S-methyl cysteines was utilized, this compound was not detected in the seed extracts. No larger phytochelatin oligomers (<7) were detected within the samples by high resolution MS. Based on its low relative abundance, S-methylhomophytochelatin does not appear to constitute a major sink for S-methylcysteine.

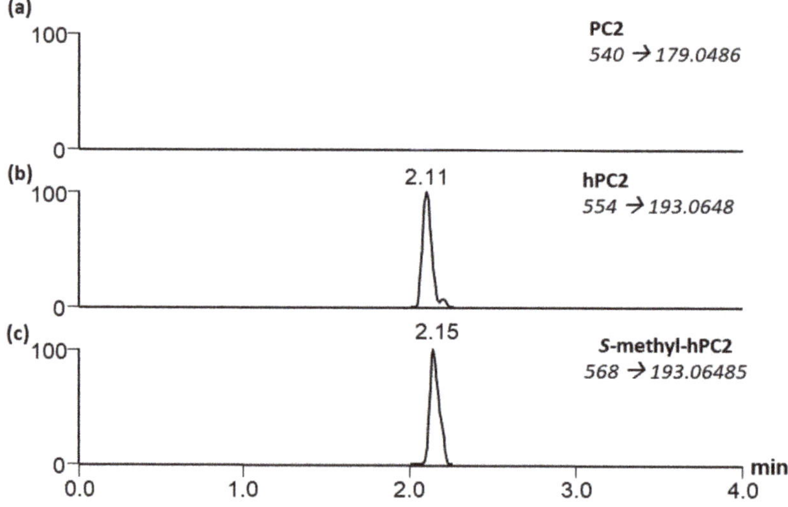


**Figure 6.** MS/MS chromatograms for monitoring (**a**) phytochelatin-2 (PC2); (**b**) homophytochelatin-2 (hPC2) and (**c**) S-methylhomophytochelatin-2 with a single S-methylcysteine (S-methyl-hPC2).

## 3. Discussion

### *3.1. Most of the S-Methylcysteine Accumulates as γ-Glutamyl-S-methylcysteine in P. vulgaris Seed*

The present study revisited the quantification of γ-glutamyl-S-methylcysteine in seed. Taylor et al. [20] reported that γ-glutamyl-S-methylcysteine accounted for only approximately 35% of total S-methylcysteine. Here, the final levels of γ-glutamyl-S-methylcysteine were about 3-fold higher than previously reported [14,35], in line with the concentration of total S-methylcysteine, measured after acid hydrolysis of mature seed flour, ranging from 14 to 35 nmol per mg seed weight [15,17,20]. We conclude that γ-glutamyl-S-methylcysteine accounts for most of the S-methylcysteine accumulated in mature seed. Giada et al. [14] determined that the average concentration of γ-glutamyl-S-methylcysteine was equal to 11 nmol per mg, with free S-methylcysteine accounting for the balance of the remaining 20%

of total S-methylcysteine measured after acid hydrolysis. In the present study, a lower concentration of free S-methylcysteine was measured. The higher levels of free S-methylcysteine measured by Giada et al. [12] may have been due to partial hydrolysis of the dipeptide during extraction, or in vivo, such as during seed storage. The difference in γ-glutamyl-S-methylcysteine levels between BAT93 and SARC1 suggests that there may be substantial genetic variability for the concentration of this metabolite (Table 1 and Figure 2). In future, it may be useful to quantify γ-glutamyl-S-methylcysteine or total S-methylcysteine in a wide range of *P. vulgaris* accessions or cultivars.

### *3.2. The Concentration of Homoglutathione or S-Methylhomoglutathione does not Appear to Limit the Accumulation of γ-Glutamyl-S-Methylcysteine*

S-Methylhomoglutathione concentration measured in mature seed of BAT93 in the present study was slightly higher than previously reported [18] and similar to that in *Vigna radiata* seeds [13]. Quantification of homoglutathione and S-methylhomoglutathione at different time points during seed development in SARC1 and SMARC1N-PN1 did not reveal major fluctuations with respect to the different levels of accumulation of the end-product, γ-glutamyl-S-methylcysteine (Figure 2). This is in contrast with the cysteine precursor, O-acetylserine, which was shown to be depleted in SMARC1N-PN1, in relation with the higher concentration of total cysteine in this genotype [27]. In BAT93, the decrease in S-methylhomoglutathione at early stages of seed development paralleled a rapid increase in γ-glutamyl-S-methylcysteine levels (Table 1).

### *3.3. Transcript Expression of BSAS4;1 and hGS is Correlated with the Accumulation of Free S-Methylcysteine and γ-Glutamyl-S-methylcysteine, Respectively*

The product of *BSAS4;1* is presumed to be directly involved in the formation of free S-methylcysteine. The developmental correlation between *BSAS4;1* transcript levels and free S-methylcysteine concentration, observed in either SARC1 or SMARC1N-PN1 (Figures 2 and 3), takes its meaning in this context. The positive genotypic correlation, observed at an early time point, when S-methylcysteine is at its highest levels, further implicates *BSAS4;1* as a plausible candidate gene (Figures 2 and 3). The higher levels of *hGS* transcript in SARC1 as compared with SMARC1N-PN1 at three out of four developmental stages (Figure 3) supports the idea that flux through homoglutathione synthetase may control, at least in part, γ-glutamyl-S-methylcysteine accumulation.

### *3.4. Mitochondrial Localization of PCS2 Prevents the Accumulation of S-methylated Phytochelatins*

The presence of S-methylhomoglutathione throughout seed development raises the question of whether S-methylphytochelatins can accumulate in P. vulgaris. Analysis of *PCS2*, the major phytochelatin synthase gene expressed in seed, revealed a polymorphism in BAT93 which would result in alternative translation initiation at a downstream start codon (Figure 4). Determination of subcellular localization by confocal microscopy demonstrated that both variants are targeted to mitochondria (Figure 5). Information on the subcellular localization of plant phytochelatin synthases is relatively limited. Arabidopsis PCS1 was shown to be present in the cytosol [48], and rice PCS1 and PCS2 were also reported to be cytosolic [49]. In the yeast *Schizosaccharomyces pombe*, phytochelatin synthase is localized in mitochondria [50]. This is logical, given that heavy metal toxicity targets mitochondrial respiration [51]. A systematic analysis of S-methylated phytochelatins in seed extracts found S-methylhomophytochelatin with a single S-methylcysteine, at low levels (Figure 6 and Table S1). The present results suggest that the localization of PCS2 in mitochondria ensures that the formation of phytochelatins does not interfere with the accumulation of γ-glutamyl-S-methycysteine.

In conclusion, our results indicate that most of the S-methylcysteine accumulates as γ-glutamyl-S-methylcysteine in mature seed. Analysis of metabolite profiles and quantitative RT-PCR data for transcripts of candidate genes of S-methylcysteine metabolism supports the hypothesis that expression of *BSAS4;1*, encoding the major cytosolic synthase in seed, may regulate the accumulation of free S-methylcysteine, whereas expression of *hGS* may regulate the accumulation of γ-glutamyl-*S*-methylcysteine, through the provision of homoglutathione and *S*-methylhomoglutathione. The mitochondrial localization of PCS2, the major phytochelatin synthase expressed in seed, is a likely explanation for the lack of substantial accumulation of *S*-methylated homophytochelatins.

## 4. Materials and Methods

### *4.1. Plant Materials and Growth Conditions*

Three genotypes of the common bean (*Phaseolus vulgaris* L.), SARC1, SMARC1N-PN1 and BAT93 were used to evaluate sulfur metabolite profiles. Seeds were sown in small trays containing vermiculite. Ten-day-old seedlings were transplanted to a bigger pot measuring 17 cm × 20 cm containing sand, perlite, and vermiculite in a 2:1:1 ratio. SARC1 and SMARC1N-PN1 plants were grown under sulfate sufficient conditions as described in previous works [43,52]. The sulfate solution included 0.2 mM $K_2SO_4$ and 1.8 mM $MgSO_4$ with other nutrients as follows: 4.5 mM $Ca(NO_3)_2$, 1.7 mM $K_2HPO_4$, 4 μM $MnSO_4.H_2O$, 5 μM $H_3BO_3$, 10 μM FeEDTA, 0.25 μM $CuSO_4.5H_2O$, 1 μM $ZnSO_4.7H_2O$, and 0.2 μM $Na_2MoO_4.2H_2O$. The BAT93 genotype was grown as previously described [35]. The pots used in the study were placed in a randomized block design with 15 plants per genotype. Each sample consisted of three biological replicates. A biological replicate consisted of a pool of 10 seeds collected randomly from 15 plants per genotype. Plants were grown in a greenhouse with 16 h light (300–400 μmol photons $m^{-2}$ $s^{-1}$) and 8 h dark with a temperature cycling between 18 and 24 °C.

### *4.2. Extraction and Quantification of Free Amino Acids*

The frozen seeds in liquid nitrogen were ground to a fine powder using a mortar and pestle. Replicate samples consisted of independent pools of ten seeds of which 100 mg were used for extraction. Sample extraction was carried out in ethanol/water (70:30), optimal for sulfur containing γ-glutamyl dipeptides [20]. Standards for γ-glutamylcysteine, γ-glutamyl-*S*-methylcysteine, *S*-methylhomoglutathione, homoglutathione and *S*-methylhomophytochelatin-2 with two *S*-methylcysteines were from Bachem (Torrance, CA, USA). Homophytochelatin-2 and phytochelatin-3 standards were from Anaspec (Fremont, CA, USA). *S*-Methylglutathione was from Millipore Sigma (Oakville, ON, Canada). Quantification of free amino acids was performed after derivatization with *O*-phthalaldehyde and mercaptopropionic acid using an Agilent 1260 Infinity HPLC system (Mississauga, ON, Canada) as described in Jafari et al. [53].

### *4.3. RNA Extraction and Quantitative RT-PCR*

Sequence analyses were performed using *Phaseolus vulgaris* v2.1, DOE-JGI and USDA-NIFA (http://phytozome.jgi.doe.gov/). Total RNA from the developmental seed samples was extracted using a modified lithium chloride method [54]. RNA was quantified with a NanoDrop 1000 (Thermo Fisher Scientific, https://www.thermofisher.com) and its quality evaluated from $A_{260}/A_{280}$ ratio. DNase I was used to remove DNA contamination that may have happened during the RNA extraction (Thermo Fisher Scientific). RNA quality was evaluated prior to cDNA synthesis by using gel electrophoresis on a 1% (w/v) agarose gel. cDNA synthesis and PCR procedures were performed using Thermoscript RT-PCR System (Thermo Fisher Scientific) and SsoFast EvaGreen Supermix (Bio-Rad Laboratories, Mississauga, ON, Canada), respectively. Primers used for RT-qPCR are listed in Table 2. Reactions contained 2 μL of cDNA diluted 5-fold, primers at a concentration of 0.5 μM and 5 μL SsoFast EvaGreen Supermix in a final volume of 10 μL. Samples were run in Hard-Shell 96-well clear PCR plates (Bio-Rad Laboratories). In each plate, three biological replicates, with three technical replicates per biological replicate and controls without template were run for developmental seed samples. The PCR program consisted in an initial step of 2 min at 95 °C, followed by 35 cycles of 30 s at 94 °C and 30 s at 60 °C. CFX Maestro software was used to analyze the RT-qPCR data. After completion of the reactions, the threshold fluorescence ($C_q$) value for each reaction was calculated. All data were normalized using the expression of the ubiquitin reference gene. The specificity of primer pairs was confirmed by melt curve analysis in comparison with controls without template. PCR efficiency was calculated from a standard curve of $C_q$ versus the logarithm of starting template quantity. Each assay was optimized so that the efficiency ranged between 98% and 108%, with a coefficient of determination ($R^2$) > 0.98.

**Table 2.** Primer sequences used for RT-qPCR.

| Gene | Accession Number | Forward Primer Sequence (5′–3′) | Reverse Primer Sequence (5′–3′) |
|---|---|---|---|
| *BSAS4;1* | Phvul.007G185200.2 | GCGGCTGATGGTGGTTATATTT | CACCAGTTCCTATCCCTGCA |
| *GCL1* | Phvul.002G289200.1 | TGCATTACCAGCACTTTGGG | ACATCTGTCTTTCTTCTGGGGT |
| *hGS* | Phvul.006G094500.1 | CCGCAAAGAGAAGGAGGAGG | TGCTGGAAAAGTGGCTGGAA |
| *PCS2* | Phvul.001G162700.1 | TTCTCTGGGAGGCAATGAGC | ACCTTCATCTCTACAACTCACAGT |
| *Ubiquitin* | Phvul.007G05260.1 | ACAGCTGGAGGATGAAAGGA | GTCCGAACTCTCCACCTCAA |

### 4.4. Cloning of *PCS2* for Subcellular Localization

The coding sequence of *PCS2* was PCR amplified using attB1 and attB2 site-containing Gateway primers from SARC1 cDNA. For the full-length version, primers were, PVPCS2-ATTB1YFPF, 5′-GGGGACAAGTTTGTACAAAAAAGCAGGCTTCATGTGCATGGCGAACCCAG-3′ and PVPCS2-ATTB1YFPR, 5′-GGGGACCACTTTGTACAAGAAAGCTGGGTTCCGCTGCACAGTCCAGATTGCT-3′. For the short variant using the alternative downstream start codon, the forward primer was PCS2-ATTB1YFPF, 5′-GGGGACAAGTTTGTACAAAAAAGCAGGCTTCATGGAAGCCTTCTTCAAG C-3′. The PCR products were inserted into the entry vector pDONR-Zeo (Thermo Fisher Scientific). The integrity of the PCR fragments was confirmed with Sanger sequencing. Using Gateway technology, LR recombination reactions were performed with entry clones pDONR-Zeo-PCS2fl and pDONR-Zeo-PCS2tr and destination vector pEarleyGate101 [55]. The final expression constructs (pEG101-PCS2fl and pEG101-PCS2tr) were transformed into *Agrobacterium tumefaciens* strain GV3101.

### 4.5. Plant Transformation and Confocal Observations

*Nicotiana benthamiana* plants were grown for 6 weeks and used for transient expression. Plants were grown in a growth chamber at 22 °C with a 16 h photoperiod (110 µmol photons $m^{-2}$ $s^{-1}$). Plants were always watered with the water soluble fertilizer (N:P:K = 20:8:20) at 0.25 g/L (Plant Products, Brampton, ON, Canada). *Agrobacterium tumefaciens* cultures were grown to an optical density at 600 nm ($OD_{600}$) of 0.5–0.8, and collected by centrifugation at 3,000 × *g* for 30 min. The pellets were resuspended in Gamborg's solution (3.2 g/L Gamborg's B5 medium and vitamins, 20 g/L sucrose, 10 mM 2-(*N*-morpholino)ethanesulfonic acid pH 5.6, 200 µM acetosyringone) to a final $OD_{600}$ of 1, followed by incubation at room temperature (21 °C) with gentle agitation for 1 h. For co-infiltration, the mitochondrial protein (cytochrome oxidase with CFP) was mixed with YFP construct (pEG101-PCS2) in a 1:1 ratio [56]. The suspension was used for co-infiltration of the abaxial side of the leaf with a 1 mL syringe. Three to four days post infiltration, the abaxial epidermis of the leaves was observed using an OLYMPUS FV1200 Laser Scanning Microscope (https://www.olympus-lifescience.com/). A 60 × water immersion objective was used at excitation wavelengths of 514 and 458 nm. The fluorescence signals were detected using an emission spectra of 530–560 nm for YFP and 470-500 nm for CFP. Sequential Scan Tool, which records fluorescence in a sequential fashion, was used for studying co-localization of PCS2 with marker protein.

### 4.6. Analysis of S-methylated Phytochelatins

Extraction of phytochelatins from mature seed tissue was performed according to a published method with slight modifications [57]. The mature seed samples were ground with a Kleco ball mill (Garcia Machine, Visalia, CA, USA). One hundred mg of powder was mixed with 1 mL of cold (4 °C) 100 mM dithiothreitol in a polypropylene centrifuge tube. The suspension was vortexed for 1 min, and then sonicated for 5 min at room temperature. The extract was precipitated by centrifugation at 15,000× *g* at 4 °C for 20 min. The supernatants were transferred to a polypropylene centrifuge tube and the centrifugation repeated one more time. The supernatants were filtered with PTFE syringe filters (0.22 µm) into 2 mL amber glass HPLC vials.

The extracts were screened using a Q-Exactive Quadrupole Orbitrap mass spectrometer (Thermo Fisher Scientific), coupled to an Agilent 1290 high-performance liquid chromatography (HPLC) system with a Zorbax Eclipse Plus RRHD C18 column maintained at 35 °C (2.1 × 50 mm, 1.8 µm; Agilent). Mobile phase A (0.1% formic acid in LC-MS grade $H_2O$, Thermo Fisher Scientific) began at 100% and was held for 1.25 min. Mobile phase B (0.1% formic acid in LC-MS grade acetonitrile, Thermo Fisher Scientific) was then increased to 50% over 1.75 min, and 100% over 0.5 min. Mobile phase B was maintained at 100% for 1.5 min and returned to 0% over 0.5 min. The following heated electrospray ionization (HESI) parameters were used in positive ionization mode: spray voltage, 3.9 kV; capillary temperature, 250 °C; probe heater temperature, 450 °C; sheath gas, 30 arbitrary units; auxiliary gas, 8 arbitrary units; and S-Lens RF level, 60%. High resolution, full MS was used to detect any possible phytochelatins and homophytochelatins (PC2-7) by accurate mass (Table S1), while MS/MS scans performed concurrently monitored PC2 (*m/z* 540 → 179.0486), homoPC2 (*m/z* 554 → 193.0648) and *S*-methyl-homoPC2 (*m/z* 568 → 193.0648) all at normalized collision energies (NCE) of 24. The full MS scans were performed at 70,000 resolution over a mass range of 100–2000 *m/z*; automatic gain control (AGC) target and maximum injection time (max IT) was 3 × 106 and 256 msecond respectively. The MS/MS scans were performed at 17,500 resolution AGC target and max IT were 3 × 106 and 64 msecond respectively. Data were analyzed and all theoretical masses were calculated with Xcalibur™ software.

### 4.7. Statistical Analysis

The experiments were carried out as a factorial in a completely randomized experimental design with three replications. *t*-Test was performed using IBM SPSS® software (Armonk, NY, USA).

**Supplementary Materials:** Table S1. Monitoring of phytochelatin ions in MS/MS data.

**Author Contributions:** E.S.-R., A.P. and J.R. performed the experiments. J.J. and F.M. designed the research. M.S., M.M. and F.M. supervised E.S.-R. E.S.-R., A.P., J.R. and F.M. analyzed the data. E.S.-R., J.R. and F.M. wrote the manuscript.

**Funding:** This research was funded by Agriculture and Agri-Food Canada (project no. J-001331). E.S.-R. received funding from the Ministry of Science, Research and Technology of Iran. J.J. was the recipient of the Dr. René Roth Memorial Award during her Ph. D. studies in the Department of Biology of the University of Western Ontario.

**Acknowledgments:** We thank Mark Bernards for acting as J.J.'s co-supervisor during her Ph. D. program and Alex Molnar for help with figures.

**Conflicts of Interest:** The authors declare no conflict of interest.

## References

1. Campos-Vega, R.; Reynoso-Camacho, R.; Pedraza-Aboytes, G.; Acosta-Gallegos, J.; Guzman-Maldonado, S.; Paredes-Lopez, O.; Oomah, B.; Loarca-Piña, G. Chemical composition and in vitro polysaccharide fermentation of different beans (*Phaseolus vulgaris* L.). *J. Food Sci.* **2009**, *74*, T59–T65. [CrossRef] [PubMed]
2. Mojica, L.; de Mejía, E.G. Characterization and comparison of protein and peptide profiles and their biological activities of improved common bean cultivars (*Phaseolus vulgaris* L.) from Mexico and Brazil. *Plant Foods Hum. Nutr.* **2015**, *70*, 105–112. [CrossRef] [PubMed]
3. Ganesan, K.; Xu, B. Polyphenol-rich dry common beans (*Phaseolus vulgaris* L.) and their health benefits. *Int. J. Mol. Sci.* **2017**, *18*, 2331. [CrossRef]
4. Chen, Y.; Zhang, H.; Liu, R.; Mats, L.; Zhu, H.; Pauls, K.P.; Deng, Z.; Tsao, R. Antioxidant and anti-inflammatory polyphenols and peptides of common bean (*Phaseolus vulgaris* L.) milk and yogurt in Caco-2 and HT-29 cell models. *J. Funct. Foods* **2019**, *53*, 125–135. [CrossRef]
5. Chen, P.X.; Zhang, H.; Marcone, M.F.; Pauls, K.P.; Liu, R.; Tang, Y.; Zhang, B.; Renaud, J.B.; Tsao, R. Anti-inflammatory effects of phenolic-rich cranberry bean (*Phaseolus vulgaris* L.) extracts and enhanced cellular antioxidant enzyme activities in Caco-2 cells. *J. Funct. Foods* **2017**, *38*, 675–685. [CrossRef]

6. Nguyen, H.C.; Hoefgen, R.; Hesse, H. Improving the nutritive value of rice seeds: Elevation of cysteine and methionine contents in rice plants by ectopic expression of a bacterial serine acetyltransferase. *J. Exp. Bot.* **2012**, *63*, 5991–6001. [CrossRef]
7. Xiang, X.; Wu, Y.; Planta, J.; Messing, J.; Leustek, T. Overexpression of serine acetyltransferase in maize leaves increases seed-specific methionine-rich zeins. *Plant Biotechnol. J.* **2017**. [CrossRef]
8. Kim, W.S.; Chronis, D.; Juergens, M.; Schroeder, A.C.; Hyun, S.W.; Jez, J.M.; Krishnan, H.B. Transgenic soybean plants overexpressing *O*-acetylserine sulfhydrylase accumulate enhanced levels of cysteine and Bowman-Birk protease inhibitor in seeds. *Planta* **2012**, *235*, 13–23. [CrossRef]
9. Galili, G.; Amir, R. Fortifying plants with the essential amino acids lysine and methionine to improve nutritional quality. *Plant Biotechnol. J.* **2013**, *11*, 211–222. [CrossRef]
10. Chiaiese, P.; Ohkama-Ohtsu, N.; Molvig, L.; Godfree, R.; Dove, H.; Hocart, C.; Fujiwara, T.; Higgins, T.J.V.; Tabe, L.M. Sulphur and nitrogen nutrition influence the response of chickpea seeds to an added, transgenic sink for organic sulphur. *J. Exp. Bot.* **2004**, *55*, 1889–1901. [CrossRef]
11. Zhang, Y.; Schernthaner, J.; Labbé, N.; Hefford, M.A.; Zhao, J.; Simmonds, D.H. Improved protein quality in transgenic soybean expressing a de novo synthetic protein, MB-16. *Transgenic Res.* **2014**, *23*, 455–467. [CrossRef] [PubMed]
12. Van, K.; McHale, L.K. Meta-analyses of QTLs associated with protein and oil contents and compositions in soybean [*Glycine max* (L.) Merr.] seed. *Int. J. Mol. Sci.* **2017**, *18*, 1180. [CrossRef]
13. Kasai, T.; Shiroshita, Y.; Sakamura, S. γ-Glutamyl peptides of *Vigna radiata* seeds. *Phytochemistry* **1986**, *25*, 679–682. [CrossRef]
14. Giada, M.D.L.R.; Miranda, M.T.M.; Marquez, U.M.L. Sulphur γ-glutamyl peptides in mature seeds of common beans (*Phaseolus vulgaris* L.). *Food Chem.* **1998**, *61*, 177–184. [CrossRef]
15. Baldi, G.; Salamini, F. Variability of essential amino acid content in seeds of 22 *Phaseolus* species. *Theor. Appl. Genet.* **1973**, *43*, 75–78. [CrossRef] [PubMed]
16. Li, C.J.; Brownson, D.M.; Mabry, T.J.; Perera, C.; Bell, E.A. Nonprotein amino acids from seeds of *Cycas circinalis* and *Phaseolus vulgaris*. *Phytochemistry* **1996**, *42*, 443–445. [CrossRef]
17. Evans, I.M.; Boulter, D. *S*-Methyl-L-cysteine content of various legume meals. *Plant Foods Hum. Nutr.* **1975**, *24*, 257–261. [CrossRef]
18. Liao, D.; Cram, D.; Sharpe, A.G.; Marsolais, F. Transcriptome profiling identifies candidate genes associated with the accumulation of distinct sulfur γ-glutamyl dipeptides in *Phaseolus vulgaris* and *Vigna mungo* seeds. *Front. Plant Sci.* **2013**, *4*, 60. [CrossRef] [PubMed]
19. Osborn, T.; Hartweck, L.; Harmsen, R.; Vogelzang, R.; Kmiecik, K.; Bliss, F. Registration of *Phaseolus vulgaris* genetic stocks with altered seed protein compositions. *Crop Sci.* **2003**, *43*, 1570–1572. [CrossRef]
20. Taylor, M.; Chapman, R.; Beyaert, R.; Hernández-Sebastià, C.; Marsolais, F. Seed storage protein deficiency improves sulfur amino acid content in common bean (*Phaseolus vulgaris* L.): Redirection of sulfur from γ-glutamyl-*S*-methyl-cysteine. *J. Agric. Food Chem.* **2008**, *56*, 5647–5654. [CrossRef]
21. Padovese, R.; Kina, S.M.; Barros, R.M.C.; Borelli, P.; Marquez, U.M.L. Biological importance of γ-glutamyl-*S*-methylcysteine of kidney bean (*Phaseolus vulgaris* L.). *Food Chem.* **2001**, *73*, 291–297. [CrossRef]
22. Hawkesford, M.J.; De Kok, L.J. Managing sulphur metabolism in plants. *Plant Cell Environ.* **2006**, *29*, 382–395. [CrossRef]
23. Takahashi, H.; Watanabe-Takahashi, A.; Smith, F.W.; Blake-Kalff, M.; Hawkesford, M.J.; Saito, K. The roles of three functional sulphate transporters involved in uptake and translocation of sulphate in *Arabidopsis thaliana*. *Plant J.* **2000**, *23*, 171–182. [CrossRef]
24. Hell, R.; Wirtz, M. Molecular biology, biochemistry and cellular physiology of cysteine metabolism in *Arabidopsis thaliana*. *Arabidopsis Book* **2011**, *9*, e0154. [CrossRef]
25. Smith, F.W.; Ealing, P.M.; Hawkesford, M.J.; Clarkson, D.T. Plant members of a family of sulfate transporters reveal functional subtypes. *Proc. Natl. Acad. Sci. USA* **1995**, *92*, 9373–9377. [CrossRef] [PubMed]
26. Takahashi, H.; Kopriva, S.; Giordano, M.; Saito, K.; Hell, R. Sulfur assimilation in photosynthetic organisms: Molecular functions and regulations of transporters and assimilatory enzymes. *Annu. Rev. Plant Biol.* **2011**, *62*, 157–184. [CrossRef]
27. Liao, D.; Pajak, A.; Karcz, S.R.; Chapman, B.P.; Sharpe, A.G.; Austin, R.S.; Datla, R.; Dhaubhadel, S.; Marsolais, F. Transcripts of sulphur metabolic genes are co-ordinately regulated in developing seeds of common bean lacking phaseolin and major lectins. *J. Exp. Bot.* **2012**, *63*, 6283–6295. [CrossRef] [PubMed]

28. Hell, R.; Jost, R.; Berkowitz, O.; Wirtz, M. Molecular and biochemical analysis of the enzymes of cysteine biosynthesis in the plant *Arabidopsis thaliana*. *Amino Acids* **2002**, *22*, 245–257. [CrossRef]
29. Bonner, E.R.; Cahoon, R.E.; Knapke, S.M.; Jez, J.M. Molecular basis of cysteine biosynthesis in plants: Structural and functional analysis of *O*-acetylserine sulfhydrylase from *Arabidopsis thaliana*. *J. Biol. Chem.* **2005**, *280*, 38803–38813. [CrossRef]
30. Hatzfeld, Y.; Maruyama, A.; Schmidt, A.; Noji, M.; Ishizawa, K.; Saito, K. β-Cyanoalanine synthase is a mitochondrial cysteine synthase-like protein in spinach and Arabidopsis. *Plant Physiol.* **2000**, *123*, 1163–1172. [CrossRef]
31. Hell, R.; Bergmann, L. γ-Glutamylcysteine synthetase in higher plants: Catalytic properties and subcellular localization. *Planta* **1990**, *180*, 603. [CrossRef] [PubMed]
32. Hicks, L.M.; Cahoon, R.E.; Bonner, E.R.; Rivard, R.S.; Sheffield, J.; Jez, J.M. Thiol-based regulation of redox-active glutamate-cysteine ligase from *Arabidopsis thaliana*. *Plant Cell* **2007**, *19*, 2653–2661. [CrossRef]
33. Yadav, S. Heavy metals toxicity in plants: An overview on the role of glutathione and phytochelatins in heavy metal stress tolerance of plants. *S. Afr. J. Bot.* **2010**, *76*, 167–179. [CrossRef]
34. Oven, M.; Raith, K.; Neubert, R.H.; Kutchan, T.M.; Zenk, M.H. Homo-phytochelatins are synthesized in response to cadmium in azuki beans. *Plant Physiol.* **2001**, *126*, 1275–1280. [CrossRef]
35. Yin, F.; Pajak, A.; Chapman, R.; Sharpe, A.; Huang, S.; Marsolais, F. Analysis of common bean expressed sequence tags identifies sulfur metabolic pathways active in seed and sulfur-rich proteins highly expressed in the absence of phaseolin and major lectins. *BMC Genomics* **2011**, *12*, 268. [CrossRef]
36. O'Rourke, J.A.; Iniguez, L.P.; Fu, F.; Bucciarelli, B.; Miller, S.S.; Jackson, S.A.; McClean, P.E.; Li, J.; Dai, X.; Zhao, P.X.; et al. An RNA-Seq based gene expression atlas of the common bean. *BMC Genomics* **2014**, *15*, 866. [CrossRef] [PubMed]
37. Rébeillé, F.; Jabrin, S.; Bligny, R.; Loizeau, K.; Gambonnet, B.; Van Wilder, V.; Douce, R.; Ravanel, S. Methionine catabolism in *Arabidopsis* cells is initiated by a γ-cleavage process and leads to *S*-methylcysteine and isoleucine syntheses. *Proc. Natl. Acad. Sci. USA* **2006**, *103*, 15687–15692. [CrossRef] [PubMed]
38. Yoshimoto, N.; Saito, K. Biosynthesis of *S*-alk(en)yl-L-cysteine sulfoxides in *Allium*: Retro perspective. In *Sulfur Metabolism in Higher Plants—Fundamental, Environmental and Agricultural Aspects*; De Kok, L.J., Hawkesford, M.J., Haneklaus, S.H., Schnug, E., Eds.; Springer International Publishing: Cham, Switzerland, 2017; Volume 49–60.
39. Watanabe, M.; Mochida, K.; Kato, T.; Tabata, S.; Yoshimoto, N.; Noji, M.; Saito, K. Comparative genomics and reverse genetics analysis reveal indispensable functions of the serine acetyltransferase gene family in *Arabidopsis*. *Plant Cell* **2008**, *20*, 2484–2496. [CrossRef]
40. Haas, F.H.; Heeg, C.; Queiroz, R.; Bauer, A.; Wirtz, M.; Hell, R. Mitochondrial serine acetyltransferase functions as a pacemaker of cysteine synthesis in plant cells. *Plant Physiol.* **2008**, *148*, 1055–1067. [CrossRef]
41. Walbot, V.; Clutter, M.; Sussex, I.M. Reproductive development and embryogeny in *Phaseolus*. *Phytomorphology* **1972**, *22*, 59–68.
42. Bobb, A.J.; Eiben, H.G.; Bustos, M.M. PvAlf, an embryo-specific acidic transcriptional activator enhances gene expression from phaseolin and phytohemagglutinin promoters. *Plant J.* **1995**, *8*, 331–343. [CrossRef] [PubMed]
43. Pandurangan, S.; Sandercock, M.; Beyaert, R.; Conn, K.L.; Hou, A.; Marsolais, F. Differential response to sulfur nutrition of two common bean genotypes differing in storage protein composition. *Front. Plant Sci.* **2015**, *6*, 92. [CrossRef]
44. Schmutz, J.; McClean, P.E.; Mamidi, S.; Wu, G.A.; Cannon, S.B.; Grimwood, J.; Jenkins, J.; Shu, S.; Song, Q.; Chavarro, C.; et al. A reference genome for common bean and genome-wide analysis of dual domestications. *Nat. Genet.* **2014**, *46*, 707–713. [CrossRef]
45. Vatamaniuk, O.K.; Mari, S.; Lu, Y.-P.; Rea, P.A. Mechanism of heavy metal ion activation of phytochelatin (PC) synthase: Blocked thiols are sufficient for PC synthase-catalyzed transpeptidation of glutathione and related thiol peptides. *J. Biol. Chem.* **2000**, *275*, 31451–31459. [CrossRef] [PubMed]
46. Vlasova, A.; Capella-Gutierrez, S.; Rendon-Anaya, M.; Hernandez-Onate, M.; Minoche, A.E.; Erb, I.; Camara, F.; Prieto-Barja, P.; Corvelo, A.; Sanseverino, W.; et al. Genome and transcriptome analysis of the Mesoamerican common bean and the role of gene duplications in establishing tissue and temporal specialization of genes. *Genome Biol.* **2016**, *17*, 32. [CrossRef]

47. Pandurangan, S.; Diapari, M.; Yin, F.; Munholland, S.; Perry, G.; Chapman, B.P.; Huang, S.; Sparvoli, F.; Bollini, R.; Crosby, W.; et al. Genomic analysis of storage protein deficiency in genetically related lines of common bean (*Phaseolus vulgaris*). *Front. Plant Sci.* **2016**, *7*, 389. [CrossRef]
48. Blum, R.; Meyer, K.C.; Wünschmann, J.; Lendzian, K.J.; Grill, E. Cytosolic action of phytochelatin synthase. *Plant Physiol.* **2010**, *153*, 159–169. [CrossRef] [PubMed]
49. Hayashi, S.; Kuramata, M.; Abe, T.; Takagi, H.; Ozawa, K.; Ishikawa, S. Phytochelatin synthase OsPCS1 plays a crucial role in reducing arsenic levels in rice grains. *Plant J.* **2017**, *91*, 840–848. [CrossRef]
50. Shine, A.M.; Shakya, V.P.; Idnurm, A. Phytochelatin synthase is required for tolerating metal toxicity in a basidiomycete yeast and is a conserved factor involved in metal homeostasis in fungi. *Fungal Biol. Biotechnol.* **2015**, *2*, 3. [CrossRef]
51. Mendoza-Cózatl, D.G.; Jobe, T.O.; Hauser, F.; Schroeder, J.I. Long-distance transport, vacuolar sequestration, tolerance, and transcriptional responses induced by cadmium and arsenic. *Curr. Opin. Plant Biol.* **2011**, *14*, 554–562. [CrossRef] [PubMed]
52. Sánchez, E.; Ruiz, J.M.; Romero, L. Proline metabolism in response to nitrogen toxicity in fruit of French Bean plants (*Phaseolus vulgaris* L. cv Strike). *Sci. Hortic.* **2002**, *93*, 225–233. [CrossRef]
53. Jafari, M.; Rajabzadeh, A.R.; Tabtabaei, S.; Marsolais, F.; Legge, R.L. Physicochemical characterization of a navy bean (*Phaseolus vulgaris*) protein fraction produced using a solvent-free method. *Food Chem.* **2016**, *208*, 35–41. [CrossRef]
54. Bruneau, L.; Chapman, R.; Marsolais, F. Co-occurrence of both L-asparaginase subtypes in *Arabidopsis*: *At3g16150* encodes a $K^+$-dependent L-asparaginase. *Planta* **2006**, *224*, 668–679. [CrossRef]
55. Earley, K.W.; Haag, J.R.; Pontes, O.; Opper, K.; Juehne, T.; Song, K.; Pikaard, C.S. Gateway-compatible vectors for plant functional genomics and proteomics. *Plant J.* **2006**, *45*, 616–629. [CrossRef]
56. Nelson, B.K.; Ca, X.; Nebenführ, A. A multicolored set of *in vivo* organelle markers for co-localization studies in Arabidopsis and other plants. *Plant J.* **2007**, *51*, 1126–1136. [CrossRef]
57. Yu, S.; Bian, Y.; Zhou, R.; Mou, R.; Chen, M.; Cao, Z. Robust method for the analysis of phytochelatins in rice by high-performance liquid chromatography coupled with electrospray tandem mass spectrometry based on polymeric column materials. *J. Sep. Sci.* **2015**, *38*, 4146–4152. [CrossRef]

*Article*

# An Exploration of the Roles of Ferric Iron Chelation-Strategy Components in the Leaves and Roots of Maize Plants

**Georgios Saridis [1], Styliani N. Chorianopoulou [2,*], Yannis E. Ventouris [2], Petros P. Sigalas [3] and Dimitris L. Bouranis [2]**

1 Botanical Institute, Cologne Biocenter, University of Cologne, D–50674 Cologne, Germany; g.saridis@uni-koeln.de

2 Plant Physiology and Morphology Laboratory, Crop Science Department, Agricultural University of Athens, 75 Iera Odos, Athens 11855, Greece; yannisventouris@gmail.com (Y.E.V.); bouranis@aua.gr (D.L.B.)

3 Rothamsted Research, West Common, Harpenden, Hertfordshire AL5 2JQ, UK; petros.sigalas@rothamsted.ac.uk

* Correspondence: s.chorianopoulou@aua.gr; Tel.: +30-210-529-4290

Received: 25 February 2019; Accepted: 16 May 2019; Published: 18 May 2019

**Abstract:** Plants have developed sophisticated mechanisms for acquiring iron from the soil. In the graminaceous species, a chelation strategy is in charge, in order to take up ferric iron from the rhizosphere. The ferric iron chelation-strategy components may also be present in the aerial plant parts. The aim of this work was to search for possible roles of those components in maize leaves. To this end, the expression patterns of ferric iron chelation-strategy components were monitored in the leaves and roots of mycorrhizal and non-mycorrhizal sulfur-deprived maize plants, both before and after sulfate supply. The two levels of sulfur supply were chosen due to the strong impact of this nutrient on iron homeostasis, whilst mycorrhizal symbiosis was chosen as a treatment that forces the plant to optimize its photosynthetic efficiency, in order to feed the fungus. The results, in combination with the findings of our previous works, suggest a role for the aforementioned components in ferric chelation and/or unloading from the xylem vessels to the aerial plant parts. It is proposed that the gene expression of the DMA exporter *ZmTOM1* can be used as an early indicator for the establishment of a mycorrhizal symbiotic relationship in maize.

**Keywords:** arbuscular mycorrhizal symbiosis; iron deficiency; iron homeostasis; sulfur deficiency; sulfur interactions; *Zea mays*

## 1. Introduction

Iron is an essential element for plant growth and productivity. It participates in electron transfer reactions and is central to the function of heme- and Fe–S cluster-requiring enzymes; thus, iron is required for various cellular processes, including respiration, photosynthesis, sulfur assimilation, and nitrogen fixation [1]. Although soil contains abundant iron, its bioavailability is extremely low. Under aerobic conditions and in the physiological pH range, iron is mainly present as oxidized ferric (hydr)oxides, which are sparingly soluble and not available to plants [2]. Therefore, plants have developed sophisticated and tightly regulated mechanisms for acquiring iron from soil. Non-graminaceous plants reduce ferric iron to the more soluble ferrous form at the root surface, by exporting a number of metabolites (including organic acids, phenolics, flavonoids and flavins), inducing the expression of ferric-chelate reductase, and they transport the resulting ferrous ions across the root plasma membrane. In contrast, graminaceous plants use a chelation strategy to take up iron from their rhizosphere. They secrete phytosiderophore compounds from their roots, phytosiderophores solubilize iron in the soil, and then the roots take up the resulting ferric iron–phytosiderophore complexes [1]. Maize in particular biosynthesizes only the 2′-deoxymugineic acid (DMA) via DMA synthase (ZmDMAS1). DMA is secreted into the rhizosphere via a specific exporter (ZmTOM1), and then the ferric iron-DMA complex is taken up by the ZmYS1 transporter [3].

Once iron has entered the root symplast, it is transported through the root cortex in the form of ferrous iron–nicotianamine complexes. After entering the root pericycle, it can be loaded into the xylem for transport to the shoot's sink tissues. The dominant form of iron in the xylem sap of non-graminaceous plants is ferric iron–citrate, whereas in graminaceous plants the bound iron forms may be ferric iron–citrate, along with ferric iron–phytosiderophores [4]. An important sink tissue for iron is the leaves, where it re-enters the symplast, is reduced to the ferrous form, and is again found as ferrous iron–nicotianamine.

During this "route of iron" within a graminaceous plant, it seems that ferric iron chelation-strategy components may also be present in the aerial plant parts. The expression of the DMA synthase gene in the leaves of rice and barley suggests a possible role(s) of DMA in leaf iron homeostasis, under various growth conditions and developmental stages. In rice shoot tissue, no expression of *OsDMAS1* was detected in the leaves of iron-sufficient plants, whereas under iron-deficient conditions, the *OsDMAS1* gene was specifically expressed in vascular bundles of the leaves [3].

Therefore, a plausible question arises as regards the role (or roles), if any, of those components in maize leaves. In a previous work of our group it was demonstrated that arbuscular mycorrhizal (AM) symbiosis impeded the expected iron deprivation responses in sulfur-deprived maize plants [5]. Sulfur deficiency has a negative impact on the iron homeostasis of all plants, due to the fact that the primary precursor of nicotianamine (i.e., the primary iron-chelating compound into the symplasm) is the sulfur-containing amino acid methionine. Nicotianamine is synthesized via the enzyme nicotianamine synthase (NAS), which uses S-adenosyl-methionine as a substrate molecule. For the graminaceous plants in particular, sulfur deficiency also exerts an impact on the iron uptake, since nicotianamine is the precursor compound of phytosiderophores. In this study, the gene expression patterns of ferric iron chelation-strategy components were monitored in the leaves (*ZmDMAS1*, *ZmTOM1* and *ZmYS1*) and roots (*ZmDMAS1* and *ZmTOM1*) of mycorrhizal and non-mycorrhizal sulfur-deprived maize plants, both before and after (24 and 48 h) sulfate supply.

## 2. Results

### 2.1. Expression Levels of ZmDMAS1 in the Leaves

No significant difference in the expression levels of *ZmDMAS1* in the leaves of non-mycorrhizal (NM) plants was monitored during the long sulfur-deficient period of time. The sulfur supply on day 60 resulted in a transient upregulation of this gene 24 h after this supply, whilst afterwards the expression levels reverted to the previous values (Figure 1a,c). The leaves of mycorrhizal (M) plants showed a differential response, both during the sulfur-deprived as well as after the sulfur-repletion period. The expression levels were constantly increased during the first 60 days of the study. The addition of sulfate had no significant effect on the expression values, which remained unaffected on days 61 and 62 (Figure 1b,d).

### 2.2. Expression Levels of ZmTOM1 in the Leaves

A prolonged sulfur deprivation resulted in a strong overexpression of *ZmTOM1* in the leaves of NM plants on day 60. The influx of sulfate caused a further overexpression 24 h after the addition of sulfur. The second day after the sulfur supply, the expression levels decreased significantly (Figure 2a,c). *ZmTOM1* was also strongly upregulated in the leaves of M plants on day 60, but in this case, it was already upregulated from day 45. The sulfur supply induced a strong overexpression on day 61, followed by an equally strong downregulation on day 62 (Figure 2b,d).

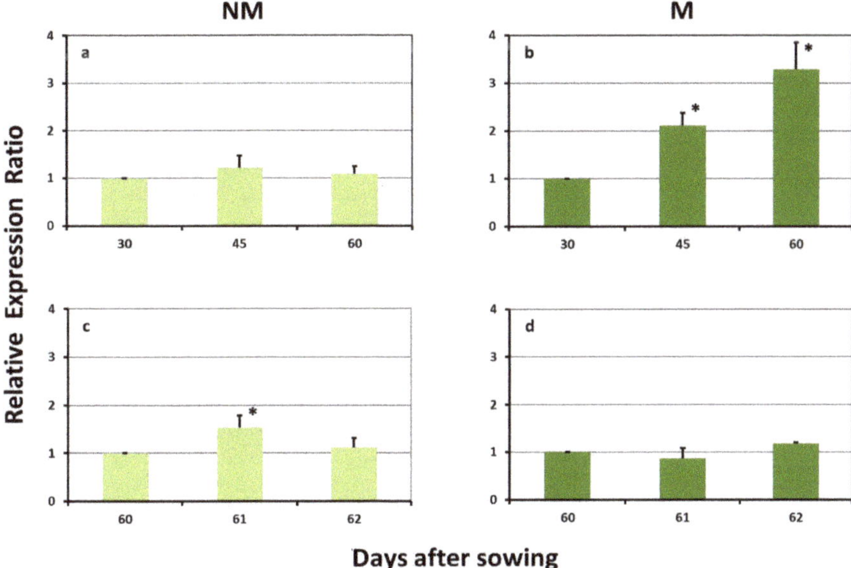


**Figure 1.** Expression of *ZmDMAS1* in the leaves of non-mycorrhizal (NM; **a**,**c**) and mycorrhizal (M; **b**,**d**) maize plants, (**a**,**b**) before and (**c**,**d**) after the sulfur supply, relative to the expression of ubiquitin. Bars show the mean of the biological replicates ± SE, asterisk (*) indicates the statistically significant difference between the sampling and the respective control at $p < 0.05$.

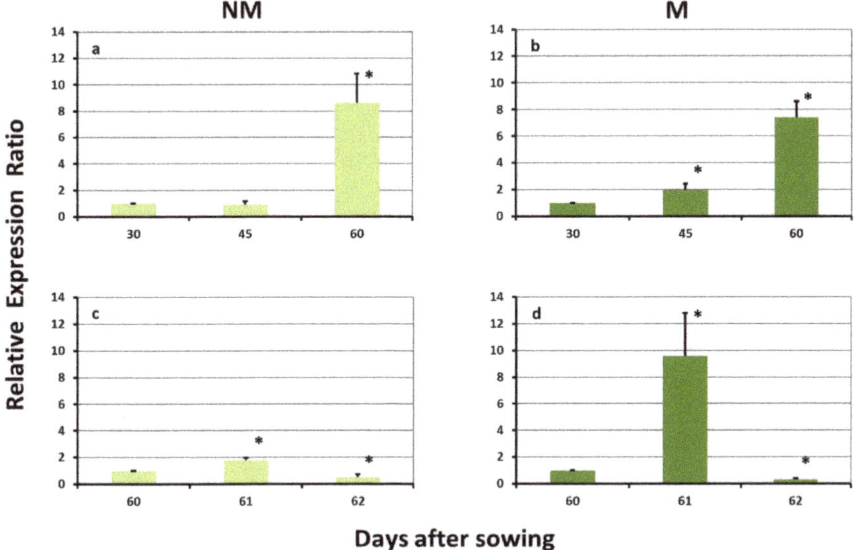


**Figure 2.** Expression of *ZmTOM1* in the leaves of non-mycorrhizal (NM; **a**,**c**) and mycorrhizal (M; **b**,**d**) maize plants, (**a**,**b**) before and (**c**,**d**) after the sulfur supply, relative to the expression of ubiquitin. Bars show the mean of the biological replicates ± SE, asterisk (*) indicates the statistically significant difference between the sampling and the respective control at $p < 0.05$.

### 2.3. Expression Levels of ZmYS1 in the Leaves

*ZmYS1* was strongly upregulated in the leaves of NM plants on day 60, before sulfur supply. The sulfur supply resulted in a transient downregulation on day 61, followed by a strong overexpression 2 days after the influx of sulfate (Figure 3a,c). The expression levels of this gene in the leaves of M plants increased constantly during the sulfur-deprived period and presented a strong overexpression on both sampling days. The addition of sulfate resulted in a strong downregulation on day 61, whilst the expression levels also remained reduced on day 62 (Figure 3b,d).

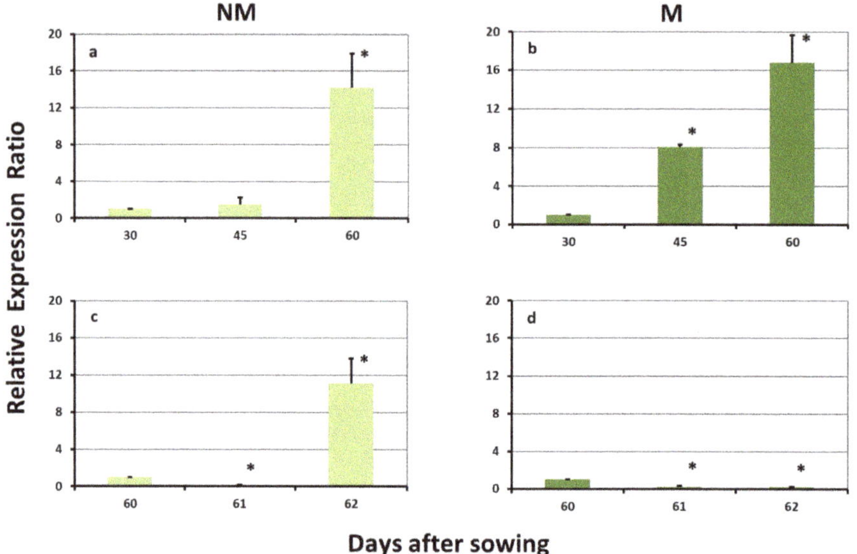


**Figure 3.** Expression of *ZmYS1* in the leaves of non-mycorrhizal (NM; **a**,**c**) and mycorrhizal (M; **b**,**d**) maize plants, (**a**,**b**) before and (**c**,**d**) after the sulfur supply, relative to the expression of ubiquitin. Bars show the mean of the biological replicates ± SE, asterisk (*) indicates the statistically significant difference between the sampling and the respective control at $p < 0.05$.

### 2.4. Expression Levels of ZmDMAS1 in the Roots

No significant difference in the expression levels of *ZmDMAS1* was monitored in the roots of NM plants during the sulfur deprivation treatment. The sulfur supply induced a strong downregulation of the expression, followed by an overexpression 2 days after the sulfate influx (Figure 4a,c). No significant difference in the expression levels of *ZmDMAS1* during sulfur deprivation was also observed in the roots of M plants. The addition of sulfate caused a strong downregulation of this gene expression on day 61. The same reduced levels were also observed on day 62 (Figure 4b,d).

### 2.5. Expression Levels of ZmTOM1 in the Roots

The expression levels of *ZmTOM1* in the roots of NM plants were significantly downregulated on day 45 and remained reduced until day 60. The sulfur supply induced a further downregulation on day 61, whilst the next day the expression was strongly upregulated (Figure 5a,c). In the roots of M plants, the respective expression levels of *ZmTOM1* were constantly downregulated, both before as well as after the sulfur supply (Figure 5b,d).

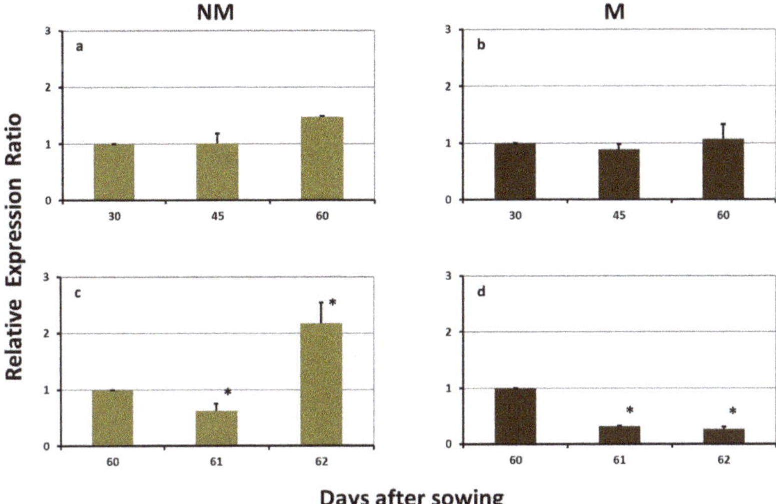


**Figure 4.** Expression of *ZmDMAS1* in the roots of non-mycorrhizal (**NM**; **a**,**c**) and mycorrhizal (**M**; **b**,**d**) maize plants, (**a**,**b**) before and (**c**,**d**) after the sulfur supply, relative to the expression of ubiquitin. Bars show the mean of the biological replicates ± SE, asterisk (*) indicates the statistically significant difference between the sampling and the respective control at $p < 0.05$.

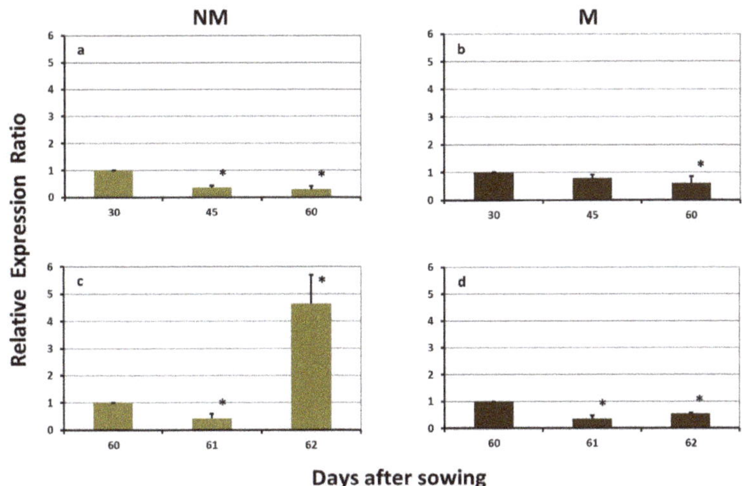


**Figure 5.** Expression of *ZmTOM1* in the roots of non-mycorrhizal (NM; **a**,**c**) and mycorrhizal (M; **b**,**d**) maize plants, (**a**,**b**) before and (**c**,**d**) after the sulfur supply, relative to the expression of ubiquitin. Bars show the mean of the biological replicates ± SE, asterisk (*) indicates the statistically significant difference between the sampling and the respective control at $p < 0.05$.

### 2.6. Data Meta-Analysis

The data on all the iron homeostasis components examined in this study, as well as in our previous works [5,6] were combined and further analyzed, and the comparative analysis is depicted in Table 1. In this analysis, the relative expression ratios of *ZmNAS3*, *ZmNAS1*, *ZmDMAS1*, *ZmTOM1* and *ZmYS1* in the leaves and roots of M plants were calculated using the respective values of NM plants as control. Even on day 30, the ferric iron chelation components were differentially regulated in the leaves of

M plants, although there was no difference observed in the M roots on that day, except for a strong downregulation of *ZmTOM1*. As a matter of fact, *ZmTOM1* was the sole iron homeostasis component permanently downregulated in the roots and upregulated in the leaves of M plants throughout the experiment. On the other hand, all of the examined genes were downregulated in the M roots after the addition of sulfate to the nutrient solution.

## 3. Discussion

### 3.1. Roles of Ferric Iron Chelation Components in Maize Leaves

Following its entry in the root central cylinder, iron is transported via the xylem vessels to be utilized in the aerial parts. The dominant bound iron forms in the xylem sap of graminaceous plants were found to be of two types: ferric iron-citrate, along with various ferric iron-phytosiderophores; DMA and citrate were present in large concentrations in the xylem sap from rice and maize [4]. The iron deficiency-inducible genes of barley have been found to be expressed almost exclusively in roots, whereas many iron deficiency-inducible genes in rice were expressed in both roots and shoots [3]. In the shoot tissues, the *OsDMAS1* promoter-GUS analysis showed an expression in vascular bundles, specifically under iron-deficient conditions. No GUS expression was detected in the leaves of iron-sufficient plants, whereas under iron-deficient conditions, a GUS activity was detected in the phloem sieve tubes and companion cells, as well as in the xylem parenchyma cells of the large vascular bundles [3]. A similar study assessing the tissue specific localization of *OsTOM1* showed the same GUS reporter gene expression pattern in the vascular bundles of the leaves of iron-deficient plants [7]. In the case of YS1, the expression of *HvYS1* was only specifically induced by iron deficiency in barley roots, whereas *ZmYS1* was expressed in maize in the leaf blades and sheaths, crown, and seminal roots [8,9]. The fact that ZmYS1 was expressed in the leaves led to the hypothesis that it must have additional functions that are not related to iron uptake from the soil. The expression of ZmYS1 was only found in the leaves of iron deficient maize plants, which are thus producing DMA. In those iron-starved plants, the ferric iron-DMA substrate for ZmYS1 could arrive to the leaves following transport from the roots through the xylem.

In this study, M maize plants presented an early response in regulating iron homeostasis in the young expanding leaves (Table 1). The young expanding leaves were chosen as strong sinks of iron, whilst AM symbiosis was chosen as a treatment that forces the plant to optimize its photosynthetic efficiency, in order to feed the fungus. Ferric iron chelation components were differentially regulated in the leaves of M plants even on day 30, when the AM symbiosis was not yet visibly established. The roots of the maize plants were colonized by the fungus on day 45. In view of this, on day 30 the plant was already prepared for the upcoming symbiosis, and its efforts to maintain an efficient iron homeostasis in the young leaves was evident throughout the experimental period (Table 1). However, it seems that the genes related to iron uptake from the roots do have a role in the aerial plant parts. More specifically, it is hypothesized that the enzymes ZmNAS1, ZmDMAS1, ZmTOM1 and ZmYS1 have a role in the ferric iron chelation and/or unloading from the xylem vessels.

It is hypothesized that DMA synthesized in the xylem parenchyma cells (through *ZmNAS1* and *ZmDMAS1*) is secreted to the xylem vessels (through *ZmTOM1*). The xylem sap in maize may contain ferric iron-citrate and ferric iron-DMA coming up from the roots. The secreted DMA chelates ferric iron delivered there in the form of ferric iron-citrate, because the ferric iron-DMA chelate has a higher stability than the ferric iron-citrate [10]. The resulting ferric iron-DMA, as well as the root-originated ferric iron-DMA in the xylem, may then enter the neighboring xylem parenchyma cells via the ZmYS1 transporter. The chelated iron may then be transported throughout the leaf and plant body via YSL transporters, for an efficient iron remobilization (Figure 6).

**Table 1.** Relative expression ratios of *ZmNAS3*, *ZmNAS1*, *ZmDMAS1*, *ZmTOM1* and *ZmYS1* in the leaves and roots of mycorrhizal plants, before (days 30, 45, and 60) and after (days 61 and 62) the sulfur supply, relative to the expression of ubiquitin. Non-mycorrhizal plants were used accordingly as a control for the calculation of the relative expression ratios. The values show the mean of the biological replicates ± SE. Table cells highlighted with dark grey indicate an upregulation, and those in light grey indicate a downregulation of the respective gene, when the difference between the sampling and the respective control was statistically significant at $p < 0.05$.

| | | Days after Sowing | | | | |
|---|---|---|---|---|---|---|
| | Gene | 30 | 45 | 60 | 61 | 62 |
| Leaves | *ZmNAS3* | 0.91 ± 0.34 | 1.28 ± 0.12 | 5.32 ± 1.25 | 2.85 ± 0.16 | 0.45 ± 0.11 |
| | *ZmNAS1* | 2.01 ± 0.27 | 2.69 ± 0.43 | 1.13 ± 0.26 | 1.19 ± 0.20 | 0.55 ± 0.03 |
| | *ZmDMAS1* | 0.61 ± 0.16 | 1.37 ± 0.21 | 1.80 ± 0.09 | 0.47 ± 0.03 | 1.27 ± 0.15 |
| | *ZmTOM1* | 1.99 ± 0.12 | 9.04 ± 1.68 | 4.12 ± 1.09 | 13.76 ± 3.54 | 3.31 ± 0.36 |
| | *ZmYS1* | 0.11 ± 0.06 | 1.76 ± 0.21 | 0.35 ± 0.14 | 0.28 ± 0.13 | 0.03 ± 0.01 |
| Roots | *ZmNAS3* | 0.83 ± 0.13 | 1.94 ± 0.11 | 0.99 ± 0.07 | 1.33 ± 0.31 | 0.67 ± 0.12 |
| | *ZmNAS1* | 0.96 ± 0.22 | 0.73 ± 0.08 | 0.63 ± 0.05 | 1.76 ± 0.12 | 0.41 ± 0.07 |
| | *ZmDMAS1* | 1.12 ± 0.18 | 1.10 ± 0.25 | 0.85 ± 0.20 | 0.48 ± 0.09 | 0.41 ± 0.15 |
| | *ZmTOM1* | 0.18 ± 0.10 | 0.38 ± 0.06 | 0.48 ± 0.14 | 0.45 ± 0.16 | 0.12 ± 0.02 |
| | *ZmYS1* | 1.56 ± 0.31 | 0.78 ± 0.04 | 1.36 ± 0.19 | 1.05 ± 0.08 | 0.12 ± 0.08 |

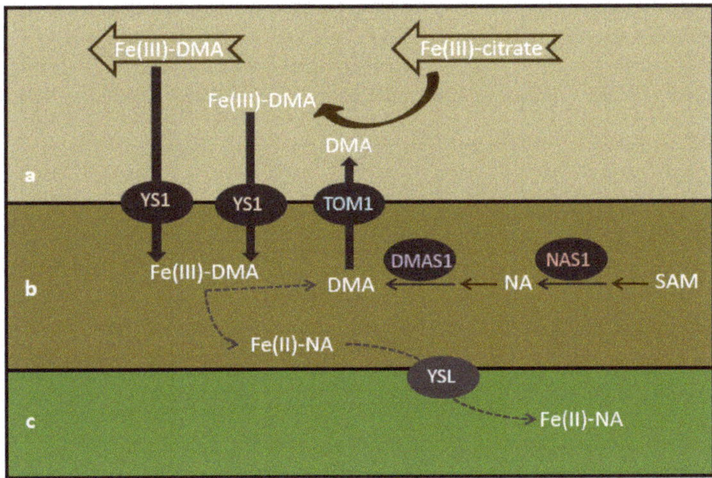


**Figure 6.** Conceptual model of iron xylem transport in young maize leaves, xylem-to-phloem iron exchange and iron re-translocation to younger tissues. The processes are considered to take place in the large vein of a maize leaf: (**a**) xylem vessel, (**b**) xylem parenchyma cell, and (**c**) phloem. DMA: deoxymugineic acid, DMAS: DMA synthase, NA: nicotianamine, NAS: nicotianamine synthase, SAM: S-adenosyl-methionine, TOM: DMA efflux transporter, YS: yellow stripe transporter, and YSL: yellow stripe like transporter.

## *3.2. Arbuscular Mycorrhizal Fungus Provides Iron to M Plants*

The expression patterns of the examined genes were completely different among the various treatments, i.e., NM vs. M under sulfur deprivation and NM vs. M after sulfate supply. This fact holds true for the roots as well as for the leaves of the examined maize plants. This can be explained if we consider the hypothesis that NM plants suffer from iron deficiency throughout the treatment, whilst mycorrhizal colonization prevented iron deprivation responses due to a symbiotic iron uptake pathway [5].

The expression profile patterns of the examined genes in the roots of M and NM maize plants, both before as well as after sulfur supply, supported the previously suggested hypothesis that the fungus provides iron to the plants [5]. Despite the facts that: (a) before the addition of sulfur the concentration of total Fe in the shoots of NM plants decreased from day 45 to day 60, whilst the corresponding concentrations in M plants remained at the same levels and (b) five days after the addition of sulfur the concentration of total iron in M plants was higher than the corresponding concentration in NM plants [5], none of the genes related to iron uptake was overexpressed in the roots of M plants, neither during the sulfur deprivation period, nor after the supply of sulfur. Given this, in the M plants iron was probably delivered to the central cortex through the fungal arbuscules. Consequently, the M plants did not need to excrete DMA to their rhizosphere and *ZmTOM1* was strongly downregulated (Table 1, Figures 4 and 5, [5]).

Twenty four hours after the sulfur supply, a general transient downregulation of all the genes related to iron uptake was monitored in the roots of both M and NM plants (Figures 4 and 5, [5,6]). Sulfate may act as a signal molecule regulating the expression of the iron uptake related genes in roots. It is suggested that the "massive entry" of inorganic sulfur into the root cells creates a transient "shortage of organic sulfur" condition, which is required for the DMA synthesis and that during this period of time the plants downregulate the whole DMA biosynthesis pathway, until organic sulfur will be again available.

Two days after the addition of sulfur to the nutrient solution, a diverse response was observed: the iron-uptake-related genes were upregulated in the roots of NM plants, whilst their expression in the M roots remained stable and downregulated from day 61 (Table 1, Figures 4 and 5, [5]).

Moreover, *ZmTOM1* was the only gene permanently overexpressed in the leaves and downregulated in the roots of M plants relative to NM ones, throughout the long-term experiment, even on day 30 when the M plants were not actually colonized by the fungus yet (Table 1). Therefore, the expression of *ZmTOM1* could be an early indicator of a mycorrhizal symbiotic relationship.

## 4. Materials and Methods

### 4.1. Plant Material and Growth Conditions

Mycorrhizal (M) and non-mycorrhizal (NM) maize plants were grown in pots with sterile river sand and practically insoluble $FePO_4$ (500 mg per pot) in a long-term experiment, as previously described [5]. The plants were watered with a nutrient solution deprived of iron and sulfur and containing a minimum phosphorus concentration, in order to enhance the establishment of the symbiotic relationship with the AM fungus *Rhizophagus irregularis*. Iron was provided to plants throughout the experiment in the sparingly soluble form of $FePO_4$. After a 60-day period of sulfur deprivation, sulfur was provided to the plants in the form of sulfate (2.5 mM $CaSO_4$ $2H_2O$ and 1 mM $MgSO_4$ $7H_2O$).

### 4.2. Plant Samplings

The samplings were performed on days 30, 45, 60, 61 (24 h after the sulfur supply) and 62 (48 h after the sulfur supply) after sowing and 3 h after the onset of light. The sampling of day 60 took place before the addition of sulfur. The lateral roots, as well as two young expanding leaves, were immediately frozen in liquid nitrogen and stored at −80 °C until use. In each experiment, plant material from at least three biological replicates per treatment and sampling day was used [5].

### 4.3. Gene Expression Analysis

The gene expression analysis was conducted by means of Real-Time RT-PCR, as previously described [5]. The oligonucleotide primers used for RT-qPCR were as previously referred to [6]. The efficiency of each Real-Time RT-PCR reaction was calculated using the LinRegPCR software [11]. The following mathematical formula, from [12], was used for the calculations of the relative expression ratios of the target genes (*ZmDMAS1*, *ZmTOM1* and *ZmYS1*), whilst ubiquitin (*ZmUBQ*) was used as the reference gene:

$$ratio = \frac{\text{Etarget}^{\Delta CP\text{target}(control-sample)}}{\text{Eref}^{\Delta CP\text{ref}(control-sample)}} \quad (1)$$

The samples of day 30 (for the samplings before the sulfur supply) or day 60 (for the samplings after the sulfur supply) of the respective treatment (NM or M) were used as control, unless otherwise specified.

### 4.4. Data Meta-Analysis

Data from [5,6] were combined and further analyzed in order to make a comparative analysis, taking into consideration the data of all of the examined iron homeostasis components. The results of this analysis are depicted in Table 1.

### 4.5. Statistical Analysis

The experiment was performed two times under the same conditions and during two distinct time periods: autumn 2013 and spring 2014. The data were analyzed by t-test variance analysis with a two-tailed distribution and two-sample unequal variance to determine the significance of differences among the samplings.

## 5. Conclusions

It is suggested that the components related to iron uptake from the roots have a role in the ferric iron chelation and/or unloading from the xylem vessels in the aerial plant parts. Moreover, the expression of the gene codes for the DMA exporter *ZmTOM1* could be an early indicator for the establishment of a mycorrhizal symbiotic relationship.

**Author Contributions:** G.S., S.N.C. and D.L.B. conceived and designed the experiment, elaborated the research questions, analyzed the data, and wrote the article; G.S., P.P.S. and Y.E.V. carried out the gene expression analysis. All authors have read and approved the manuscript.

**Funding:** This research received no external funding.

**Conflicts of Interest:** The authors declare no conflict of interest.

## References

1. Connorton, J.M.; Balk, J.; Rodriguez-Celma, J. Iron homeostasis in plants—A brief overview. *Metallomics* **2017**, *9*, 813–823. [CrossRef] [PubMed]
2. Li, Y.; Yu, S.; Strong, J.; Wang, H. Are the biogeochemical cycles of carbon, nitrogen, sulfur, and phosphorus driven by the "$Fe^{III}$-$Fe^{II}$ redox wheel" in dynamic redox environments? *J. Soils Sediments* **2012**, *12*, 683–693. [CrossRef]
3. Bashir, K.; Inoue, H.; Nagasaka, S.; Takahashi, M.; Nakanishi, H.; Mori, S.; Nishizawa, N.K. Cloning and characterization of deoxymugineic acid synthase genes from graminaceous plants. *J. Biol. Chem.* **2006**, *281*, 32395–32402. [CrossRef] [PubMed]
4. Ariga, T.; Hazama, K.; Yanagisawa, S.; Yoneyama, T. Chemical forms of iron in xylem sap from graminaceous and non-graminaceous plants. *Soil Sci. Plant Nutr.* **2014**, *60*, 460–469. [CrossRef]
5. Chorianopoulou, S.N.; Saridis, Y.I.; Dimou, M.; Katinakis, P.; Bouranis, D.L. Arbuscular mycorrhizal symbiosis alters the expression patterns of three key iron homeostasis genes, *ZmNAS1*, *ZmNAS3*, and *ZmYS1*, in S deprived maize plants. *Front. Plant Sci.* **2015**, *6*, 257. [CrossRef] [PubMed]
6. Saridis, G.I.; Chorianopoulou, S.N.; Katinakis, P.; Bouranis, D.L. Evidence for regulation of the iron uptake pathway by sulfate supply in S-deprived maize plants. In *Sulfur Metabolism in Higher Plants—Fundamental, Environmental and Agricultural Aspects, Proceedings of the International Plant Sulfur Workshop, Goslar, Germany, 1–4 September 2015*; De Kok, L.J., Hawkesford, M.J., Haneklaus, S.H., Schnug, E., Eds.; Springer International Publishing AG.: Berlin/Heidelberg, Germany, 2017; pp. 175–180.

7. Nozoye, T.; Nagasaka, S.; Kobayashi, T.; Takahashi, M.; Sato, Y.; Sato, Y.; Uozumi, N.; Nakanishi, H.; Nishizawa, N.K. Phytosiderophore efflux transporters are crucial for iron acquisition in graminaceous plants. *J. Biol. Chem.* **2011**, *86*, 5446–5454. [CrossRef] [PubMed]
8. Roberts, L.A.; Pierson, A.J.; Panaviene, Z.; Walker, E.L. Yellow Stripe1. Expanded roles for the maize iron-phytosiderophore transporter. *Plant Physiol.* **2004**, *135*, 112–120. [CrossRef] [PubMed]
9. Ueno, D.; Yamaji, N.; Ma, J.F. Further characterization of ferric—Phytosiderophore transporters ZmYS1 and HvYS1 in maize and barley. *J. Exp. Bot.* **2009**, *60*, 3513–3520. [CrossRef] [PubMed]
10. Fodor, F. Heavy metals competing with iron under conditions involving phytoremediation. In *Iron Nutrition in Plants and Rhizospheric Microorganisms*; Barton, L.L., Abadía, J., Eds.; Springer: Dordrecht, The Netherlands, 2006; pp. 129–151.
11. Ruijter, J.M.; Ramakers, C.; Hoogaars, W.M.; Karlen, Y.; Bakker, O.; van den Hoff, M.J.; Moorman, A.F. Amplification efficiency: Linking base line and bias in the analysis of quantitative PCR data. *Nucleic Acids Res.* **2009**, *37*, e45. [CrossRef] [PubMed]
12. Pfaffl, M.W. A new mathematical model for relative quantification in real-time RT-PCR. *Nucleic Acids Res.* **2001**, *29*, 2002–2007. [CrossRef] [PubMed]

Article

# Contribution of Root Hair Development to Sulfate Uptake in *Arabidopsis*

**Yuki Kimura [1], Tsukasa Ushiwatari [1], Akiko Suyama [1], Rumi Tominaga-Wada [2], Takuji Wada [2] and Akiko Maruyama-Nakashita [1,*]**

1 Department of Bioscience and Biotechnology, Faculty of Agriculture, Kyushu University, 744 Motooka, Nishi-ku, Fukuoka 819-0395, Japan; YukiKimura@gmail.com (Y.K.); 2BE18410P@s.kyushu-u.ac.jp (T.U.); aksuyama@nm.beppu-u.ac.jp (A.S.)

2 Graduate School of Biosphere Sciences, Hiroshima University, 1-4-4 Kagamiyama, Higashi Hiroshima, Hiroshima 739-8528, Japan; rtomi@hiroshima-u.ac.jp (R.T.-W.); twada86@hiroshima-u.ac.jp (T.W.)

* Correspondence: amaru@agr.kyushu-u.ac.jp; Tel.: +81-92-802-4712

Received: 26 February 2019; Accepted: 17 April 2019; Published: 19 April 2019

**Abstract:** Root hairs often contribute to nutrient uptake from environments, but the contribution varies among nutrients. In *Arabidopsis*, two high-affinity sulfate transporters, SULTR1;1 and SULTR1;2, are responsible for sulfate uptake by roots. Their increased expression under sulfur deficiency (−S) stimulates sulfate uptake. Inspired by the higher and lower expression, respectively, of *SULTR1;1* in mutants with more (*werwolf* [*wer*]) and fewer (*caprice* [*cpc*]) root hairs, we examined the contribution of root hairs to sulfate uptake. Sulfate uptake rates were similar among plant lines under both sulfur sufficiency (+S) and −S. Under −S, the expression of *SULTR1;1* and *SULTR1;2* was negatively correlated with the number of root hairs. These results suggest that both −S-induced *SULTR* expression and sulfate uptake rates were independent of the number of root hairs. In addition, we observed (1) a negative correlation between primary root lengths and number of root hairs and (2) a greater number of root hairs under −S than under +S. These observations suggested that under both +S and −S, sulfate uptake was influenced by the root biomass rather than the number of root hairs.

**Keywords:** *Arabidopsis thaliana*; root hair; sulfate transporter; sulfate uptake; CPC; CAPRICE; WER; WERWOLF

---

## 1. Introduction

Plant nutrients are generally absorbed from the roots in appropriate chemical forms. From the root surface to cortex, nutrients are transported via the symplasmic or apoplastic routes, but have to take the symplasmic route when they have reached the endodermis. At the endodermis, nutrients are subsequently transported to the xylem parenchyma cells and loaded into the xylem stream for further transport to shoot tissues [1,2]. The root surface cell layers, epidermis, and cortex all play important roles in nutrient uptake by expressing high-affinity transporters for most essential nutrients [2–5]. It has been suggested that increasing the root surface area exposed to the soil environments enables root hairs to also contribute to nutrient uptake [6–8]. Root hairs have been shown to increase the uptake of some nutrients, including phosphate and ammonium [8–10], but not others, such as iron and silicon [11,12]. Root hair growth is stimulated by the deficiency of phosphorous, manganese, iron, and some other nutrients in *Arabidopsis*, whereas deficiencies of other nutrients fail to produce a similar phenotype [3,9,13–17]. Thus, the importance of root hairs in nutrient uptake varies among nutrients.

Plants absorb sulfate as the main source of sulfur through the activity of sulfate transporters (SULTRs) [5,18]. Sulfate taken up from the soil is transported to the shoots and primarily metabolized through reductive assimilation to form cysteine (Cys). Then, methionine and glutathione (GSH) are synthesized from Cys [5,18–20]. In *Arabidopsis*, two high-affinity sulfate transporters, SULTR1;1 and

SULTR1;2, which are localized to the root epidermis and cortex, are responsible for the initial uptake of sulfate from the soil environment [21–25]. The transcript levels of both *SULTR1;1* and *SULTR1;2* are increased by sulfur deficiency (−S), which increases the sulfate uptake rate [21–25]. The −S-induced the expression of *SULTR1;1* and *SULTR1;2* is controlled by the promoter activities of their 5′-upstream regions [26–29].

Although SULTR1;1 and SULTR1;2 both contribute to sulfate uptake from the roots and their expression is stimulated by −S, they differ with regard to several features related to transcriptional regulation. When sulfate is sufficiently supplied, the transcript levels of *SULTR1;2* are higher than those of *SULTR1;1* are [24,25]. Under–S conditions, the transcript levels of *SULTR1;2* are still higher than those of *SULTR1;1*, but the rate of induction by −S is less in *SULTR1;2* than that in *SULTR1;1* [23–25,30,31]. The sulfate uptake rate and the levels of Cys and GSH in *SULTR1;1* and *SULTR1;2* knockout lines indicate that SULTR1;2 is the main contributor to sulfate uptake under sulfur sufficiency (+S) and −S conditions [23–25,30]. In addition, the tissue specificity of *SULTR1;1* and *SULTR1;2* expression in root surface cell layers is not identical: *SULTR1;1* is mainly expressed in the epidermis, including root hairs, whereas *SULTR1;2* is mainly expressed in the cortex [21,24,26].

The genetic control of root hair development has been extensively studied in *Arabidopsis* [6,7,9,32–35]. Root epidermal cells are separated into root hair and non-hair cells in *Arabidopsis*. Many transcription factors function in the early events of epidermal cell differentiation. In non-hair cells, a core transcriptional complex consisting of TRANSPARENT TESTA GLABRA1 (TTG1, WD40 repeat protein, [36]), GLABRA3 (GL3, bHLH transcription factor, [37]), ENHANCER of GLABRA3 (EGL3, bHLH transcription factor, [38]), and WERWOLF (WER, R2R3 MYB transcription factor, [39]) promotes the expression of the transcription factor GLABRA2 (GL2, homeobox-leucine zipper protein, [40]) which negatively regulates root hair cell-specific genes and positively regulates non-hair cell specific genes to determine non-hair cell fate. In root hair cells, a different protein complex, comprising R3 MYB transcription factors CAPRICE (CPC, [41]), TRIPTYCHON (TRY, [42]), and ENHANCER of TRY and CPC1 (ETC1, [43,44]), inhibits the association of the WER protein with the transcriptional complex described above, thereby repressing GL2 [45,46]. Therefore, given their functions in root hair development, mutation in WER results in an increase in the number of root hairs and mutation in CPC results in a decrease in the number of root hairs.

In addition to these transcription factors, which function as cell fate determinants, many other factors involved in root hair growth have been identified [35]. Among them, a basic Helix-Loop-Helix (bHLH) transcription factor, ROOT HAIR DEFECTIVE 6 (RHD6), plays key roles in determining root hair identity under the control of CPC and GL2 [35,47–49]. RHD6 promotes the expression of other bHLH transcription factors including RHD6-LIKE4 (RSL4), which positively regulates root hair growth [50]. The expression of RSL4 is also induced by ethylene through direct induction by the transcription factors ETHYLENE INSENSITIVE 3 (EIN3) and its homolog EIN3-LIKE1 (EIL1) [17]. EIN3 and EIL1 physically interact with RHD6 to synergistically induce *RSL4* expression [17]. As phosphate starvation increases the level of EIN3 protein [16], a phosphate starvation-responsive increase in the number of root hairs would also be mediated by RSL4 and the MYB transcription factor PHOSPHATE STARVATION RESPONSE 1 (PHR1) which plays a central role in phosphate starvation responses [51,52].

In the present study, we investigated the role of root hairs in sulfate uptake using *werwolf* (*wer*) and *caprice* (*cpc*) *Arabidopsis* mutants, as CPC and WER control epidermal cell fate upstream of the signal transduction system of root hair development [6,7,9,32–35]. Our results indicated that the fate of root epidermal cells influenced the expression of *SULTR1;1* and *SULTR1;2*. However, the –S-induced expression of *SULTR1;1* and *SULTR1;2*, and sulfate uptake were independent of root epidermal cell development. Overall, the sulfate uptake rate was likely influenced by the total root biomass rather than the number of root hairs under both +S and –S conditions.

## 2. Results

### 2.1. *SULTR1;1 and SULTR1;2 Were Differentially Expressed in the Root Epidermal Cells under* −*S*

The difference in tissue-specific expression between *SULTR1;1* and *SULTR1;2* in the root surface was investigated by assessing green fluorescent protein (GFP) fluorescence in the root elongation zones of $P_{SULTR1;1}$*-GFP* [26] and $P_{SULTR1;2}$*-GFP* [27] plants. When the plants were grown under +S sulfur sufficient (S1500) conditions, GFP fluorescence was not detected in either plant line (data not shown). Under −S sulfur deficient (S15) conditions, GFP fluorescence was clearly visible in root elongation zones (Figure 1). In $P_{SULTR1;1}$*-GFP* plants, GFP fluorescence was mainly detected in root hair cells (Figure 1a–c), whereas it was detected in non-hair cells in $P_{SULTR1;2}$*-GFP* plants (Figure 1g–i). Interestingly, GFP fluorescence was detected in a striped pattern in the root hair cells and non-hair cells in $P_{SULTR1;1}$*-GFP* and $P_{SULTR1;2}$*-GFP* plants, respectively, suggesting that they were differentially expressed in these cell lines and played different roles in sulfate uptake from the root surface.

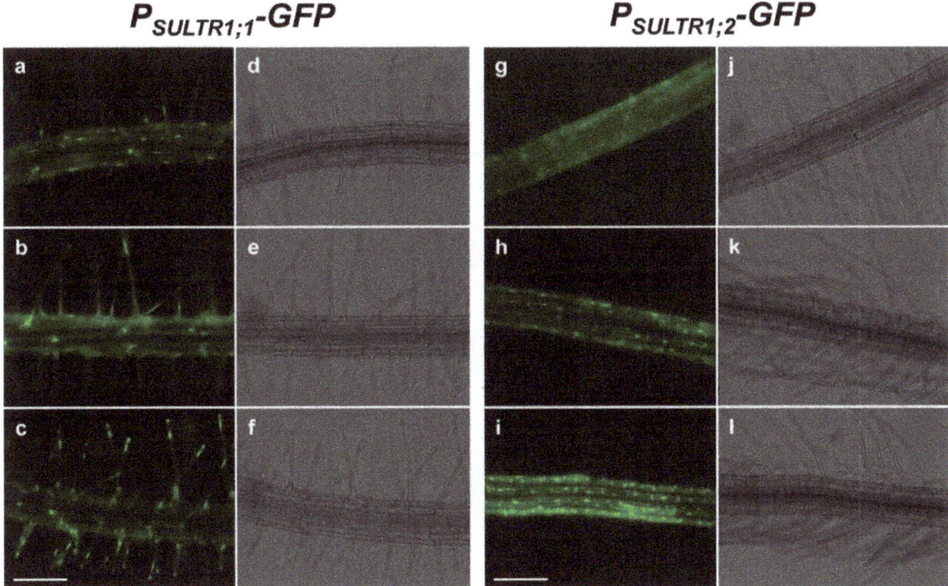


**Figure 1.** Expression of green fluorescent protein (GFP) in the root elongation zones of $P_{SULTR1;1}$*-GFP* plants (**a**–**f**, [26]) and $P_{SULTR1;2}$*-GFP* plants (**g**–**l**, [27]). Three $T_2$ generation lines carrying each construct were grown for 7 d on agar media supplied with 15 μM sulfate and observed using fluorescent microscopy (EVOS FL Auto 2). GFP images (**a**–**c**,**g**–**i**) and bright field images (**d**–**f**,**j**–**l**) are presented. Scale bars, 100 μm.

### 2.2. *Primary Root Length and Numbers of Root Hairs Compensated for Each Other in wer and cpc*

To assess the relationship between sulfate uptake and root hair development, root phenotypes of *wer*, *cpc* and wild-type Columbia plants (WT) under S1500 and S15 conditions were observed (Figure 2). The number of root hairs 5 mm from the root tip were counted in plants grown under S1500 and S15 conditions (Figure 2a–c). The number of root hairs was significantly higher under S15 conditions than under S1500 conditions (Figure 2c), suggesting that plants increased the number of root hairs for more efficient sulfate uptake under –S. Under both sulfur conditions, *wer* had more root hairs than the WT, and *cpc* tended to have less root hairs than the WT, as reported previously [39,41] (Figure 2c).

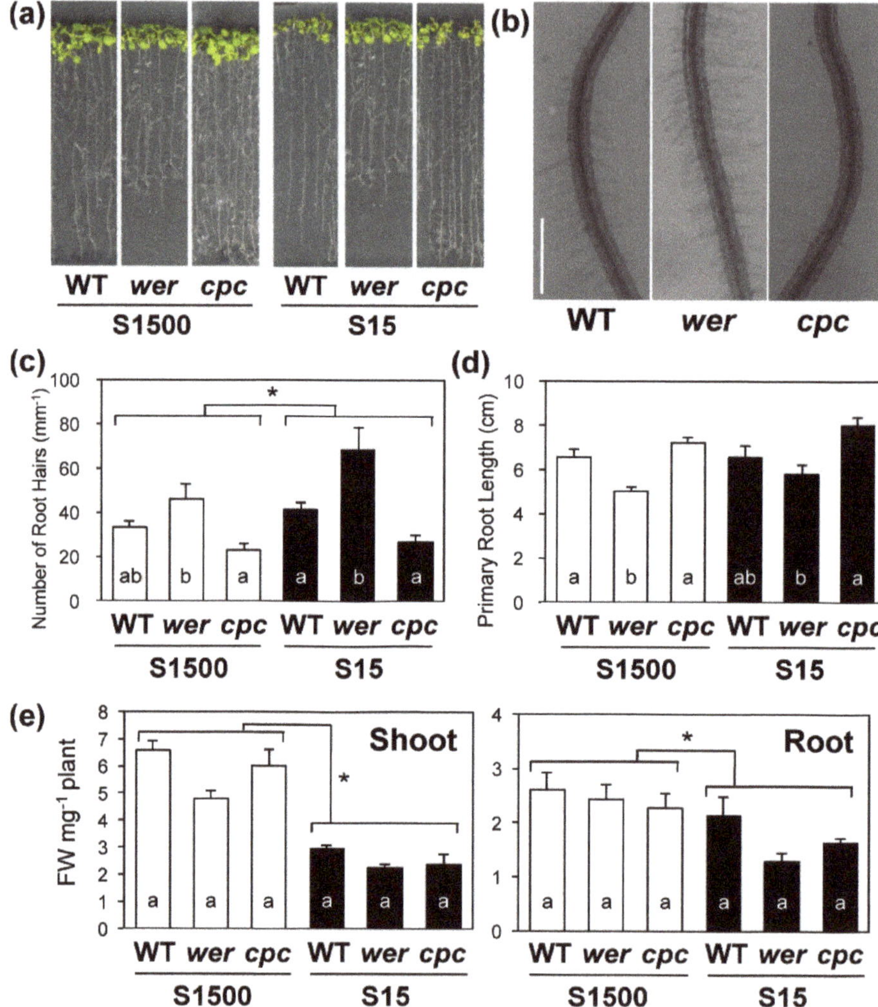


**Figure 2.** Plant growth and root development of wild-type (WT), and *werwolf* (*wer*), and *caprice* (*cpc*) mutants grown under sulfur-sufficient and -deficient conditions. Plants were vertically grown for 10 d on agar media supplied with 1500 μM sulfate (sulfur sufficient, S1500, white bar) or 15 μM sulfate (sulfur-deficient, S15, black bar). (**a**) Images of plants on their media. (**b**) Typical root image of plants grown under the S15 condition. Scale bar, 500 μm. (**c**) Number of root hairs in root elongation zone. Number of root hairs 5 mm from root tip was counted. (**d**) Primary root lengths. (**e**) Shoot and root fresh weights (FW) per plant. Values and error bars indicate means ± SEM [n = 5 in (**c**), n = 8 to 12 in (**d**), and n = 3 in (**e**)]. Two-way ANOVA was used to detect the effects of sulfur conditions, genotypes, and their interactions. Asterisks indicate significant differences between the S1500 and S15 conditions (two-way analysis of vatiance [ANOVA], * $P < 0.05$). As all comparisons in (**c**–**e**) did not detect interaction, the Tukey–Kramer test was used to analyze significant differences among mutants and the WT under the same growth condition. Different letters indicate significant differences ($P < 0.05$).

Primary root lengths were not influenced by sulfate concentrations (Figure 2d). Interestingly, primary root lengths were influenced in the opposite direction as the number of root hairs in the two mutants as compared to the WT. That is, primary root lengths tended to be shorter in *wer* and longer in *cpc* than they were in WT under both the sulfur conditions. Significant negative correlations between the number of root hairs and primary root lengths were detected using Sperman's Rank correlation coefficient (R = 0.047, correlation constant = −0.89). These phenotypes suggested that plants adjusted their root biomass by balancing the inverse relationship between primary root lengths and root hair densities.

Both shoot and root fresh weights were decreased under –S conditions (Figure 2e). In *wer*, shoot fresh weights under S1500 and S15 conditions tended to be less than those in the WT (Figure 2e). When plants were grown under the S15 condition, the root fresh weights of *wer* and *cpc* tended to be less than those in the WT (Figure 2e).

### 2.3. Regulatory Gene Expression in Root Hair Development Was Consistent with Root Hair Numbers under +S but Not –S Conditions

The number of root hairs was increased under –S conditions and, therefore, we analyzed the expression of transcription factors *CPC, WER, GL2, RHD6,* and *RSL4,* which regulate root hair development (Figure 3).

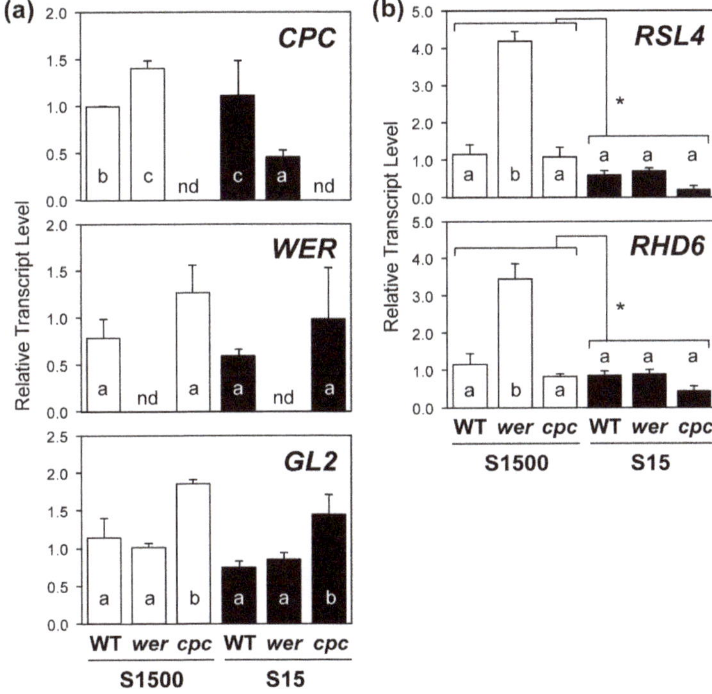


**Figure 3.** Expression of transcription factors regulating root hair development in wild-type (WT), *werwolf (wer)*, and *caprice (cpc)* mutants grown under S1500 and S15 conditions. (**a**) Relative transcript levels of *CPC, WER,* and *GL2,* and (**b**) those of *RHD6,* and *RSL4,* in root tissues were determined using quantitative reverse transcription-polymerase chain reaction (qRT-PCR). Plants were vertically grown for 10 d on S1500 (white bars) and S15 (black bars) agar media. Values and error bars indicate means ± SEM (n = 3). Two-way analysis of variance (ANOVA) was used to detect the effects of sulfur conditions, genotypes, and their interactions. In (**a**), all comparisons did not detect interactions, so the Tukey–Kramer test was applied to the three genotypes under the same growth conditions. In (**b**), interactions between sulfur conditions and genotypes were detected, so the Tukey–Kramer test was applied to all six experimental conditions. Different letters indicate significant differences ($P < 0.05$).

The expression of *CPC* in *wer* was higher under the S1500 condition but lower under the S15 condition than that in the WT, whereas the expression of *WER* was not influenced in *cpc* (Figure 3a). The expression of *GL2* was higher in *cpc* under both conditions than that in the WT.

The transcript levels of root hair-specific transcription factors, *RHD6* and *RSL4*, were higher in *wer* under the S1500 condition than they were in WT but were similar among the genotypes under S15 (Figure 3b). Their transcript levels tended to be lower in *cpc* under both conditions than they were in the WT, which was likely due to the increased expression of *GL2* in *cpc* as GL2 suppressed the expression of *RHD6* [49]. With regard to the expression of *RHD6* and *RSL4*, an interaction was found between sulfur conditions and genotypes using a two-way ANOVA, suggesting their expressions were differentially influenced by the disruption of WER or the number of root hairs under +S and −S.

### 2.4. Sulfate Uptake Was Constant among WT, wer, and cpc

We analyzed the transcript levels of *SULTR1;1* and *SULTR1;2* in the roots and the sulfate uptake activity in WT, *wer*, and *cpc* plants grown under S1500 and S15 conditions (Figure 4). Sulfur deficiency increased the transcript levels of *SULTR1;1* and *SULTR1;2* in all plant lines (Figure 4a). Under the S1500 condition, transcript levels of *SULTR1;2* were similar between the mutants and WT, whereas those of *SULTR1;1* were higher in *wer* and lower in *cpc* than they were in WT (Figure 3a), which is consistent with previous reports [53,54]. These results indicated that the expression of *SULTR1;1* was correlated with the number of root hairs when there was sufficient sulfate for plant growth (Figure 2b,c and Figure 4a).

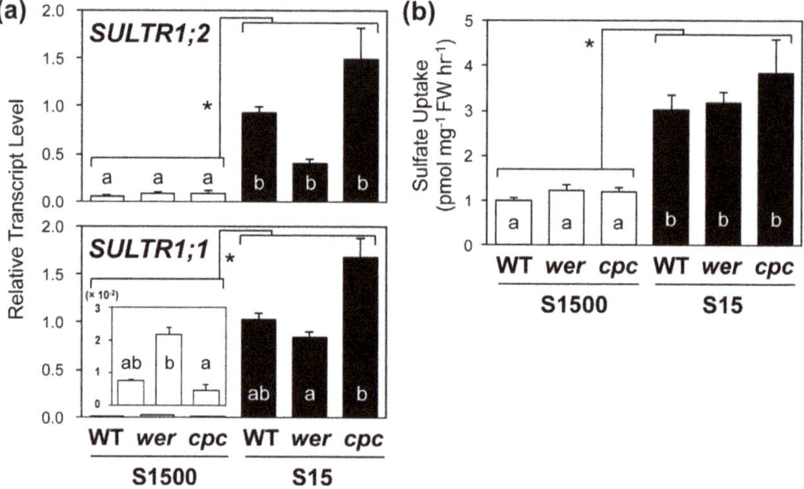


**Figure 4.** Transcript levels of *SULTR1;1* and *SULTR1;2*, and sulfate uptake activity in wild-type (WT), *werwolf* (*wer*), and *caprice* (*cpc*) mutants grown under S1500 and S15 conditions. (**a**) Relative transcript levels of *SULTR1;1* and *SULTR1;2* in root tissues determined using quantitative reverse transcription-polymerase chain reaction (qRT-PCR). (**b**) Sulfate uptake activity of plants. Absolute values of [$^{35}$S] sulfate uptake rates are presented. Plants were vertically grown for 10 d on S1500 (white bars) and S15 (black bars) agar media. Values and error bars indicate means ± SEM [n = 3 in (**a**), n = 6 in (**b**)]. Two-way analysis of variance (ANOVA) was used to detect the effects of sulfur conditions, genotypes, and their interactions. Asterisks indicate significant differences between S1500 and S15 conditions (two-way ANOVA; * $P < 0.05$). In (**a**), interactions between sulfur conditions and genotypes were detected and, so the Tukey–Kramer test was applied to all six experimental conditions. In (**b**), no interaction was detected and, so the Tukey–Kramer test was applied to the three genotypes grown under the same condition. Different letters indicate significant differences ($P < 0.05$).

However, under –S, transcript levels of *SULTR1;1* were lower in *wer* than they were in *cpc*, indicating a negative correlation with the number of root hairs (Figure 2b,c and Figure 4a). Supporting this observation, interactive effects between sulfur conditions and genotypes on the expression of *SULTR1;1* and *SULTR1;2* were detected using a two-way ANOVA, which indicates that the effects of the number of root hairs on their expression would be switched from positive to negative with environmental changes from +S to –S.

The sulfate uptake rate was stimulated under the S15 condition, which was consistent with previous reports [23–25,27,55,56] (Figure 4b). Despite the differences in *SULTR1;1* expression between the plant lines, sulfate uptake rates were relatively constant between the two lines under both S1500 and S15 conditions (Figure 4b).

The levels of sulfur-containing metabolites, sulfate, Cys, GSH and total sulfur in the precipitate after extraction with 10 mM HCl were decreased under the S15 condition in all plant lines (Figure 5). Similar to the sulfate uptake rate, the content of sulfate, Cys, and GSH in shoots was generally not influenced by mutations in *WER* and *CPC* under both S1500 and S15 conditions, except for slightly higher sulfur levels in the precipitate of *wer* under the S15 condition than in the WT.

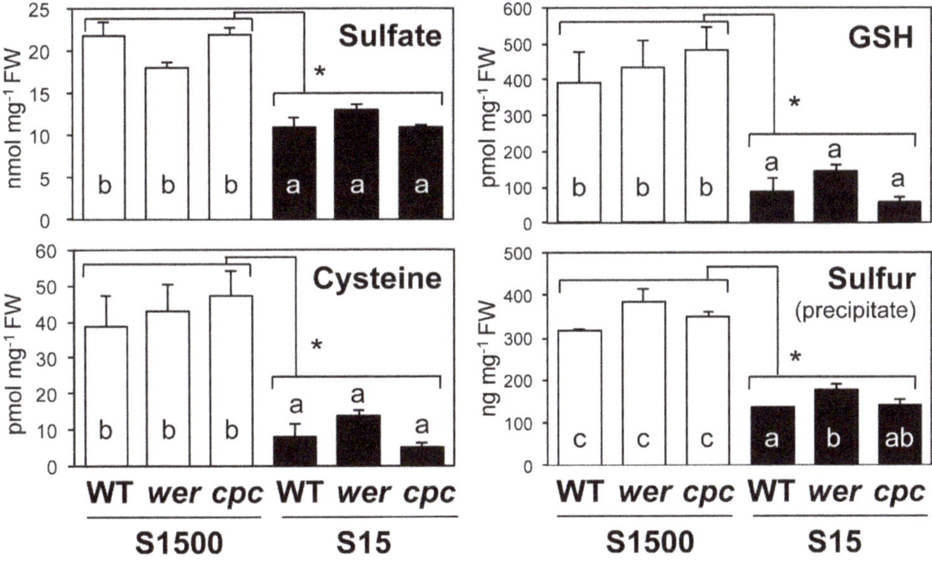


**Figure 5.** Accumulation of sulfate, cysteine (Cys), glutathione (GSH), and sulfur in shoots of wild-type (WT), *werwolf* (*wer*), and *caprice* (*cpc*) mutants grown under S1500 and S15 conditions. Plants were vertically grown for 10 d on S1500 (white bars) and S15 (black bars) agar media. Sulfate content in shoot tissues was determined using ion chromatography. Cys and GSH content of shoot tissues was analyzed using a high-performance liquid chromatography (HPLC)-fluorescent detection system, after labeling thiol bases with monobromobimane. Total sulfur content in precipitates was analyzed using inductively coupled plasma-mass spectroscopy (ICP-MS) after nitric acid digestion. Values and error bars indicate means ± SEM (n = 3). Two-way analysis of variance (ANOVA) was used to detect effects of sulfur conditions, genotypes, and their interactions. Asterisks indicate significant differences between S1500 and S15 conditions (two-way ANOVA; * $P < 0.05$). As all comparisons did not detect interactions, the Tukey–Kramer test was applied to detect significant differences among genotypes grown under similar conditions. Different letters indicate significant differences ($P < 0.05$).

The sum of sulfur levels analyzed in this study was converted to sulfur content in shoot per plant (Figure 6). Concentrations of sulfate, Cys, and GSH (Figure 5) were converted to sulfur content, and added to the sulfur content of the precipitate, and then the total was multiplied by the shoot FW per plant (Figure 2e). The sum of sulfur levels was lower in *wer* under the S1500 condition than it was in the WT and was similar among the genotypes under the S15 condition (Figure 6). Interaction between sulfur conditions and genotypes was detected using a two-way ANOVA, suggesting that sulfur levels were differentially controlled by the number of root hairs under +S and −S conditions.

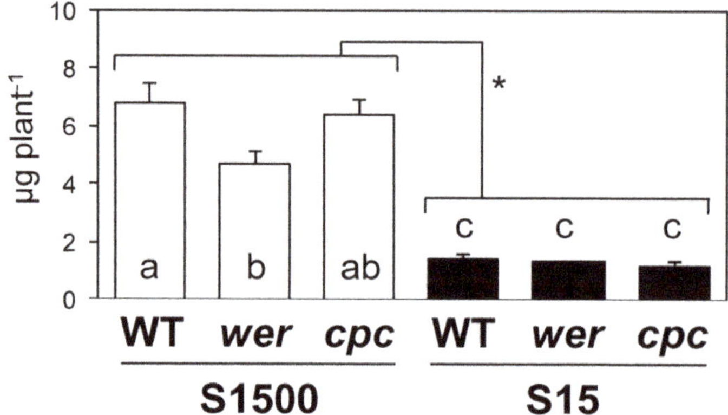


**Figure 6.** Sum of sulfur levels in shoots of WT, *wer*, and *cpc* mutants grown under S1500 and S15 conditions. Plants were vertically grown for 10 d on S1500 (white bars) and S15 (black bars) agar media. Sulfate, Cys, GSH, and sulfur contents in precipitate of shoot tissues were calculated to μg sulfur per plant. The values and error bars indicate means ± SEM (n = 3). As two-way ANOVA detected the interaction between sulfur conditions and genotypes, the Tukey–Kramer test was applied to all six experimental conditions. Different letters mean significant differences ($P < 0.05$). Asterisks indicate significant differences between S1500 and S15 conditions (2-way ANOVA; * $P < 0.05$).

## 3. Discussion

The mechanisms by which cellular differentiation influences the cellular biochemical or physiological functions and, vice versa, how cellular function influences cell fate, have both long been topics of discussion. In the case of sulfate uptake, two high-affinity sulfate transporters, SULTR1;1 and SULTR1;2, show different spatial expression patterns in root epidermal cells [21,24,26] (Figure 1). Specifically, we found that the expression of *SULTR1;1* was induced preferentially in the root-hair cells in a −S condition [57], whereas *SULTR1;2* was preferentially expressed in non-hair cells (Figure 1). Additionally, the expression of *SULTR1;1* was reported to be higher in *wer* and lower in *cpc* than in WT, whereas the expression of *SULTR1;2* was not influenced in the mutants [53,54]. Based on this information, we postulated that a decrease in the number of root hairs would reduce sulfate uptake activity. However, an increase or decrease in the number of root hairs did not influence the sulfate uptake rate in plants under either +S or −S conditions (Figure 4b). These results indicated that the number of root hairs does not contribute to the sulfate uptake rate, at least in the comparison among WT, *wer* and *cpc*.

Although we observed no difference in sulfate uptake between the WT and mutants, the contribution of root hairs to sulfate uptake cannot be completely ruled out because there was a negative correlation between the number of root hairs and primary root length [58] (Figure 2). The negative correlation suggested that there might be a mechanism to maintain a constant root biomass. A similar negative correlation was observed when plants were exposed to phosphate-deficient conditions, ethylene, or both [16,59], suggesting the involvement of general homeostatic mechanisms. The molecular machinery coordinating the negative interaction between root hair development and primary root length would be an interesting subject to investigate as a potential tool to improve the understanding of nutrient acquisition by plants. The S15 condition also induced a slight but significant increase in the number of root hairs (Figure 2c), which could contribute to the increase in sulfate uptake rate under –S conditions. These results suggested that the sulfate uptake rate was influenced by the total root biomass rather than the number of root hairs under both +S and –S conditions.

The expression of both *SULTRs* was correlated with the number of root hairs negatively under the S15 condition, but positively under the S1500 condition (Figures 2c and 4a). A similar tendency was observed for the expression of *RHD6* and *RSL4* (Figure 3b). Under the S1500 condition, the expression levels of *RHD6* and *RSL4* were higher in *wer* and tended to be lower in *cpc* than they were in the WT, which was consistent with the number of root hairs (Figures 2c and 3b). However, under the S15 condition, the expression of *RHD6* and *RSL4* was similar among the plant lines, whereas the number of root hairs varied in the mutants in a manner similar to that under S1500 condition (Figures 2c and 3b). Although it is not clear whether the expression of *RHD6*, *RSL4*, and *SULTRs* was regulated by the same mechanism under –S conditions, these results suggested that another mechanism likely increased root hair development under –S conditions.

Furthermore, phosphate deficiency induces *RSL4* expression, probably by increasing the stability of EIN3 and EIL1, which both bind to cognate binding sites in the promoter of *RSL4* [16,17]. In another root hair-less line, NR23, several events are induced by phosphate deficiency, including increased the expression of phosphate transporter genes and secretion of acid phosphatases and organic acids, compared to the WT [8]. This observation also supports the differing contribution of root hairs to the uptake of phosphate and sulfate, which varies with the chemical forms of nutrients in the soil [3,4].

Although there were negative or positive correlations between the expression of *SULTRs* and the number of root hairs under both +S and –S conditions, the sulfate uptake rate did not fluctuate among the plant lines (Figure 4). This finding suggests that the transcript levels of *SULTR1;1* and *SULTR1;2* were not the only determinants of sulfate uptake rate. Further studies of the relationship between root architecture and sulfate uptake may shed light on the as yet unknown determinants of this process.

## 4. Materials and Methods

### 4.1. Plant Materials and Growth Conditions

*Arabidopsis thaliana* plants, ecotype "Columbia" (Col-0), were used as the wild-type (WT) while *cpc* [41] (*cpc-1*) and *wer* [39] (*wer-1*) mutants were obtained from the Arabidopsis Biological Resource Center (ABRC). Plants were grown at 22 °C under continuous light (40 $\mu$mol $m^{-1}$ $s^{-1}$) conditions on mineral nutrient media containing 1% sucrose [60,61]. For the preparation of the agar medium, agar was washed twice with 1 L de-ionized water and vacuum filtered to remove the sulfate. Sulfur sufficient (S1500) agar medium was supplemented with 1500 $\mu$M $MgSO_4$. Sulfur-deficient (S15) agar medium was supplemented with 15 $\mu$M $MgSO_4$ and Mg concentration was adjusted to 1500 $\mu$M by adding $MgCl_2$. After the indicated shown in each figure, the shoot and root tissues were harvested separately, rinsed with distilled water, and subsequently subjected to various analyses.

### 4.2. Observation of GFP Fluorescence

The tissue-specific expression of GFP in $P_{SULTR1;1}$-*GFP* and $P_{SULTR1;2}$-*GFP* transgenic plants was visualized in whole mounts of 7-day-old plants using a fluorescent microscope system (EVOS FL Auto 2 Imaging System) equipped with the EVOS Light Cube, GFP (Ex: 470/22, Em: 525/50) (Thermo Fisher Scientific, USA).

### 4.3. Observation of Root Development

The primary root lengths of plants were analyzed using images captured with a STAGE2000-BG system (AMZ System Science, Japan). The number of root hairs in 5 mm from the root tip was analyzed from the image captured using a CCD camera (WRAYCAM G500, WRAYMER, Japan) connected to a stereoscopic microscope (SW-700TD, WRAYMER). The free software package, ImageJ [62,63] was used for the analysis.

### 4.4. Quantitative Real-Time RT-PCR Analysis

RNA preparation and RT were performed as reported previously [27,54,55]. Real-time PCR was carried out using a SYBR Green Perfect real-time kit (Takara, Japan) and Thermal Cycler Dice real-time system (Takara) using the gene-specific primers for CPC, CPC-F (5′-GGATGTATAAACTCGTTGG CGACAG-3′) and CPC-R (5′-GCCGTGTTTCATAAGCCAATATCTC-3′) [64]; for WER, WER-F (5′-TGGTAATAGGTATAACTTCATTTGC-3′) and WER-R (5′-TTTGATTCCGAGTTTCTTACTAAGG ATG-3′); for GL2, GL2-F (5′-TCGGATCACTGAGACCACAA-3′) and GL2-R (5′-GTGTATCCCGG AACCAGTGT-3′) [64]; for RHD6, RHD6-real-F (5′-TGATTTGGTGACAATGCTTGA-3′) and RHD6-real-R (5′-GGAGAGAATGGCATCAATGG-3′) [49]; for RSL4, RSL4_q-PCR f new (5′-AACCTT GTGCCAAACGGGAC-3′) and RSL4_q-PCR r new (5′-CCAGGCCGTTGTAAGCCAAT-3′) [17]; and, for *SULTR1;2*, SULTR1;2-1854F (5′-GGATCCAGAGATGGCTACATGA-3′) and SULTR1;2-1956R (5′-TCGATGT CCGTAACAGGTGAC-3′) [27]; for *SULTR1;1*, SULTR1;1-625F (5′-GCCATCACAA TCGCTCTCCAA-3′) and SULTR1;1-750R (5′-TTGCCAATTCCACCCATGC-3′) [30]; and for ubiquitin, UBQ2-144F (5′-CCAAG ATCCAGGACAAAGAAGGA-3′) and UBQ2-372R (5′-TGGAGACGAGC ATAACACTTGC-3′) [30]. The relative mRNA levels were calculated using ubiquitin2 as an internal standard.

### 4.5. Sulfate Uptake Assay

Plants were vertically grown for 10 days on S1500 and S15 agar media. The roots were submerged in S1500 medium containing 15 μM [$^{35}$S] sodium sulfate (American Radiolabeled Chemicals, USA) and incubated for 1 h. Washing and measurement were carried out as described previously [25,27,54,55,65,66].

### 4.6. Measurement of Sulfate, Cysteine and Glutathione, and Total Sulfur Levels

Plant tissues were frozen in liquid nitrogen and homogenized in 5 volumes of 10 mM HCl using a Tissue Lyser (Retsch, Germany). After homogenization, the cell debris was removed by centrifugation, and the supernatant was subsequently analyzed.

For sulfate measurement, the extracts were diluted 100-fold with extra pure water and analyzed using ion chromatography (IC-2001, TOSOH, Japan). Using serial 30-μL injections, the diluted extracts were separated at 40 °C using a TSK SuperIC-AZ column (TOSOH) at a flow rate of 0.8 mL min$^{-1}$ with an eluent containing 1.9 mM $NaHCO_3$ and 3.2 mM $Na_2CO_3$. Anion mixture standard solution 1 (Wako Pure Chemicals, Japan) was used as a standard.

Cys and GSH contents were determined using monobromobimane (Invitrogen, USA) labeling of the thiols after reduction of the plant extracts with dithiothreitol (DTT). The labeled products were then separated using HPLC (JASCO, Japan) using the TSKgel ODS-120T column (150 × 4.6 mm, TOSOH) and detected using a fluorescence detector, FP-920 (JASCO), monitoring for fluorescence of thiol-bimane adducts at 482 nm under excitation at 390 nm. Cys and GSH (Nacalai Tesque, Japan) were used as standards.

The total sulfur content was determined using inductively coupled plasma-mass spectroscopy (ICP-MS; Agilent7700, Agilent Technologies, USA). The precipitates obtained from the extraction described above were digested in 200 μL $HNO_3$ at 95 °C for 30 min and then 115 °C for 90 min. After cooling to room temperature, the digested samples were diluted to 1 mL with extra pure water, filtered using 0.45 μm filters (DISMIC-03CP, ADVANTEC, Japan). The filtered samples were diluted 10 times with a solution consisting of 0.1 M $HNO_3$ and 10 μg L$^{-1}$ gallium (KANTO CHEMICAL, Japan) as an internal standard, before being subjected to ICP-MS. Quantification was performed using the standard curve obtained via serial dilutions of the sulfur standard solution (KANTO CHEMICAL).

### 4.7. Statistical Analysis

Two-way ANOVA was used to detect the effects of sulfur conditions, genotypes, and their interactions (Figures 2–6). Significant differences between S1500 and S15 conditions detected using a two-way ANOVA are indicated with asterisks ($P < 0.05$). Furthermore, where an interaction was detected between sulfur conditions and genotypes, the Tukey–Kramer test was applied to all experimental conditions (Figures 3 and 4a). In addition, where interaction was not detected between the values, the Tukey–Kramer test was used to analyze all genotypes grown under the same growth condition. Significant differences detected by the Tukey–Kramer test were shown with different letters ($P < 0.05$). Statcel4 software (OMS Publishing Inc., Tokyo, Japan) was used for all statistical analysis with the Microsoft Excel program.

**Author Contributions:** A.M.N., R.T.W. and T.W. designed the research. Y.K., T.U., A.S. and A.M.N. performed the experiments and analyzed the data. A.M.N. wrote the manuscript.

**Funding:** This work was supported by JSPS KAKENHI Grant Number 24380040 and 17H03785 (for A.M.N.).

**Acknowledgments:** We thank Yuki Mori (Kyushu University, Japan) for instructing the total sulfur analysis with ICP-MS. We thank Yasuo Niwa (University of Shizuoka, Japan) for providing the sGFP(S65T) vector. We thank Chihiro Iwaya and Yukiko Okuo for the technical support.

**Conflicts of Interest:** The authors declare no conflict of interest.

## References

1. White, P. Ion uptake mechanisms of individual cells and roots: Short-distance transport. In *Mineral Nutrition of Higher Plants*, 3rd ed.; Marschner, P., Ed.; Academic Press: Oxford, UK, 2012; pp. 7–48.
2. Patrick, J.W.; Tyerman, S.D.; van Bel, A.J.E. Long-Distance Transport. In *Biochemistry & Molecular Biology of Plants*, 2nd ed.; Buchana, B.B., Gruissem, W., Jones, R.L., Eds.; Wiley Blackwell: Oxford, UK, 2015; pp. 658–710.
3. Lopez-Bucio, J.; Cruz-Ramirez, A.; Herrera-Estrella, L. The role of nutrient availability in regulating root architecture. *Curr. Opin. Plant Biol.* **2003**, *6*, 280–287. [CrossRef]
4. Delhaize, E.; Schachtman, D.; Kochian, L.; Ryan, P.R. Mineral nutrient acquisition, transport, and utilization. In *Biochemistry & Molecular Biology of Plants*, 2nd ed.; Buchana, B.B., Gruissem, W., Jones, R.L., Eds.; Wiley Blackwell: Oxford, UK, 2015; pp. 1101–1131.
5. Long, S.R.; Kahn, M.; Seefeldt, L.; Tsay, Y.F.; Kopriva, S. Nitrogen and Sulfur. In *Biochemistry & Molecular Biology of Plants*, 2nd ed.; Buchana, B.B., Gruissem, W., Jones, R.L., Eds.; Wiley Blackwell: Oxford, UK, 2015; pp. 746–768.
6. Schiefelbein, J.; Kwak, S.-H.; Wieckowski, Y.; Barron, C.; Bruex, A. The gene regulatory network for root epidermal cell-type pattern formation in *Arabidopsis*. *J. Exp. Bot.* **2009**, *60*, 1515–1521. [CrossRef] [PubMed]
7. Libault, M.; Brechenmacher, L.; Cheng, J.; Xu, D.; Stacey, G. Root hair systems biology. *Trends Plant Sci.* **2010**, *15*, 641–650. [CrossRef] [PubMed]
8. Tanaka, N.; Kato, M.; Tomioka, R.; Kurata, R.; Fukao, Y.; Aoyama, T.; Maeshima, M. Characteristics of a root hair-less line of *Arabidopsis thaliana* under physiological stresses. *J. Exp. Bot.* **2014**, *65*, 1497–1512. [CrossRef] [PubMed]
9. Gilroy, S.; Jones, D.L. Through form to function: Root hair development and nutrient uptake. *Trends Plant Sci.* **2000**, *5*, 56–60. [CrossRef]
10. Gahoonia, T.S.; Nielsen, N.E.; Joshi, P.A.; Jahoor, A. A root hairless barley mutant for elucidating genetic of root hairs and phosphorus uptake. *Plant Soil* **2001**, *235*, 211–219. [CrossRef]
11. Schmidt, W.; Tittel, J.; Schikora, A. Role of hormones in the induction of iron deficiency responses in Arabidopsis roots. *Plant Physiol.* **2000**, *122*, 1109–1118. [CrossRef]
12. Ma, J.F.; Goto, S.; Tamai, K.; Ichii, M. Role of root hairs and lateral roots in silicon uptake by rice. *Plant Phys.* **2001**, *127*, 1773–1780. [CrossRef]
13. Ma, Z.; Bielenberg, D.G.; Brown, K.M.; Lynch, J.P. Regulation of root hair density by phosphorus availability in *Arabidopsis thaliana*. *Plant Cell Environ.* **2001**, *24*, 459–467. [CrossRef]
14. Mueller, M.; Schmidt, W. Environmentally induced plasticity of root hair development in Arabidopsis. *Plant Phys.* **2004**, *134*, 409–419. [CrossRef] [PubMed]
15. Yang, T.J.W.; Perry, P.J.; Ciani, S.; Pandian, S.; Schmidt, W. Manganese deficiency alters the patterning and development of root hairs in *Arabidopsis*. *J. Exp. Bot.* **2008**, *59*, 3453–3464. [CrossRef] [PubMed]
16. Song, L.; Yu, H.; Dong, J.; Che, X.; Jiao, Y.; Liu, D. The molecular mechanism of ethylene-mediated root hair development induced by phosphate starvation. *PLoS Genet.* **2016**, *12*, e1006194. [CrossRef]

17. Feng, Y.; Xu, P.; Li, B.; Li, P.; Wen, X.; An, F.; Gong, Y.; Xin, Y.; Zhu, Z.; Wang, Y.; et al. Ethylene promotes root hair growth through coordinated EIN3/EIL1 and RHD6/RSL1 activity in *Arabidopsis*. *Proc. Natl. Acad. Sci. USA* **2017**, *114*, 13834–13839. [CrossRef]
18. Takahashi, H.; Kopriva, S.; Giordano, M.; Saito, K.; Hell, R. Sulfur assimilation in photosynthetic organisms: Molecular functions and regulations of transporters and assimilatory enzymes. *Annu. Rev. Plant Biol.* **2011**, *62*, 157–184. [CrossRef]
19. Leustek, T.; Martin, M.N.; Bick, J.; Davies, J.P. Pathways and regulation of sulfur metabolism revealed through molecular genetic studies. *Annu. Rev. Plant Physiol. Plant Mol. Biol.* **2000**, *51*, 141–166. [CrossRef]
20. Saito, K. Sulfur assimilatory metabolism. The long and smelling road. *Plant Physiol.* **2004**, *136*, 2443–2450. [CrossRef]
21. Takahashi, H.; Watanabe-Takahashi, A.; Smith, F.W.; Blake-Kalff, M.; Hawkesford, M.J.; Saito, K. The roles of three functional sulfate transporters involved in uptake and translocation of sulfate in *Arabidopsis thaliana*. *Plant J.* **2000**, *23*, 171–182. [CrossRef] [PubMed]
22. Vidmar, J.J.; Tagmount, A.; Cathala, N.; Touraine, B.; Davidian, J.C.E. Cloning and characterization of a root specific high-affinity sulfate transporter from *Arabidopsis thaliana*. *FEBS Lett.* **2000**, *475*, 65–69. [CrossRef]
23. Shibagaki, N.; Rose, A.; McDermott, J.P.; Fujiwara, T.; Hayashi, H.; Yoneyama, T.; Davies, J.P. Selenate-resistant mutants of *Arabidopsis thaliana* identify *Sultr1;2*, a sulfate transporter required for efficient transport of sulfate into roots. *Plant J.* **2002**, *29*, 475–486. [CrossRef] [PubMed]
24. Yoshimoto, N.; Takahashi, H.; Smith, F.W.; Yamaya, T.; Saito, K. Two distinct high-affinity sulfate transporters with different inducibilities mediate uptake of sulfate in *Arabidopsis* roots. *Plant J.* **2002**, *29*, 465–473. [CrossRef] [PubMed]
25. Yoshimoto, N.; Inoue, E.; Watanabe-Takahashi, A.; Saito, K.; Takahashi, H. Posttranscriptional regulation of high-affinity sulfate transporters in Arabidopsis by sulfur nutrition. *Plant Physiol.* **2007**, *145*, 378–388. [CrossRef] [PubMed]
26. Maruyama-Nakashita, A.; Nakamura, Y.; Watanabe-Takahashi, A.; Yamaya, T.; Takahashi, H. Induction of *SULTR1;1* sulfate transporter in *Arabidopsis* roots involves protein phosphorylation/dephosphorylation circuit for transcriptional regulation. *Plant Cell Physiol.* **2004**, *45*, 340–345. [CrossRef]
27. Maruyama-Nakashita, A.; Nakamura, Y.; Yamaya, T.; Takahashi, H. A novel regulatory pathway of sulfate uptake in *Arabidopsis* roots: Implication of CRE1/WOL/AHK4-mediated cytokinin-dependent regulation. *Plant J.* **2004**, *38*, 779–789. [CrossRef]
28. Maruyama-Nakashita, A.; Nakamura, Y.; Watanabe, A.; Inoue, E.; Yamaya, T.; Takahashi, H. Identification of a novel *cis*-acting element conferring sulfur deficiency response in *Arabidopsis* roots. *Plant J.* **2005**, *42*, 305–314. [CrossRef] [PubMed]
29. Maruyama-Nakashita, A. Metabolic changes sustain the plant life in low-sulfur environments. *Curr. Opin. Plant Biol.* **2017**, *39*, 144–151. [CrossRef]
30. Maruyama-Nakashita, A.; Inoue, E.; Watanabe-Takahashi, A.; Yamaya, T.; Takahashi, H. Transcriptome profiling of sulfur-responsive genes in Arabidopsis reveals global effect on sulfur nutrition on multiple metabolic pathways. *Plant Physiol.* **2003**, *132*, 597–605. [CrossRef]
31. Rouached, H.; Wirtz, M.; Alary, R.; Hell, R.; Arpat, A.B.; Davidian, J.C.; Fourcroy, P.; Berthomieu, P. Differential regulation of the expression of two high-affinity sulfate transporters, SULTR1.1 and SULTR1.2, in Arabidopsis. *Plant Physiol.* **2008**, *147*, 897–911. [CrossRef]
32. Grebe, M. The patterning of epidermal hairs in *Arabidopsis*—updated. *Curr. Opin. Plant Biol.* **2012**, *15*, 31–37. [CrossRef]
33. Salazar-Henao, J.E.; Veélez-Bermuédez, I.C.; Schmidt, W. The regulation and plasticity of root hair patterning and morphogenesis. *Development* **2016**, *143*, 1848–1858. [CrossRef]
34. Ishida, T.; Kurata, T.; Okada, K.; Wada, T. A genetic regulatory network in the development of trichomes and root hairs. *Annu. Rev. Plant Biol.* **2008**, *59*, 365–386. [CrossRef] [PubMed]
35. Shibata, M.; Sugimoto, K. A gene regulatory network for root hair development. *J. Plant Res.* **2019**. [CrossRef] [PubMed]
36. Walker, A.R.; Davison, P.A.; Bolognesi-Winfield, A.C.; James, C.M.; Srinivasan, N.; Blundell, T.L.; Esch, J.J.; Marks, M.D.; Gray, J.C. The *TRANSPARENT TESTA GLABRA1* Locus, which regulates trichome differentiation and anthocyanin biosynthesis in *Arabidopsis*, encodes a WD40 repeat protein. *Plant Cell* **1999**, *11*, 1337–1349. [CrossRef]

37. Payne, C.T.; Zhang, F.; Lloyd, A.M. *GL3* encodes a bHLH protein that regulates trichome development in *Arabidopsis* through interaction with GL1 and TTG1. *Genetics* **2000**, *156*, 1349–1362.
38. Bernhardt, C.; Lee, M.M.; Gonzalez, A.; Zhang, F.; Lloyd, A.; Schiefelbein, J. The bHLH genes *GLABRA3* (*GL3*) and *ENHANCER OF GLABRA3* (*EGL3*) specify epidermal cell fate in the *Arabidopsis* root. *Development* **2003**, *130*, 6431–6439. [CrossRef]
39. Lee, M.M.; Schiefelbein, J. WERWOLF, a MYB-related protein in *Arabidopsis*, is a position-dependent regulator of epidermal cell patterning. *Cell* **1999**, *99*, 473–483. [CrossRef]
40. Rerie, W.G.; Feldmann, K.A.; Marks, M.D. The GLABRA2 gene encodes a homeo domain protein required for normal trichome development in *Arabidopsis*. *Genes Dev.* **1994**, *8*, 1388–1399. [CrossRef] [PubMed]
41. Wada, T.; Tachibana, T.; Shimura, Y.; Okada, K. Epidermal cell differentiation in *Arabidopsis* determined by a Myb-homolog, CPC. *Science* **1997**, *277*, 1113–1116. [CrossRef] [PubMed]
42. Schellmann, S.; Schnittger, A.; Kirik, V.; Wada, T.; Okada, K.; Beermann, A.; Thumfahrt, J.; Jürgens, G.; Hülskamp, M. *TRIPTYCHON* and *CAPRICE* mediate lateral inhibition during trichome and root hair patterning in *Arabidopsis*. *EMBO J.* **2002**, *21*, 5036–5046. [CrossRef] [PubMed]
43. Kirik, V.; Simon, M.; Huelskamp, M.; Schiefelbein, J. The ENHANCER OF TRY AND CPC1 gene acts redundantly with TRIPTYCHON and CAPRICE in trichome and root hair cell patterning in Arabidopsis. *Dev. Biol.* **2004**, *268*, 506–513. [CrossRef]
44. Tominaga, R.; Iwata, M.; Sano, R.; Okada, K.; Wada, T. *Arabidopsis CAPRICE-like Myb 3* (*CPL3*) controls endoreduplication and flowering development in addition to trichome and root-hair formation. *Development* **2008**, *135*, 1335–1345. [CrossRef]
45. Tominaga, R.; Iwata, M.; Okada, K.; Wada, T. Functional analysis of the epidermal-specific MYB genes *CAPRICE* and *WEREWOLF* in *Arabidopsis*. *Plant Cell* **2007**, *19*, 2264–2277. [CrossRef] [PubMed]
46. Song, S.K.; Ryu, K.H.; Kang, Y.H.; Song, J.H.; Cho, Y.H.; Yoo, S.D.; Schiefelbein, J.; Lee, M.M. Cell fate in the *Arabidopsis* root epidermis is determined by competition between WEREWOLF and CAPRICE. *Plant Physiol.* **2011**, *157*, 1196–1208. [CrossRef]
47. Masucci, J.D.; Schiefelbein, J.W. The rhd6 mutation of Arabidopsis thaliana alters root-hair initiation through an auxin- and ethylene-associated process. *Plant Physiol.* **1994**, *106*, 1335–1346. [CrossRef]
48. Menand, B.; Yi, K.; Jouannic, S.; Hoffmann, L.; Ryan, E.; Linstead, P.; Schaefer, D.G.; Dolan, L. An ancient mechanism controls the development of cells with a rooting function in land plants. *Science* **2007**, *316*, 1477–1480. [CrossRef] [PubMed]
49. Lin, Q.; Ohashi, Y.; Kato, M.; Tsuge, T.; Gu, H.; Qu, L.-J.; Aoyama, T. GLABRA2 directly suppresses basic helix-loop-helix transcription factor genes with diverse functions in root hair development. *Plant Cell* **2015**, *27*, 2894–2906. [CrossRef]
50. Yi, K.; Menand, B.; Bell, E.; Dolan, L. A basic helix-loop-helix transcription factor controls cell growth and size in root hairs. *Nat. Genet.* **2010**, *42*, 264–267. [CrossRef]
51. Rubio, V.; Linhares, F.; Solano, R.; Martin, A.C.; Iglesias, J.; Leyva, A.; Paz-Ares, J. A conserved MYB transcription factor involved in phosphate starvation signaling both in vascular plants and in unicellular algae. *Genes Dev.* **2001**, *15*, 2122–2133. [CrossRef] [PubMed]
52. Bustos, R.; Castrillo, G.; Linhares, F.; Puga, M.I.; Rubio, V.; Pérez-Pérez, J.; Solano, R.; Leyva, A.; Paz-Ares, J. A central regulatory system largely controls transcriptional activation and repression responses to phosphate starvation in *Arabidopsis*. *PLoS Genet.* **2010**, *6*, e1001102. [CrossRef]
53. Bruex, A.; Kainkaryam, R.M.; Wieckowski, Y.; Kang, Y.H.; Bernhardt, C.; Xia, Y.; Zheng, X.; Wang, J.Y.; Lee, M.M.; Benfey, P.; et al. A gene regulatory network for root epidermis cell differentiation in *Arabidopsis*. *PLoS Genet.* **2012**, *8*, e1002446. [CrossRef] [PubMed]
54. Simon, M.; Bruex, A.; Kainkaryam, R.M.; Zheng, X.; Huang, L.; Woolf, P.J.; Schiefelbein, J. Tissue-specific profiling reveals transcriptome alterations in *Arabidopsis* mutants lacking morphological phenotypes. *Plant Cell* **2013**, *25*, 3175–3185. [CrossRef] [PubMed]
55. Maruyama-Nakashita, A.; Nakamura, Y.; Tohge, T.; Saito, K.; Takahashi, H. *Arabidopsis* SLIM1 is a central transcriptional regulator of plant sulfur response and metabolism. *Plant Cell* **2006**, *18*, 3235–3251. [CrossRef]
56. Maruyama-Nakashita, A.; Watanabe-Takahashi, A.; Inoue, E.; Yamaya, T.; Saito, K.; Takahashi, H. Sulfur-responsive elements in the 3′-nontranscribed intergenic region are essential for the induction of SULFATE TRANSPORTER 2;1 gene expression in Arabidopsis roots under sulfur deficiency. *Plant Cell* **2015**, *27*, 1279–1296. [CrossRef] [PubMed]

57. Lan, P.; Li, W.; Lin, W.-D.; Santi, S.; Schmidt, W. Mapping gene activity of Arabidopsis root hairs. *Genome Biol.* **2013**, *14*, R67. [CrossRef]
58. Tominaga-Wada, R.; Wada, T. Relationship between root hair formation and primary root length in *Arabidopsis thaliana*. *Res. Rev. Biosci.* **2017**, *12*, 128.
59. Negi, S.; Ivanchenko, M.G.; Muday, G.K. Ethylene regulates lateral root formation and auxin transport in *Arabidopsis thaliana*. *Plant J.* **2008**, *55*, 175–187. [CrossRef]
60. Fujiwara, T.; Hirai, M.Y.; Chino, M.; Komeda, Y.; Naito, S. Effects of sulfur nutrition on expression of the soybean seed storage protein genes in transgenic petunia. *Plant Physiol.* **1992**, *99*, 263–268. [CrossRef] [PubMed]
61. Hirai, M.Y.; Fujiwara, T.; Chino, M.; Naito, S. Effects of sulfate concentrations on the expression of a soybean seed storage protein gene and its reversibility in transgenic *Arabidopsis thaliana*. *Plant Cell Physiol.* **1995**, *36*, 1331–1339. [PubMed]
62. Schneider, C.A.; Rasband, W.S.; Eliceiri, K.W. NIH image to ImageJ: 25 years of image analysis. *Nat. Methods* **2012**, *9*, 671–675. [CrossRef] [PubMed]
63. Hartig, S.M. Basic image analysis and manipulation in ImageJ. *Curr. Protoc. Mol. Biol.* **2013**, *102*, 14–15. [CrossRef]
64. Hayashi, N.; Tetsumura, T.; Sawa, S.; Wada, T.; Tominaga-Wada, R. CLE14 peptide signaling in Arabidopsis root hair cell fate determination. *Plant Biotechnol.* **2018**, *35*, 17–22. [CrossRef]
65. Kataoka, T.; Hayashi, N.; Yamaya, T.; Takahashi, H. Root-to-shoot transport of sulfate in *Arabidopsis*: Evidence for the role of SULTR3; 5 as a component of low-affinity sulfate transport system in the root vasculature. *Plant Physiol.* **2004**, *136*, 4198–4204. [CrossRef] [PubMed]
66. Yoshimoto, N.; Kataoka, T.; Maruyama-Nakashita, A.; Takahashi, H. Measurement of uptake and root-to-shoot distribution of sulfate in *Arabidopsis* seedlings. *Bio-Protocol* **2016**, *6*, e1700. [CrossRef]

*Article*

# CLE-CLAVATA1 Signaling Pathway Modulates Lateral Root Development under Sulfur Deficiency

**Wei Dong, Yinghua Wang and Hideki Takahashi ***

Department of Biochemistry and Molecular Biology, Michigan State University, East Lansing, MI 48824, USA; wdong@msu.edu (W.D.); wangyi64@msu.edu (Y.W.)
* Correspondence: htakaha@msu.edu

Received: 2 March 2019; Accepted: 17 April 2019; Published: 18 April 2019

**Abstract:** Plant root system architecture changes drastically in response to availability of macronutrients in the soil environment. Despite the importance of root sulfur (S) uptake in plant growth and reproduction, molecular mechanisms underlying root development in response to S availability have not been fully characterized. We report here on the signaling module composed of the CLAVATA3 (CLV3)/EMBRYO SURROUNDING REGION (CLE) peptide and CLAVATA1 (CLV1) leucine-rich repeat receptor kinase, which regulate lateral root (LR) development in *Arabidopsis thaliana* upon changes in S availability. The wild-type seedlings exposed to prolonged S deficiency showed a phenotype with low LR density, which was restored upon sulfate supply. In contrast, the *clv1* mutant showed a higher daily increase rate of LR density relative to the wild-type under prolonged S deficiency, which was diminished to the wild-type level upon sulfate supply, suggesting that CLV1 directs a signal to inhibit LR development under S-deficient conditions. *CLE2* and *CLE3* transcript levels decreased under S deficiency and through CLV1-mediated feedback regulations, suggesting the levels of CLE peptide signals are adjusted during the course of LR development. This study demonstrates a fine-tuned mechanism for LR development coordinately regulated by CLE-CLV1 signaling and in response to changes in S availability.

**Keywords:** *Arabidopsis thaliana*; CLE peptide; CLAVATA1; root system architecture; small signaling peptide; sulfate; sulfur

## 1. Introduction

Plant roots optimize nutrient uptake capacity by altering the root system architecture (RSA) in the soil environment [1]. Changes in nutrient availabilities have a distinct effect on RSA depending on the nutrient types and the amount supplied or locally concentrated in the soil environment [2]. The macronutrient sulfur (S), in the form of sulfate, is a mobile resource found in deeper soil profiles [1]. It is proposed that a combination of a thick and deep primary root (PR) with few and long lateral roots (LRs) can improve the uptake of S [3]. To achieve the adjustment of RSA, individual root traits can be regulated independently in response to changes in nutrient availabilities and patterns of nutrient distributions in the soil environment [4]. Among all root traits, LRs are phenotypically evaluated by length, total numbers, and density, which is considered a major determinant of RSA [5]. Changes in sulfate availability have a variable effect on LR development. For example, several studies have demonstrated that S deficiency leads to reduction in LR length [2,6–9] and LR number or density [2,7,9]. However, active growth of LR appears to be another response to sulfate starvation as it is described with longer LR length [10,11] and higher LR number or density [11–13].

Studies on functional characterization of small signaling peptides (SSPs) reveal that several distinct groups of SSPs play important roles in plant root development [14–24]. Nitrogen (N)-responsive C-TERMINALLY ENCODED PEPTIDEs (CEPs) are a group of SSPs known to be functional as negative regulators of LR development under N-limited conditions [14–17], while they have also been shown to be involved in long-distance regulation of N uptake [18]. A few distinct members of the CLAVATA3 (CLV3)/EMBRYO SURROUNDING REGION (CLE) family are also well characterized as SSPs regulating LR development in *Arabidopsis thaliana* (*A. thaliana*) [19,24]. *CLE1*, *-3*, *-4*, and *-7* are characterized particularly in relation to N nutritional responses as they are expressed in roots under N deficient conditions [19]. The LR phenotypes depicted in our previous study therefore highlight the *CLE3* gene expression enhanced in roots as a potential mechanism suppressing LR development under low N supply or in response to systemic N demand signals [19]. *CLE2* and *CLE3* also demonstrate positive responses of gene expression after N resupply to N-starved seedlings, as their transcript levels are shown to dramatically increase in response to nitrate and ammonium, respectively [19,25]. *CLE1*, *-2*, *-3*, *-4*, and *-7* are predominantly expressed in pericycle cells in roots, while *CLE1* and *CLE5* promoters are also found to be active in epidermal cells of the primary root tip [19]. These partially overlapping expression patterns, environmental responses and feedback regulation suggest functional redundancy of these N-responsive *CLE* genes in LR development. In addition to mechanisms characterized in relation to the N status, *CLE* genes are also known to be transcriptionally modulated by changes in availability of other macronutrients including S, phosphorus (P) and potassium (K) [24,26], as well as perturbation of cellular status caused by phytohormones and environmental stimuli [24,27]. More specifically to responses to S in roots, *CLE12* and *CLE2* are known to be up- and down-regulated, respectively, by S deprivation [24]. Thus, *CLE2* seems to be controlled by the S-responsive pathways in addition to being up-regulated by resupply of N [19]. In contrast, the S-responsive regulation of *CLE3* gene expression has not been studied despite its roles in LR development documented in relation to responses to N nutrition. CLE2 peptide has been shown to physically bind to the CLE receptor CLAVATA1 (CLV1) [28], and CLE3 requires CLV1 to transmit signals to modulate LR development [19]. Based on these aspects of nutrient-responsive regulation of *CLE2* and *CLE3* gene expression and their specific relationship with CLV1, we focused on investigating the effect of S on the CLE-CLV1 signaling pathway and LR development.

Here, we report the CLE-CLV1 signaling pathway is associated with S-responsive mechanisms modulating LR development in *A. thaliana*. The results shown in this study suggest a link between the morphological responses of LRs and the CLE-CLV1 signaling pathway, as well as S-responsive regulations of *CLE2* and *CLE3* genes in *A. thaliana* seedlings exposed to prolonged S deficiency.

## 2. Results

### 2.1. CLAVATA1 Controls Lateral Root Development under S Deficiency

To investigate the effect of S supply on root development, the wild-type *A. thaliana* (accession Columbia-0 (Col-0)) were germinated and precultured on a –S (15 μM sulfate) or +S (1500 μM sulfate) medium for 7 days. The seedlings were then transferred to the medium with the same concentration of sulfate, or from the –S preculture to the +S medium, or from the +S preculture to the –S medium, and grown for 3 days to validate the effect of S starvation and S replenishment (Figure 1). The most significant changes in root morphology at Day 10 were the decrease in length and number of LRs after long-term limitation of S (Figure 1a; plants transferred from –S to –S) compared to the recovery of the roots observed in response to supply of sulfate (Figure 1b; plants transferred from –S to +S). The PR growth was slightly enhanced when the seedlings were transferred from the –S preculture to the +S medium (Figure 1a,b). In contrast, the seedlings from the +S preculture medium transferred either to the +S or –S medium showed no significant changes in the root morphological phenotypes (Figure 1c,d).

Based on these observations, we hypothesized the CLE-CLV1 signaling pathway, which has been shown to regulate LR development under N-starved conditions [19], would be involved in pathways modulating root growth under S-limited conditions. To test this hypothesis, we focused on studying the effect of *clv1* mutations on root growth under conditions where the differences in root morphology were most significant. As mentioned, the most significant morphological changes were observed for the –S-precultured seedlings transferred to either the –S or +S medium (Figure 1a,b). The effect of S supply on changes in root morphology was recorded at Day 7 before the transfer and during the 3 consecutive days in the post-transfer growth period until Day 10 (Figures 2–4). The results indicated that changes in S conditions had little influence on primary root growth (Figure 2). The *clv1-101* mutant had slightly shorter PR values compared to its background wild-type accession (Col-0), regardless of changes in S conditions; however, the same phenotype was not observed in the *clv1-4* mutant in comparison with its background wild-type accession Landsberg *erecta* (L*er*) (Figure 2).

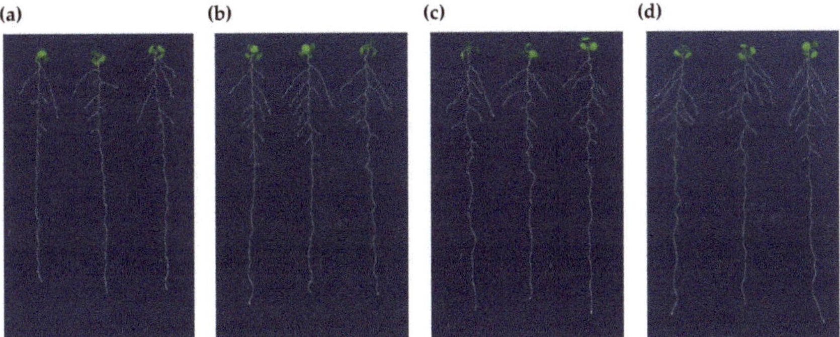


**Figure 1.** Effect of sulfur (S) supply on root phenotypes of *Arabidopsis thaliana* (*A. thaliana*) seedlings. The wild-type Columbia-0 (Col-0) seedlings were grown vertically on –S (15 μM sulfate) or +S (1500 μM sulfate) medium for 7 days and transferred to –S or +S medium in a reciprocal manner to be grown subsequently for 3 days. The scanned images of seedlings on Day 10 are shown. The images show the phenotypes of representative seedlings transferred from (**a**) –S to –S, (**b**) –S to +S, (**c**) +S to +S, and (**d**) +S to –S, respectively.

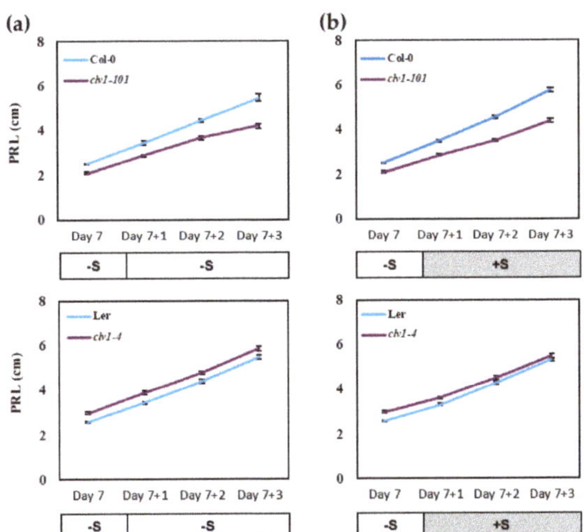


**Figure 2.** Effect of S supply on primary root growth. Wild-type (Col-0 and L*er*) and *clv1* mutant lines (*clv1-101* and *clv1-4*) were grown vertically on –S (15 μM sulfate) medium for 7 days and then transferred to (**a**) –S medium or (**b**) +S medium to be grown subsequently for 3 days. Primary root length (PRL) was measured before (Day 7) and after the transfer (Day 7+1, 7+2, and 7+3) as indicated on each graph. Values show means (± SE) of 24 individual plants per treatment. White and dark grey bars labeled –S (15 μM sulfate) and +S (1500 μM sulfate) below the graph represent the S conditions before the transfer and during the treatment.

As for the phenotypes associated with LR development, the LR density (LRD; total lateral root number divided by primary root length ($cm^{-1}$)) was maintained at low levels in the wild-type Col-0 seedlings during the period of S limitation which was extended for 3 days after the transfer, while it was restored upon sulfate supply and became higher during the time course (Figure 3). The LRD showed a trend of linear increase over the time course, for which the slope value can be calculated based on linear regression. The slope value of the linear regression line is expressed as the number of LR developed in one cm unit length of PR per day, indicating the daily increase rate of LRD. Thus, it provides a quantitative measure for the assessment of incremental changes in LR development demonstrated over time as a part of the RSA phenotype (Figure 1). The slope values were 0.36 on +S and 0.20 on –S medium, respectively (Figure 3a,b), suggesting that LR emergence which gave rise to visible LR happened 1.8-fold more frequently in Col-0 roots transferred to sulfate-supplied conditions compared to those exposed to prolonged S deficiency. These estimations were generally consistent with the visible phenotypes (Figure 1). Similar changes were observed in the wild-type *Ler* seedling; however, the differences between the slope values (0.41 on +S and 0.35 on –S medium, respectively) were not as significant as those estimated for Col-0 (Figure 3a,b).

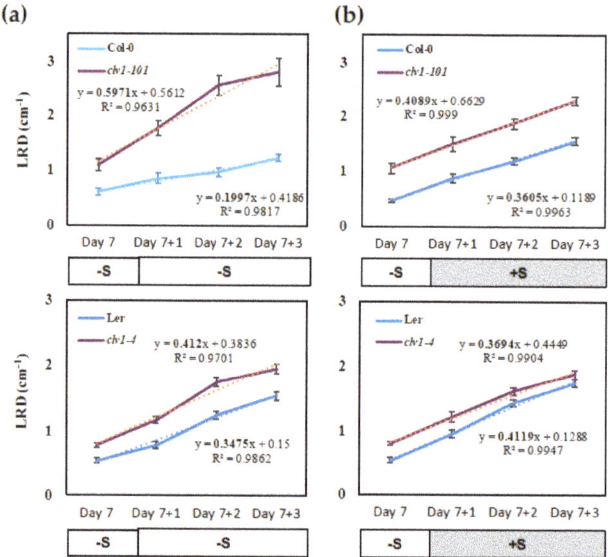


**Figure 3.** Effect of S supply on lateral root density (LRD). Wild-type (Col-0 and *Ler*) and *clv1* mutant lines (*clv1-101* and *clv1-4*) were grown vertically on –S (15 μM sulfate) medium for 7 days and then transferred to (**a**) –S medium or (**b**) +S medium to be grown subsequently for 3 days. LRD was calculated at each time point based on the number of emerged lateral roots (LRs) and the length of the primary root (PR) of one seedling. Values represent means (± SE) of 24 individual plants per treatment. The equations for the linear repression and the R-squared values are indicated on each graph.

Consistent with our previous findings [19], LRD was constantly higher in the *clv1* mutants than in the wild-type (Figure 3). LRD increased significantly in the *clv1* mutants during the period of prolonged S deficiency (Figure 3a; transfer from –S to –S) compared to those transferred to the sulfate-supplied medium (Figure 3b; transfer from –S to +S). In contrast, the wild-type seedlings showed a lower increase rate of LRD on the –S medium compared to those transferred to the +S medium. Based on the slope values of the linear regression, the daily increase rate of LRD was estimated to be 3-fold greater in the *clv1-101* mutant than in Col-0 under S-deficient conditions (Figure 3a). In contrast, upon sulfate supply, the slope value representing the increase rate of LRD was diminished in *clv1-101* though increased in Col-0, apparently converging to the same level (Figure 3b). Similar trends were found when the *clv1-4* mutant was compared with L*er*, while the differences were not so obvious as those shown with the *clv1-101* mutant and Col-0. These results indicated that CLV1 is a signaling component which is necessary for regulation of LR development under S-deficient conditions.

Lateral root branching density (BD) is an alternative measure of LR density, which is calculated by the number of emerged LRs divided by the length of the branching zone (the distance from shoot base to the youngest emerged LR) [29]. BD was calculated 2 and 3 days after the transfer of seedlings to –S or +S medium from 7 days of preculture on the –S medium (Supplemental Figure S1). Consistent with the results shown for LRD (Figure 3), the wild-type plants exposed to prolonged S deficiency showed low BD, while this branching phenotype was recovered upon sulfate supply (Supplemental Figure S1). BD was significantly higher in the *clv1* mutants compared to the wild-type under prolonged S deficiency, while it was lowered after S supplementation. These changes in BD implicate that LR primordia located between the emerged LRs are partially arrested or delayed in progression in the wild-type under the prolonged S deficiency, while their growth can be recovered by the S supplementation. These results also indicate that CLV1 is involved in the regulatory pathway that inhibits the growth of LR primordia, as shown by an increase in BD in the *clv1* mutants relative to the wild-type (Supplemental Figure S1), which confirms our previous findings providing evidence for its essential role in regulating developmental stage progression of LR primordia [19]. Our present findings suggest that long-term S deficiency signals to the CLV1-directed pathway to modulate LR development as demonstrated by changes in LRD and BD.

The total LR length density (TLRLD; total lateral root length divided by primary root length (cm $cm^{-1}$)) was also calculated based on the measurement of the length of the entire LR developed in the root system (Figure 4). The *clv1* mutants showed a significant increase in TLRLD during the period of prolonged S limitation, where an enhanced response to –S was identified in comparison with +S. In contrast, the wild-type showed a more significant increase in TLRLD after sulfate supplementation. These opposing effects of S limitation and sulfate supply on TLRLD in the *clv1* mutants and the wild-type were consistent with those identified for LRD and BD (Figure 3 and Supplemental Figure S1). These results further suggested the essential role of CLV1 in regulation of LR development under S deficiency. As shown in Figure 1, the effect of +S to +S and +S to –S transfer on root morphology was examined simultaneously in this study. In contrast to the results obtained from the –S to –S and the –S to +S transfer experiments (Figures 3 and 4), the LRD and TLRLD of the +S-precultured seedlings changed only slightly in response to S supplementation and S removal (Supplemental Figures S2 and S3). The LRD and TLRLD were higher in the *clv1* mutants than the wild-type as expected, but there was no substantial effect of S in contrast to the phenotypes identified in roots transferred from the –S preculture to –S and +S conditions (Figures 3 and 4).

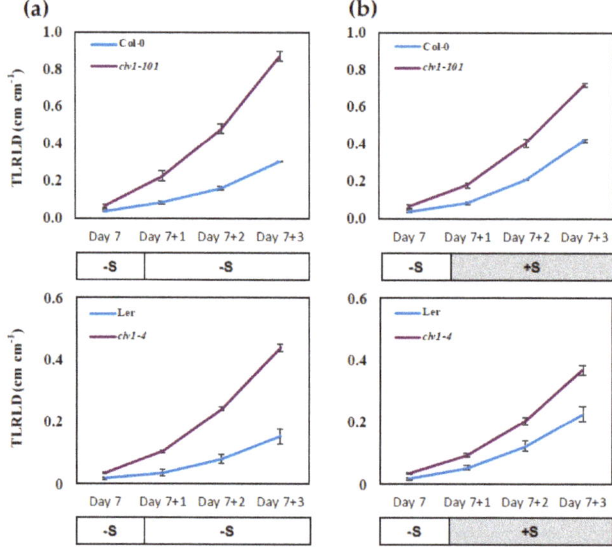


**Figure 4.** Effect of S supply on total lateral root length density (TLRLD). Wild-type (Col-0 and *Ler*) and *clv1* mutant lines (*clv1-101* and *clv1-4*) were grown vertically on –S (15 µM sulfate) medium for 7 days and then transferred to (**a**) –S medium or (**b**) +S medium to be grown subsequently for 3 days. TLRLD was calculated at each time point based on the lengths of the entire LR in one seedling and the length of the PR. Values show means (± SE) of 24 individual plants per treatment.

### *2.2. Regulation of CLE2 and CLE3 Gene Expression under S Deficiency*

To investigate the responses of *CLE* gene expression to the alteration of S nutritional status, the wild-type and *clv1* mutant lines were grown with different S supplies as indicated above for the root phenotype analysis. The root RNAs were extracted at Day 10 (i.e., Day 7+3) for the gene expression analysis. Among the *CLE* genes, *CLE2* and *CLE3* were selected for gene expression analysis, because *CLE2* was previously reported as a S-responsive *CLE* gene [24], and *CLE3* was shown to be significantly up-regulated by N deficiency and inhibits LR development through acting on the CLV1 signaling pathway [19]. *SULTR1;1* encodes a high-affinity sulfate transporter which is highly regulated in response to sulfur deficiency (–S) in the epidermis and cortex of *A. thaliana* roots [30,31]. Therefore, *SULTR1;1* was used as an indicator for tracking the changes in S status. The results indicated prolonged S deficiency (transfer from –S to –S) caused repression, although the sulfate replenishment (transfer from –S to +S) allowed induction of *CLE2* and *CLE3* gene expression in both wild-type and *clv1* mutant lines ($P < 0.05$) (Figure 5a,b). *CLE3* showed greater responses (2–3 fold) to S than *CLE2* (1.2–1.5 fold).

To investigate the effect of S on CLV1-mediated feedback regulation, the *clv1*/wild-type ratios of the *CLE2* and *CLE3* mRNA levels were calculated and compared between the prolonged –S (transfer from –S to –S) and the sulfate supplied (transfer from –S to +S) conditions. The results indicated both the *clv1-101*/Col-0 and *clv1-4*/*Ler* ratios of the *CLE3* mRNA levels under the prolonged –S conditions were higher compared to those estimated for roots transferred to +S medium, suggesting that prolonged S deficiency activates a pathway downstream of CLV1 to feedback regulate *CLE3*. In contrast to *CLE2* and *CLE3*, the *CLV1* expression levels did not change significantly by perturbation of S supply (Figure 5c). *SULTR1;1* was upregulated in roots exposed to prolonged S deficiency while repressed upon sulfate replenishment, showing typical patterns of its S-responsive gene expression, which verified that the S conditions tested in this study were appropriate (Figure 5d).

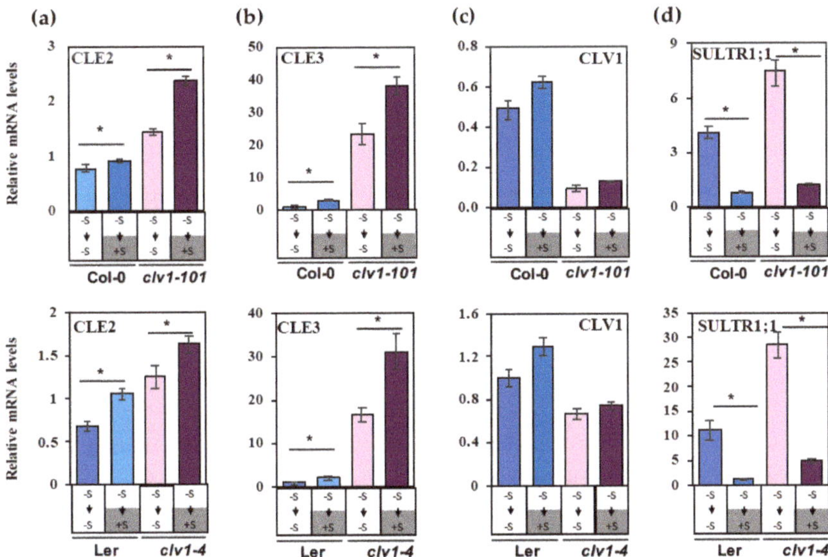


**Figure 5.** Regulation of *CLE2* and *CLE3* transcript levels by S deprivation and sulfate supply. Wild-type (Col-0 and L*er*) and *clv1* mutant lines (*clv1-101* and *clv1-4*) were grown vertically on –S (15 μM sulfate) medium for 7 days and then transferred to –S medium or +S medium to be grown subsequently for 3 days. The mRNA levels of (**a**) *CLE2*, (**b**) *CLE3*, (**c**) *CLV1* and (**d**) *SULTR1;1* in roots at Day 10 (i.e., Day 7+3) were determined by real-time PCR. Roots of wild-type plants grown on the +S medium for 10 days were used as reference samples for relative quantifications. Actin 2 and Ef1α were used as internal controls. Mean values (±SE) of 4 biological replicates with 8 plants for each replicate were calculated using two internal controls. Asterisks indicate statistically significant differences ($P < 0.05$) between gene expression on –S and +S treatment. The S conditions and the order of transfers are shown by white and dark grey bars and arrows below the graph.

## 3. Discussion

The results shown in this study indicate that the CLE-CLV1 signaling pathway is involved in regulation of LR development under prolonged S deficiency. The CLE-CLV1 signaling module has direct impact on LR development and physiological responses associated with changes in RSA. The proposed model describes three steps of the S-dependent signals and their coordinated actions controlling the CLE-CLV1-dependent pathway, extending our knowledge on S nutrient signaling mechanisms involved in regulation of RSA (Figure 6).

The inhibition of LR development occurs in the wild-type plants exposed to prolonged S deficiency, while this inhibition is abolished in the *clv1* mutants (Figures 3 and 4). In contrast, transferring the seedlings to the S-sufficient medium leads to a recovery of LR development in the wild-type, but rather a diminished response in the *clv1* mutants. The increase rate of LRD calculated based on linear regression provides additional information allowing for a quantitative interpretation of the LR phenotypes, as it is shown to be altered in response to S conditions and different among the genotypes being tested in this study. As described in Figure 3, the daily increase rate of LRD was greater in the *clv1* mutant than in the wild-type during the period of prolonged S deficiency, but estimated to be similar between the *clv1* mutant and the wild-type after the sulfate supply. These findings suggest that CLV1 is a signaling component acting on pathways negatively controlling LR development. Regulatory components expressed downstream of CLV1 may be activated under S-deficient conditions (Figure 6). The number of the emerged and visible LR was counted in experiments performed in this study. Thus, we assume that an increase in LRD after the transfer of precultured seedlings to the new medium corresponds to the number of newly emerged LR, and it is associated with developmental stage transition of LR primordia—as has been shown previously [19]. The inhibition of LR development under S deficiency and its CLV1 dependency may reflect these developmental mechanisms previously shown with relevance to the N starvation responses [19].

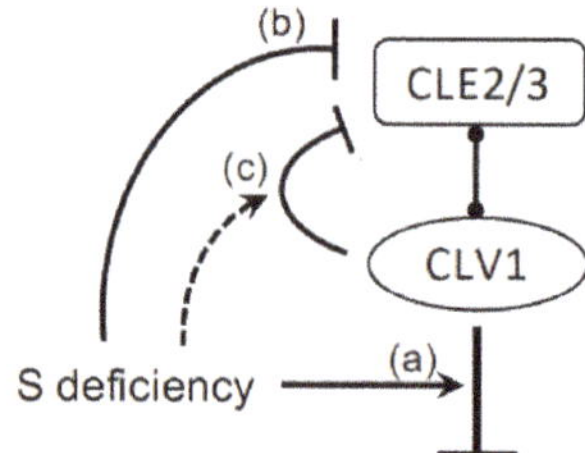


**Figure 6.** Schematic model of signaling pathway controlling lateral root development under S-deficient conditions. S availability affects the activity of the signaling module composed of CLAVATA3 (CLV3)/EMBRYO SURROUNDING REGION (CLE) peptides and CLAVATA1 (CLV1) leucine-rich repeat receptor kinase in roots. The model proposes three potential pathways involved in regulation of LR development under S-deficient conditions. Regulatory components expressed downstream of CLV1 become active under S deficiency and inhibit LR development (**a**). *CLE2* and *CLE3* can be repressed under S deficiency directly or partially through the CLV1-mediated feedback mechanism (at least for CLE3) to reduce the input of CLE signals (**b**,**c**).

*CLE2* and *CLE3* transcript levels decrease under S deficiency in both the wild-type and the *clv1* mutant lines (Figure 5). These transcript profiles suggest that *CLE2* and *CLE3* are repressed under S-deficient conditions in a CLV1-independent manner. In addition, changes in the *clv1*/wild-type ratios of the *CLE3* transcript levels indicate that the CLV1-dependent feedback control mechanism may be strengthened under S-deficient conditions (Figure 5). However, the *CLE2* and *CLE3* transcript levels being reduced in the wild-type plants under S-deficient conditions had no positive impact on promoting LR development. As shown in our model, components expressed downstream of CLV1 are suggested to be more crucial and directly involved in regulation of LR development (Figure 6). The *CLE2* and *CLE3* transcript repression occurring under S-deficient conditions and partially in conjunction with the CLV1-mediated feedback pathway may be considered mechanisms that counteractively reduce the amplitude of the input CLE signals (Figure 6). *CLE2* and *CLE3* are expressed in the pericycle, while CLV1 is expressed in the phloem companion cells [19]. *CLE2* expression is also found in the vascular tissue at the base of developing LR primordia [19,32]. It is suggested that these CLE peptides are SSPs secreted from the pericycle and diffused toward the phloem companion cells to bind to the receptor CLV1, and the CLV1-downstream components expressed in the phloem possibly carry the information to regulate development of LR primordia in long distance via trafficking through the phloem connection [19,33]. The CLE3-CLV1 ligand–receptor relationship and the potential long-distance effect of this signaling module are supported by evidence showing strong inhibition of LR development in transgenic lines with *CLE3* gene overexpression driven by its own promoter in the wild type background and an apparent insensitive response to the same transgene overexpression observed in the *clv1* mutant [19]. Our previous findings provide further implication that additional CLV1-downstream signals may be present and in turn sent back to the pericycle to feedback control *CLE* gene expression [19,33]. According to these models, both CLE peptides and CLV1-downstream components are likely transported between the two different cell types, i.e., pericycle and phloem, and through the vascular system. The actual signals involved in these putative long distance pathways yet remain to be identified.

## 4. Materials and Methods

### *4.1. Plant Growth Conditions*

Two *Arabidopsis thaliana* accessions, Columbia-0 (Col-0) and Landsberg *erecta* (L*er*), and their corresponding *clv1* mutants, *clv1-101* and *clv1-4* were used in this study. Plants were grown vertically on a nutrient medium containing 1% agar and 1% sucrose as described previously in Reference [19] in growth chambers (CU-36L4; Percival Scientific, Perry, IA, USA) conditioned at 22 °C under 16h-light/8h-dark long-day cycles with 75 μmol $m^{-2}$ $s^{-1}$ light intensity. Agar was washed 6 times with 1 liter of deionized water and remaining water was poured off after each agar settlement to remove sulfate. S-replete (+S) medium contained 1500 μM $MgSO_4$ as the sulfur source. S-deficient (–S) medium contained 15 μM $MgSO_4$, and Mg concentration was adjusted to 1500 μM by adding $MgCl_2$. After 7 d preculture on +S or –S medium, the seedlings were transferred to plates with the same or different concentration of sulfate and grown for 3 d.

### *4.2. Root Analysis*

Roots on agar plates were scanned at Day 7 before transfer and the following 3 days after transfer by using a scanner (Epson Perfection V700 Photo; Seiko Epson, Suwa, Japan) at 300 dpi. The images of roots were traced and measured using ImageJ. LR number, LR length, and PR length were recorded for calculation of LR density (LRD, $cm^{-1}$) and total LR length density (TLRLD, cm $cm^{-1}$). Total LR length (TLRL, cm) is the sum of the entire lateral root length per plant.

### *4.3. Quantitative Real-Time PCR*

Roots were homogenized by using TissueLyser II (Qiagen, Hilden, Germany). Total RNAs were extracted by using E.Z.N.A. ® Plant RNA Kit (Omega Bio-Tek, Norcross, GA, USA). Turbo DNA-free kit (Invitrogen, Thermo Fisher Scientific, Waltham, MA, USA) was used for genomic DNA removal of RNA samples. First-strand complementary DNA (cDNA) was prepared from 500 ng of root total RNA by using SuperScript III First-Strand Synthesis System (Invitrogen, Thermo Fisher Scientific). Quantitative real-time PCR was performed by using SYBR Select Master Mix (Applied Biosystems, Thermo Fisher Scientific) on a QuantStudio 7 Flex Real-Time PCR System at the Research Technology Support Facility (RTSF) of Michigan State University. The primers were previously published [19,31,34].

**Supplementary Materials:** The following are available online at http://www.mdpi.com/2223-7747/8/4/103/s1, Figure S1: Effect of S supply on lateral root branching density, Figure S2: Effect of S removal on lateral root density (LRD), Figure S3: Effect of S removal on total lateral root length density (TLRLD).

**Author Contributions:** W.D., and H.T. designed the research. W.D. performed the plant vertical culture, qRT-PCR, and data analysis. Y.W. performed the root measurements using ImageJ. W.D. wrote the paper. H.T. reviewed and edited the paper.

**Funding:** The authors acknowledge funding support from the National Science Foundation (MCB-1244300, IOS-1444549) and the USDA National Institute of Food and Agriculture (Hatch Project 1002395 and 1018575).

**Acknowledgments:** The authors thank Hiroo Fukuda (University of Tokyo) for kindly providing the *clv1* mutant lines for this research.

**Conflicts of Interest:** The authors declare no conflict of interest. The funders had no role in the design of the study; in the collection, analyses, or interpretation of data; in the writing of the manuscript, or in the decision to publish the results.

## References

1. Giehl, R.F.H.; von Wirén, N. Root nutrient foraging. *Plant Physiol.* **2014**, *166*, 509–517. [CrossRef] [PubMed]
2. Gruber, B.D.; Giehl, R.F.H.; Friedel, S.; von Wirén, N. Plasticity of the Arabidopsis root system under nutrient deficiencies. *Plant Physiol.* **2013**, *163*, 161–179. [CrossRef]
3. Lynch, J.P. Steep, cheap and deep: An ideotype to optimize water and N acquisition by maize root systems. *Ann. Bot.* **2013**, *112*, 347–357. [CrossRef] [PubMed]

4. Ristova, D.; Busch, W. Natural variation of root traits: from development to nutrient uptake. *Plant Physiol.* **2014**, *166*, 518–527. [CrossRef] [PubMed]
5. Paez-Garcia, A.; Motes, C.M.; Scheible, W.-R.; Chen, R.; Blancaflor, E.B.; Monteros, M.J. Root traits and phenotyping strategies for plant improvement. *Plants* **2015**, *4*, 334–355. [CrossRef] [PubMed]
6. Gao, Y.; Tian, Q.; Zhang, W.-H. Systemic regulation of sulfur homeostasis in Medicago truncatula. *Planta* **2014**, *239*, 79–96. [CrossRef] [PubMed]
7. Gao, Y.; Li, X.; Tian, Q.-Y.; Wang, B.-L.; Zhang, W.-H. Sulfur deficiency had different effects on Medicago truncatula ecotypes A17 and R108 in terms of growth, root morphology and nutrient contents. *J. Plant Nutr.* **2016**, *39*, 301–314. [CrossRef]
8. Hubberten, H.-M.; Hesse, H.; Hoefgen, R. Lateral root growth in sulphur enriched patches. In *Sulfur Metabolism in Higher Plants*; Sirko, A., De Kok, L.J., Haneklaus, S., Hawkesford, M.J., Rennenberg, H., Saito, K., Schnug, E., Stulen, I., Eds.; Backhuys Publishers: Leiden, The Netherlands, 2009; pp. 105–108.
9. Dan, H.; Yang, G.; Zheng, Z.-L. A negative regulatory role for auxin in sulphate deficiency response in *Arabidopsis thaliana*. *Plant Mol. Biol.* **2007**, *63*, 221–235. [CrossRef]
10. Hopkins, L.; Parmar, S.; Bouranis, D.L.; Howarth, J.R.; Hawkesford, M.J. Coordinated expression of sulfate uptake and components of the sulfate assimilatory pathway in maize. *Plant Biol.* **2004**, *6*, 408–414. [CrossRef]
11. Kutz, A.; Müller, A.; Hennig, P.; Kaiser, W.M.; Piotrowski, M.; Weiler, E.W. A role for nitrilase 3 in the regulation of root morphology in sulphur-starving *Arabidopsis thaliana*. *Plant J.* **2002**, *30*, 95–106. [CrossRef]
12. Hell, R.; Hillebrand, H. Plant concepts for mineral acquisition and allocation. *Curr. Opin. Biotechnol.* **2001**, *12*, 161–168. [CrossRef]
13. López-Bucio, J.; Cruz-Ramírez, A.; Herrera-Estrella, L. The role of nutrient availability in regulating root architecture. *Curr. Opin. Plant Biol.* **2003**, *6*, 280–287. [CrossRef]
14. Mohd-Radzman, N.A.; Laffont, C.; Ivanovici, A.; Patel, N.; Reid, D.; Stougaard, J.; Frugier, F.; Imin, N.; Djordjevic, M.A. Different pathways act downstream of the CEP peptide receptor CRA2 to regulate lateral root and nodule development. *Plant Physiol.* **2016**, *171*, 2536–2548. [PubMed]
15. Roberts, I.; Smith, S.; Stes, E.; De Rybel, B.; Staes, A.; van de Cotte, B.; Njo, M.F.; Dedeyne, L.; Demol, H.; Lavenus, J.; et al. CEP5 and XIP1/CEPR1 regulate lateral root initiation in Arabidopsis. *J. Exp. Bot.* **2016**, *67*, 4889–4899. [CrossRef] [PubMed]
16. Imin, N.; Mohd-Radzman, N.A.; Ogilvie, H.A.; Djordjevic, M.A. The peptide-encoding CEP1 gene modulates lateral root and nodule numbers in Medicago truncatula. *J. Exp. Bot.* **2013**, *64*, 5395–5409. [CrossRef]
17. Delay, C.; Imin, N.; Djordjevic, M.A. CEP genes regulate root and shoot development in response to environmental cues and are specific to seed plants. *J. Exp. Bot.* **2013**, *64*, 5383–5394. [CrossRef]
18. Tabata, R.; Sumida, K.; Yoshii, T.; Ohyama, K.; Shinohara, H.; Matsubayashi, Y. Perception of root-derived peptides by shoot LRR-RKs mediates systemic N-demand signaling. *Science* **2014**, *346*, 343–346. [CrossRef] [PubMed]
19. Araya, T.; Miyamoto, M.; Wibowo, J.; Suzuki, A.; Kojima, S.; Tsuchiya, Y.N.; Sawa, S.; Fukuda, H.; von Wirén, N.; Takahashi, H. CLE-CLAVATA1 peptide-receptor signaling module regulates the expansion of plant root systems in a nitrogen-dependent manner. *Proc. Natl. Acad. Sci. USA* **2014**, *111*, 2029–2034. [CrossRef] [PubMed]
20. Cho, H.; Ryu, H.; Rho, S.; Hill, K.; Smith, S.; Audenaert, D.; Park, J.; Han, S.; Beeckman, T.; Bennett, M.J.; et al. A secreted peptide acts on BIN2-mediated phosphorylation of ARFs to potentiate auxin response during lateral root development. *Nat. Cell Biol.* **2014**, *16*, 66–76. [CrossRef]
21. Kumpf, R.P.; Shi, C.-L.; Larrieu, A.; Stø, I.M.; Butenko, M.A.; Péret, B.; Riiser, E.S.; Bennett, M.J.; Aalen, R.B. Floral organ abscission peptide IDA and its HAE/HSL2 receptors control cell separation during lateral root emergence. *Proc. Natl. Acad. Sci. USA* **2013**, *110*, 5235–5240. [CrossRef]
22. Shinohara, H.; Mori, A.; Yasue, N.; Sumida, K.; Matsubayashi, Y. Identification of three LRR-RKs involved in perception of root meristem growth factor in Arabidopsis. *Proc. Natl. Acad. Sci. USA* **2016**, *113*, 3897–3902. [CrossRef] [PubMed]
23. Yamaguchi, Y.L.; Ishida, T.; Sawa, S. CLE peptides and their signaling pathways in plant development. *J. Exp. Bot.* **2016**, *67*, 4813–4826. [CrossRef] [PubMed]
24. Czyzewicz, N.; Shi, C.-L.; Vu, L.D.; Van De Cotte, B.; Hodgman, C.; Butenko, M.A.; De Smet, I. Modulation of Arabidopsis and monocot root architecture by CLAVATA3/EMBRYO SURROUNDING REGION 26 peptide. *J. Exp. Bot.* **2015**, *66*, 5229–5243. [CrossRef]

25. Patterson, K.; Cakmak, T.; Cooper, A.; Lager, I.; Rasmusson, A.G.; Escobar, M.A. Distinct signalling pathways and transcriptome response signatures differentiate ammonium- and nitrate-supplied plants. *Plant Cell Environ.* **2010**, *33*, 1486–1501. [CrossRef] [PubMed]
26. De Bang, T.C.; Lay, K.S.; Scheible, W.-R.; Takahashi, H. Small peptide signaling pathways modulating macronutrient utilization in plants. *Curr. Opin. Plant Biol.* **2017**, *39*, 31–39. [CrossRef]
27. Wang, G.; Zhang, G.; Wu, M. CLE Peptide Signaling and Crosstalk with Phytohormones and Environmental Stimuli. *Front. Plant Sci.* **2016**, *6*, 1211. [CrossRef]
28. Ohyama, K.; Shinohara, H.; Ogawa-Ohnishi, M.; Matsubayashi, Y. A glycopeptide regulating stem cell fate in *Arabidopsis thaliana*. *Nat. Chem. Biol.* **2009**, *5*, 578–580. [CrossRef]
29. Dubrovsky, J.G.; Forde, B.G. Quantitative analysis of lateral root development: pitfalls and how to avoid them. *Plant Cell* **2012**, *24*, 4–14. [CrossRef]
30. Maruyama-Nakashita, A.; Nakamura, Y.; Tohge, T.; Saito, K.; Takahashi, H. Arabidopsis SLIM1 is a central transcriptional regulator of plant sulfur response and metabolism. *Plant Cell* **2006**, *18*, 3235–3251. [CrossRef]
31. Maruyama-Nakashita, A.; Nakamura, Y.; Watanabe-Takahashi, A.; Yamaya, T.; Takahashi, H. Induction of SULTR1;1 sulfate transporter in Arabidopsis roots involves protein phosphorylation/dephosphorylation circuit for transcriptional regulation. *Plant Cell Physiol.* **2004**, *45*, 340–345. [CrossRef]
32. Jun, J.; Fiume, E.; Roeder, A.H.K.; Meng, L.; Sharma, V.K.; Osmont, K.S.; Baker, C.; Ha, C.M.; Meyerowitz, E.M.; Feldman, L.J.; et al. Comprehensive analysis of CLE polypeptide signaling gene expression and overexpression activity in Arabidopsis. *Plant Physiol.* **2010**, *154*, 1721–1736. [CrossRef] [PubMed]
33. Araya, T.; von Wirén, N.; Takahashi, H. CLE peptides regulate lateral root development in response to nitrogen nutritional status of plants. *Plant Signal. Behav.* **2014**, *9*, e29302. [CrossRef] [PubMed]
34. Czechowski, T.; Bari, R.P.; Stitt, M.; Scheible, W.-R.; Udvardi, M.K. Real-time RT-PCR profiling of over 1400 Arabidopsis transcription factors: unprecedented sensitivity reveals novel root- and shoot-specific genes. *Plant J.* **2004**, *38*, 366–379. [CrossRef] [PubMed]

*Article*

# Sulfate-Induced Stomata Closure Requires the Canonical ABA Signal Transduction Machinery

**Hala Rajab [1,2], Muhammad Sayyar Khan [2], Mario Malagoli [3], Rüdiger Hell [1] and Markus Wirtz [1,*]**

1 Centre for Organismal Studies (COS), Heidelberg University, 69120 Heidelberg, Germany; hala.rajab@cos.uni-heidelberg.de (H.R.); ruediger.hell@cos.uni-heidelberg.de (R.H.)
2 Institute of Biotechnology and Genetic Engineering, The University of Agriculture Peshawar, 25000 Peshawar, Pakistan; sayyar@aup.edu.pk
3 Department of Agronomy, Food, Natural Resources, Animals and Environment, University of Padova, 35020 Legnaro, Italy; mario.malagoli@unipd.it
* Correspondence: markus.wirtz@cos.uni-heidelberg.de; Tel.: +49-6221-54-5334

Received: 23 November 2018; Accepted: 11 January 2019; Published: 16 January 2019

**Abstract:** Phytohormone abscisic acid (ABA) is the canonical trigger for stomatal closure upon abiotic stresses like drought. Soil-drying is known to facilitate root-to-shoot transport of sulfate. Remarkably, sulfate and sulfide—a downstream product of sulfate assimilation—have been independently shown to promote stomatal closure. For induction of stomatal closure, sulfate must be incorporated into cysteine, which triggers ABA biosynthesis by transcriptional activation of NCED3. Here, we apply reverse genetics to unravel if the canonical ABA signal transduction machinery is required for sulfate-induced stomata closure, and if cysteine biosynthesis is also mandatory for the induction of stomatal closure by the gasotransmitter sulfide. We provide genetic evidence for the importance of reactive oxygen species (ROS) production by the plasma membrane-localized NADPH oxidases, RBOHD, and RBOHF, during the sulfate-induced stomatal closure. In agreement with the established role of ROS as the second messenger of ABA-signaling, the SnRK2-type kinase OST1 and the protein phosphatase ABI1 are essential for sulfate-induced stomata closure. Finally, we show that sulfide fails to close stomata in a cysteine-biosynthesis depleted mutant. Our data support the hypothesis that the two mobile signals, sulfate and sulfide, induce stomatal closure by stimulating cysteine synthesis to trigger ABA production.

**Keywords:** sulfate; abscisic acid; stomatal closure; phytohormone synthesis; NADPH oxidase; Protein phosphatases 2C; Sucrose non-fermenting Related Kinase 2 (SnRK2); reactive oxygen species (ROS)

---

## 1. Introduction

Plants have to respond to diverse environmental challenges to optimize growth and ensure survival during stress. Regulation of the stomatal aperture is a critically controlled stress response of plants. Stomata are the gates of the plants for interaction with their environment, and various input signals such as pathogen attack, $CO_2$-concentration, light, heat, humidity, and soil water supply, need to be integrated for the optimal opening of these pores. Phytohormone abscisic acid (ABA) is a potent regulator of stomatal aperture and has been shown to transduce many abiotic and biotic input signals for stomatal closure.

The ABA signal transduction cascade is one of the best-characterized input transmission pathways of plants at the molecular level. In guard cells, ABA physically interacts with the PYRABACTIN RESISTANCE1 (PYR1) and PYR1-LIKE (PYL)-proteins or regulatory components of the ABA receptor (RCAR). Binding of ABA to PYR/PYL receptor enhances the affinity of PYR/PYL receptors for ABI1, a PROTEIN PHOSPHATASE of type 2C (PP2C). PP2Cs are inhibited after binding to the ABA-PYR/PYL receptor complex. Inactivation of PP2Cs by ABA causes activation of subclass III

Sucrose non-fermenting Related Kinase 2 (SnRK2s) [1], of which SnRK2.6 (OPEN STOMATA 1, OST1, Q940H6) is most relevant for stomatal closure. OST1 is a very potent actor since it phosphorylates several targets whose activities contribute to stomatal closure. One of these targets is the SLOW ANION CHANNEL 1 (SLAC1, [2]). SLAC1-induced current changes result in activation of outward K+ channels. The K+ efflux decreases the osmotic potential in the guard cells, followed by water export. The resulting decreased turgor of the guard cell is the physical cause for stomatal is is closure [3]. Furthermore, OST1 can phosphorylate the plasma membrane-resident β-nicotinamide adenine dinucleotide 2′-phosphate (NADPH) oxidase RESPIRATORY BURST OXIDASE HOMOLOG F (RBOHF, O48538) at $Ser^{13}$ and $Ser^{174}$, which is crucial for the regulation of RBOHF activity [4]. NADPH oxidases produce apoplastic reactive oxygen species (ROS) that are essential for ABA-induced stomatal closure. Internal and apoplastic ROS affect multiple steps during ABA signaling and act as a second messenger that plays a predominant role as an ABA signal amplifier [5].

Several studies connect the drought stress response to the assimilation of sulfur [6,7]. Drought stress regulates the sulfur assimilation pathway in an organ-specific manner and causes differential accumulation of sulfur-metabolism related compounds of the primary sulfur metabolism (e.g., glutathione) and the secondary sulfur metabolism (e.g., 3′-phosphoadenosine 5′-phosphate, PAP) [7,8]. The ROS scavenger glutathione acts in the cytosol, the plastids, and the mitochondria as a redox buffer during stress-induced accumulation of ROS. Its drought-stress induced accumulation has been interpreted as a protection mechanism to avoid over-oxidation of these compartments upon stress-induced ROS formation [9–12].

In contrast, PAP acts as a redox stress-induced retrograde signal of the chloroplast in drought and high light signaling by affecting the expression of nuclear-encoded stress-related genes [8,13]. Recently, PAP has also been shown to act as a second messenger of ABA signaling during stomatal closure that bypasses the canonical ABA signaling components ABI1 and OST1 [8]. PAP is a byproduct of sulfation reactions catalyzed by cytosolic sulfotransferases that transfer the activated sulfate of 3′-phosphoadenosine 5′-phosphosulfate (PAPS) to various compounds [14]. The cytosolic PAP is counter exchanged with the predominantly plastid-generated PAPS and degraded in the plastids by the highly redox-regulated 3′(2′),5′-bisphosphate nucleotidase (EC 3.1.3.7, SAL1) [13,15].

Remarkably, sulfate is an early xylem-borne chemical signal in maize under soil drying conditions and precedes the root-to-shoot transport of ABA and the pH increase of the xylem sap [16]. ABA transport and an increase of xylem sap pH have long been assumed to serve as a root-to-shoot signal during soil drying. In Poplar, drought-stress also increases the concentration of sulfate in the xylem by lowered sulfate xylem unloading via PtaSULFATE TRANSPORTER 3;3a (PtaSULTR3;3a) and PtaSULTR1;1, and by enhanced sulfate loading from parenchyma cells into the xylem via ALUMINIUM ACTIVATED MALATE TRANSPORTER3b (PtaALMT3b). Furthermore, sulfate has been shown to decrease relative transpiration and stomatal conductance after petiole feeding of sulfate to detached Poplar leaves [17]. The studies established soil-drying induced xylem transport of sulfate as a candidate for the root-to-shoot signal of the water status but did not uncover the signal transduction pathway for sulfate-induced stomatal closure at the molecular level (see below).

After xylem transport of sulfate to the leaves, the sulfate can be stored in the vacuole of leaf cells or transported into the plastids where it can be activated to APS by ATP sulfurylase [18]. APS is either substrate for production of PAPS by APS kinase or reduced to sulfide by subsequent action of APS REDUCTASE (APR) and SULFiTE REDUCTASE (SiR). The competition between APS reductase and APS kinase for their common substrate APS controls sulfur partitioning between the primary and secondary sulfur metabolism [19]. APR and SiR are exclusively localized in plastids and control the flux through the assimilatory sulfate reduction pathway which generates sulfide [20,21]. Three O-ACETYLSERINE-THIOL-LYASE (OAS-TL) isoforms catalyze in the plastids, the mitochondria, and the cytosol the incorporation of sulfide into cysteine, which is the precursor for all reduced sulfur-containing compounds in plants, e.g., glutathione [20–24]. The carbon backbone for incorporation of sulfide into cysteine is *O*-acetylserine (OAS), and is separately produced in all subcellular compartments with its own cysteine synthesis by five SERINE ACETYLTRANSFERASEs (SERATs, [25]). The subcellular localization of OAS-TL and SERAT implies that significant amounts of the membrane-permeable sulfide move from the plastids to the cytosol and the mitochondria for incorporation into cysteine [26,27].

Sulfide is highly toxic and efficiently incorporated into cysteine in mitochondria, which significantly contributes to the detoxification of elevated sulfide levels [28]. On the other hand, sulfide acts in humans and plants as a volatile gasotransmitter that controls various physiological responses [29,30]. In plants, sulfide represses autophagy and induces stomatal closure [31,32]. However, the mode of sulfide-induced stomatal closure is still controversially discussed. Sulfide has been suggested to (1) affect ABA receptor expression directly [33], (2) act upstream of nitric oxide (NO) to modulate ABA-dependent stomatal closure [34], (3) induce in a NO-dependent manner accumulation of 8-mercapto-cGMP for stomatal closure [35], or (4) activate SLAC1 in an OST1- dependent manner [31].

We recently showed that sulfate must be incorporated into cysteine to trigger stomata dynamics. Consequently, sulfate-induced stomata closure was impaired in mutants deficient in the synthesis of cysteine or ABA [36]. Remarkably, cysteine synthesis depleted mutants are sensitive to drought and high light stress [36–38]. Both stresses also cause PAP accumulation. Since sulfide is a downstream product of assimilatory sulfate reduction pathway and PAP formation is a result of sulfation reactions, it is important to dissect how PAP, sulfide, and sulfate control stomatal aperture.

Here, we apply reverse genetics to understand the contribution of the canonical ABA signaling machinery to sulfate-induced stomata closure and dissect the potential relevance of the sulfation byproduct PAP in this process. We found that the protein phosphatase ABI1 and the down-stream kinase OST1 are essential for sulfate-induced ROS formation in stomata and stomatal closure. Since PAP acts independently of OST1, we concluded that potential accumulation of PAP upon external sulfate administration does not significantly contribute to sulfate-induced stomatal closure. In concordance with the function of ROS as an amplifier of ABA signaling, the loss-of-function mutants for the NADPH oxidases RBOHD and RBOHF are also impaired in sulfate-induced stomatal closure. We furthermore demonstrate that sulfide-induced stomatal closure requires the presence of the major SERAT isoforms located in the cytosol, the plastids, and the mitochondria, strongly suggesting that sulfide needs to be integrated into cysteine to promote stomatal closure. We suggest that sulfate and sulfide are incorporated into cysteine to trigger ABA formation, which in turn requires canonical ABA signaling components to mediate sulfate/sulfide/cysteine-induced stomatal closure.

## 2. Results

In our previous study, we demonstrated that sulfate can close stomata and that it needs to be incorporated into cysteine for activation of ABA synthesis and accumulation of ABA in the cytosol of guard cells. However, we did not show how sulfate-induced ABA is perceived to trigger stomatal closure.

### 2.1. Sulfate-induced Stomatal Closure Requires Functional ABA Signaling

In order to provide direct evidence for the biological relevance of the ABI1 during sulfate-induced stomata closure, we challenged epidermal peels of 5-week-old soil grown wild-type and *abi1-1* plants with 15 mM sulfate for three hours. In the *abi1-1* mutant, the protein phosphatase ABI1 is constitutively active, which results in a permanent open stomata phenotype ([39,40], Figure 1a). Application of 15 mM sulfate efficiently closed the stomata of wild-type plants (Figure 1a), which supports the previously reported impact of sulfate on stomatal closure [17,36]. In contrast, stomata did not close in the *abi1-1* mutant after application of sulfate, demonstrating that sulfate-induced ABA accumulation in guard cells is not triggering stomatal closure when the PYR/PYL-ABA-ABI1 ternary complex is non-functional (Figure 1a,b).

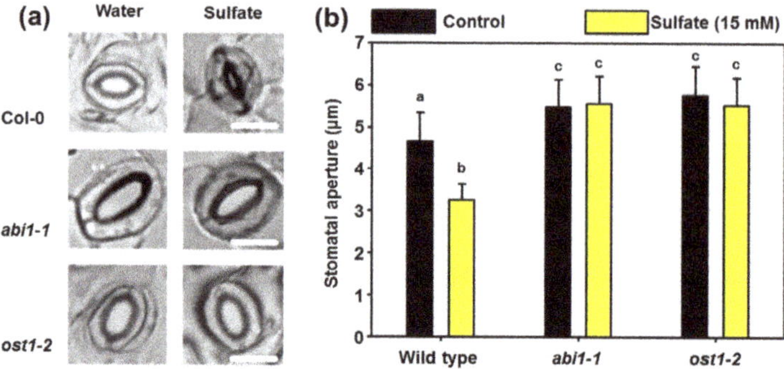


**Figure 1.** Functional ABI1 and OST1 are essential for sulfate-induced stomatal closure in Arabidopsis. (**a**) Representative guard cell in epidermal peels of 5-weeks–old soil-grown wild-type, *abi1-1*, and *ost1-2* plants that were floated on water at pH 5.5 or water supplemented with sulfate (15 mM $MgSO_4$) for three hours. (**b**) Quantification of the stomatal aperture of guard cells in epidermal peels from plants indicated in (**a**) that were treated with water (control, black) supplemented with sulfate (15 mM, yellow). Letters indicate statistically significant differences between groups determined with the one Way ANOVA test ($p < 0.05$, n = 50 stomata, derived from epidermal peels of five individual plants). Values represent means ± standard deviation (SD). Scale bar, 10 µm.

In a separate experiment, we independently confirmed the absence of sulfate responsiveness for *abi1-1* and applied ABA as a control for stomatal closure. Treatment of sulfate (15 mM) and ABA (50 µM) resulted in significant stomatal closure in the wild-type. The degree of stomatal closure was indistinguishable after sulfate and ABA application. ABA or sulfate application to epidermal peels of the *abi1-1* mutant did not affect the stomatal aperture (Figure 2).

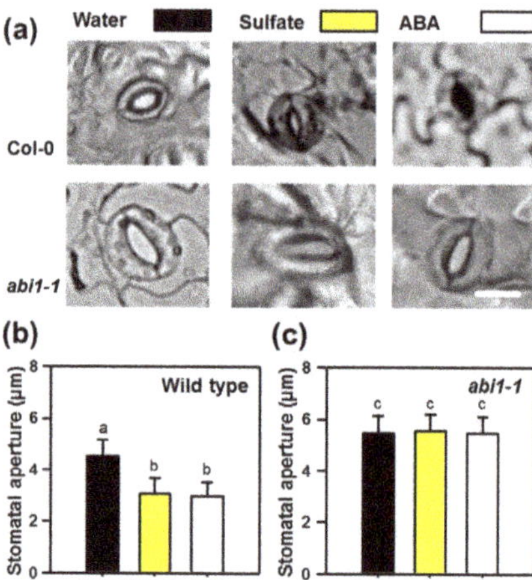


**Figure 2.** Functional ABI1 is required for stomatal closure upon ABA or sulfate treatment. (**a**) Representative guard cell in epidermal peels of 5-week–old soil-grown wild-type and *abi1-1* plants that were floated on water at pH 5.5 or water supplemented with sulfate (15 mM $MgSO_4$) or ABA (50 µM) for three hours. (**b**,**c**) Quantification of the stomatal aperture in epidermal peels of the wild-type (**b**) and the *abi1-1* mutant (**c**) that were treated with water (control, black) supplemented with sulfate (15 mM, yellow) or ABA (50 µM, white). Letters indicate statistically significant differences between groups determined with the one Way ANOVA test ($p < 0.05$, n = 50 stomata, derived from epidermal peels of five individual plants). Values represent means ± standard deviation (SD). Scale bar, 10 µm.

Binding of ABA to PYR/PYL receptors causes efficient inactivation of the PP2C named ABA INSENSITIVE1 (ABI1). The inactivation of ABI1 releases the downstream kinase OST1 from inhibition, which in turn phosphorylates multiple targets to promote ABA-induced stomatal closure. In the *ost1-2* mutant, a T-DNA insertion in the OST1 gene destabilizes the OST1 mRNA, resulting in a total loss-of-OST1 function. Like the *abi1-1* mutant, *ost1-2* displays constitutively open stomata and is sensitive to soil drying and low humidity ([41], Figure 1a). The absence of OST1 kinase activity inhibits the impact of sulfate on stomatal aperture (Figure 1a,b), which is in concordance with the previously demonstrated function of sulfate for the promotion of ABA accumulation in guard cells [36]. These results suggest that the PYR/PYL receptors sense sulfate-induced accumulation of ABA in guard cells and support the importance of the ABI1-OST1 phosphorylation relay for sulfate-induced stomatal closure.

### 2.2. ABI1 and OST1 are Essential for the Sulfate-induced Formation of ROS in Guard Cells

A known target of the ABI1-OST1 phosphorylation relay is the membrane resident NADPH oxidase RBOHF, which is essential for ABA-induced ROS production [4]. We, therefore, tested the formation of ROS in guard cells after sulfate-application to epidermal peels from the wild-type, the *abi1-1*, and the *ost1-2* mutant. Application of sulfate (15 mM) resulted in significant production of ROS in the wild type. The degree of ROS formation in response to sulfate was comparable to the formation of ROS after application of 50 µM ABA (Figure 3a,b). The guard cells in epidermal peels of *abi1-1* and *ost1-2* displayed wild-type like levels of ROS under control conditions. Both mutants of the ABI1-OST1 phosphorylation relay failed to produce ROS in response to the application of sulfate, which is in concordance with the failure to close stomata in response to sulfate. Also, ABA application did not induce ROS formation in both mutants under their applied conditions, which is in agreement with results of previous studies [5].

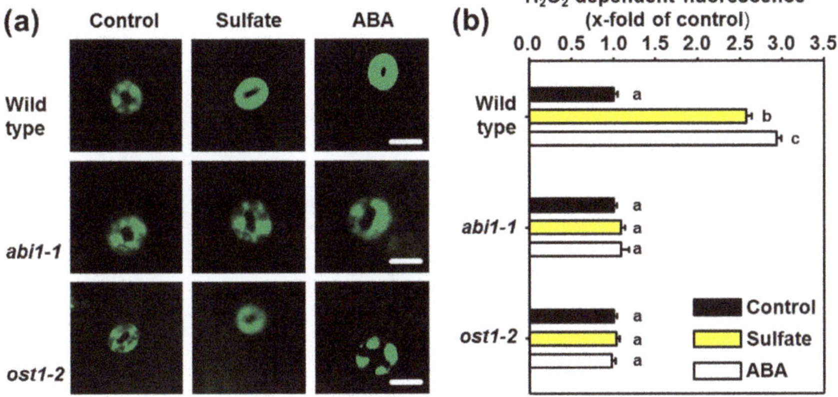


**Figure 3.** Functional ABI1 and OST1 are vital for the sulfate-induced accumulation of reactive oxygen species in guard cells. (**a**) Visualization of reactive oxygen species (ROS) with the $H_2O_2$-selective dye 2′,7′-dichlorofluorescein (H2DCF) as described in material and methods. Epidermal peels of 5-weeks–old soil-grown wild-type, *abi1-1*, and *ost1-2* plants were floated on water at pH 5.5 or water supplemented with sulfate (15 mM $MgSO_4$) or ABA (50 µM) for three hours prior staining of ROS. (**b**) Quantification of ROS staining in guard cells floated on water (control, black) supplemented with sulfate (15 mM, yellow) or ABA (50 µM, white). Letters indicate statistically significant differences between groups determined with the one way ANOVA test ($p < 0.05$, n = 50 stomata, derived from epidermal peels of five individual plants). Values represent means ± standard deviation (SD). Scale bar, 10 µm.

### 2.3. Sulfate Stimulus Activates NADPH Oxidases for Production of ROS in Guard Cells

The sulfate-induced accumulation of ROS in wild-type guard cells has also been observed in our previous study on sulfate-induced stomatal closure [36]. In this study, we demonstrated that inhibition of oxidases, like RBOH isoforms A-F, with the selective inhibitor diphenyleneiodonium prevents the sulfate-induced formation of ROS in guard cells. Co-application of sulfate and diphenyleneiodonium also impaired sulfate-induced stomatal closure, strongly suggesting that formation of ROS is vital for transduction of the sulfate stimulus. These findings support the hypothesis that sulfate induces synthesis of ABA in guard cells, which causes the formation of the ABA-PYR/PYL-ABI1 ternary complex resulting in activation of OST1. The activated OST1 potentially phosphorylates NADPH oxidases to produce the second messenger ROS. The elevated ROS levels will act as a signal amplifier to activate SLAC1, leading to stomatal closure upon sulfate stimulus [5].

In order to link the activation of OST1 and the formation of ROS upon sulfate stimulus with the membrane resident NADPH oxidases, we tested the contribution of two major isoforms of RBOHs expressed in guard cells, RBOH-D and RBOH-F, to sulfate-induced ROS formation [42]. The *rboh-D* and *rboh-F* mutant lack functional isoforms of the NADPH oxidase due to a T-DNA insertion in the respective gene. The absence of either RBOH-D or RBOH-F impaired ROS formation in guard cells upon application of ABA or sulfate for three hours to epidermal peels (Figure 4a,b). Consequently, sulfate did not affect stomatal aperture in both mutants (Figure 4c). Wild-type guard cells produced ROS and closed stomata upon sulfate application.

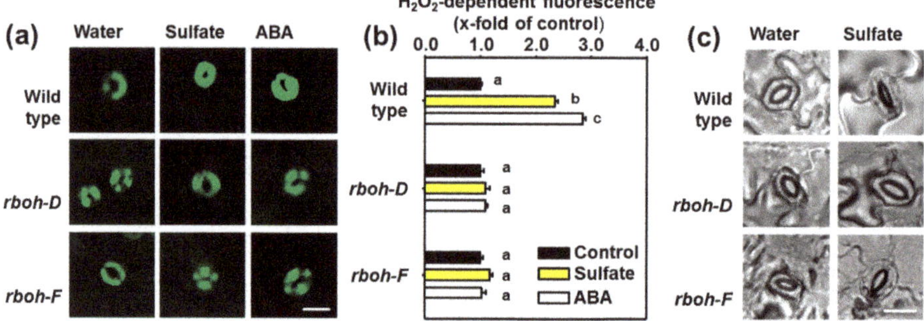


**Figure 4.** Sulfate-induced stomatal closure and ROS formation require the NADPH oxidases RBOHD, and RBOHF (**a**) Visualization of reactive oxygen species (ROS) with the $H_2O_2$-selective dye 2′,7′-dichlorofluorescein (H2DCF) as described in material and methods. Epidermal peels of 5-weeks–old soil-grown wild-type, *rboh-D*, and *rboh-F* plants were floated on water at pH 5.5 or water supplemented with sulfate (15 mM $MgSO_4$) or ABA (50 μM) for three hours prior staining of ROS. (**b**) Quantification of ROS staining in guard cells floated on water (control, black) supplemented with sulfate (15 mM, yellow) or ABA (50 μM, white). Letters indicate statistically significant differences between groups determined with the one Way ANOVA test ($p < 0.05$, n = 50 stomata, derived from epidermal peels of five individual plants). Values represent means ± standard deviation (SD). (**c**) Representative guard cells embedded in epidermal peels of the wild-type and the *rboh-D*, and *rboh-F* mutants after treatment with water or water supplemented with sulfate (15 mM). Scale bar, 10 μm.

### 2.4. Stomata of the Serat tko Mutant do not Close upon Sulfide Application

The volatile signal $H_2S$ is a downstream product of sulfate assimilation and has been shown independently to induce stomatal closure [34,43,44]. Since sulfate must be incorporated into cysteine to gain competence as an inducer of stomatal closure, we decided to test if sulfide is also incorporated into cysteine for induction of ABA biosynthesis. Sulfide is the direct sulfur-precursor of cysteine and is incorporated by the activity of OAS-TL into cysteine. Cysteine biosynthesis is not limited by OAS-TL activity, but by the formation of the carbon skeleton for cysteine, OAS. The three major SERAT isoforms SERAT1;1, SERAT2;1 and SERAT2;2 produce the bulk of OAS for the incorporation of sulfide into cysteine in the cytosol, the plastids, and the mitochondria. A SERAT triple knock-out mutant (*serat tko*) lacking these major SERAT isoforms is retarded in growth and displays significantly lowered translation due to decreased production of OAS and cysteine [45]. When we applied 100 μM sulfide (NaHS) dissolved in water at pH 5.5 to epidermal peels of the wild-type, the dissolved sulfide triggered the closure of the wild-type stomata within 90 min. The application of water at pH 5.5 for the same duration did not affect the stomatal aperture (Figure 5). Remarkably, sulfide failed to close the stomata in epidermal peels of the *serat tko* mutant. These results suggest that sulfide is not perceived by a membrane-resident receptor located at the plasma membrane, but is incorporated into cysteine to gain competence as a stomatal closure signal.

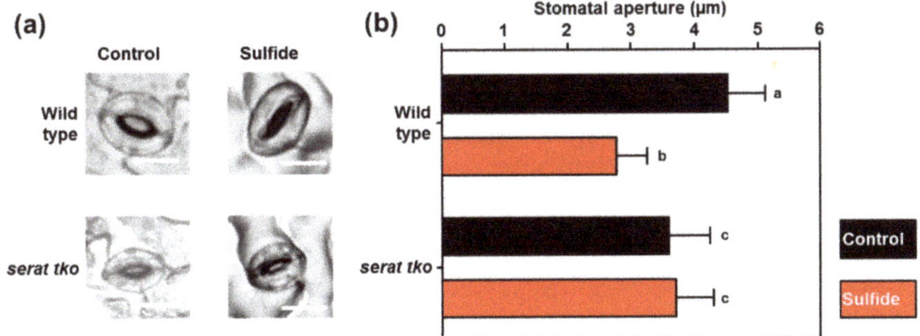


**Figure 5.** Sulfide is unable to induce stomatal closure in the OAS biosynthesis-depleted *serat tko* mutant. (**a**) Representative guard cells embedded in epidermal peels of the wild-type and the *serat tko* mutant after treatment with water at pH 5.5 (Control) or water supplemented with sulfide (100 μM NaHS) at pH 5.5 for 90 minutes. (**b**) Quantification of the stomatal aperture in epidermal peels of the wild-type and the *serat tko* mutant that were treated with water (control, black), or water supplemented with sulfide (100 μM NaHS, red). Letters indicate statistically significant differences between groups determined with the one Way ANOVA test ($p < 0.05$, n = 50 stomata, derived from epidermal peels of five individual plants). Values represent means ± standard deviation (SD). Scale bar, 10 μm.

## 3. Discussion

### *3.1. Sulfate and Sulfide are Incorporated into Cysteine to Trigger Stomatal Closure*

Stomatal closure is a dynamic process that optimizes carbon dioxide uptake with transpiration-mediated water loss. A multifaceted signaling network regulates stomatal movements and integrates diverse environmental and endogenous inputs. Some of these inputs are locally generated, e.g., pathogen-induced stomatal closure, and cause fast responses that use phosphorylation relays to control plasma-membrane resident NADPH oxidases [5]. In other cases, e.g., perceiving of the soil-water status, a distantly originated signal must travel to the guard cell and will be integrated to modulate stomatal aperture. Many of these integration processes impinge on the ABA signal transduction pathway [3,46]. However, our knowledge of the regulation of the tissue-specific ABA biosynthesis sites in response to environmental cues is currently quickly evolving, but still limited [3,46].

Characterization of the plastidic sulfate transporter SULTR3;1 uncovered for the first time a direct link between the actual sulfur supply and ABA biosynthesis in plants [47,48]. Indeed, sulfate has been shown by independent groups to accumulate in the xylem sap of drought-stress maize, Poplar, common hop and Arabidopsis plants and was supposed to act as an early signal that promotes ABA-induced stomatal closure [16,17,49,50]. The recent identification of cysteine as a trigger for ABA biosynthesis in guard cells linked the sulfate-induced stomatal closure with the biosynthesis of this factor for stomatal closure [36]. The presented findings establish ABI1 as a required transducer of the ABA signal during sulfate-induced stomatal closure (Figure 1), which is in agreement with the stimulation of ABA production by cysteine and sulfate in guard cells and rosette leaves [36].

After perception, the ABA signal is transduced via the ABI1-OST1 phosphorylation relay to multiple downstream effectors. In concordance with this canonical view on ABA perception, OST1 is also vital for sulfate-induced stomatal closure. Both results strongly support the view that sulfate stimulates ABA formation and point against a direct gating of the QUICK ANION CHANNEL 1 (QUAC1 or ALMT12) channel by sulfate, which has been hypothesized by Malcheska and co-workers as the molecular basis of sulfate-induced stomata closure. This hypothesis was based on the stimulating effect of sulfate on QUAC1 expressed in *Xenopus* oocytes and the absence of stomatal closure in the *quac1* mutant [17]. The presented findings are not in contradiction with the insensitivity of the *quac1* mutant towards sulfate as a trigger of stomatal closure [17], but offer a different interpretation of the *quac1* insensitivity towards sulfate. Like SLAC1, QUAC1 is required for ABA-induced stomatal closure and is a substrate of OST1, which can activate SLAC1 and QUAC1 by direct phosphorylation [2,51,52]. Consequently, sulfate fails to close stomata in *quac1* [17] and *slac1-1* [36], although SLAC1 is not gated by sulfate.

Sulfide is a well-established signal for stomatal closure. Like sulfate, sulfide can be incorporated into cysteine. The failure of sulfide to stimulate stomatal closure in the *serat tko* (Figure 5), which is impaired in the incorporation of sulfide into cysteine, suggests that sulfate and sulfide use the same mechanism for induction of stomatal closure. Both signals stimulate the synthesis of cysteine, which in turn promotes ABA formation. In support of this hypothesis, sulfide is known to be immediately incorporated into cysteine after short-term exposure [27], and stimulates SLAC1 by activation of OST1 [31]. Furthermore, sulfate fails to close stomata in the *sir1-1* mutant that is depleted in its capability to reduce sulfate to sulfide [36,53].

### 3.2. The Role of ROS as Second-messenger in Sulfate-induced Stomatal Closure

ROS are important signal amplifiers of the ABA signal and act downstream of ABA as second messengers during stomatal closure and systemic acquired acclimation [5,54]. Like sulfide [55], sulfate application also triggered ROS production in guard cells of the wild type (Figure 3) in an RBOH dependent manner (Figure 4). The formation of ROS is essential for sulfate-induced stomatal closure [36]. The ABA-triggered ROS production in guard cells is OST1 dependent [41], which can directly phosphorylate plasma membrane-resident NADPH oxidases, like RBOH-F [4]. In concordance with the established scheme for ABA-induced ROS formation, sulfate-induced ROS production was diminished in *abi1-1*, *ost1-2* and loss-of-function mutants for NADPH oxidases (RBOH-D and RBOH-F). Remarkably, ABA-triggered formation of ROS depends on ABI1 but not on ABI2 [56]. Indeed, *abi2-1* mutants accumulated ROS in response to sulfate application [36]. Thus, the here presented data strongly suggest that the sulfate-induced ROS formation is a consequence of sulfate-promoted ABA formation, which in turn stimulates membrane resident NADPH oxidases via the classical ABA signal transduction cascade.

### 3.3. Contribution of Cytosolic Sulfation Reactions Releasing PAP to Close Stomata in Response to Sulfate

PAP is a side-product of sulfation reactions that use PAPs as sulfate donors [14]. PAP is also a potent inducer of stomatal closure [57]. Consequently, one could have hypothesized that sulfate-triggered PAP accumulation might contribute to sulfate-induced stomatal closure. Remarkably, PAP bypasses the canonical ABA-induced stomatal closure pathway and is independent of ABI1 and OST1 [57]. In contrast, sulfate-induced stomatal closure requires functional ABI1 and OST1 (Figures 1–3) and its downstream targets (RBOH-F, Figure 4). Our findings exclude a significant contribution of PAP during sulfate-induced stomatal closure. Accumulation of PAP upon sulfate application is also highly questionable, since PAP levels are strongly regulated by the PAP degrading enzyme, SAL1 [6,8]. SAL1 is highly regulated by environmental stimuli like drought and high-light stress and tightly controls PAP level, which is the basis for the well-established retrograde signaling function of PAP [8,13].

In conclusion, our results uncover that sulfate-induced formation of the stomatal closure signal, ABA, is transduced by ABI1, which has been previously shown to physically interact with the PYR/PYL receptor in an ABA-dependent manner. The canonical ABI1-OST1 phosphorylation relay is essential for the activation of plasma-membrane resident NADPH oxidases of the RBOH type. These RBOHs produce the ABA signal amplifier ROS, which is mandatory for sulfate-induced stomatal closure. The failure of sulfide to close stomata in the *serat tko* mutant supports the view that sulfate and sulfide must be incorporated into cysteine to gain competences as stomatal closure signals due to the stimulation of ABA production.

## 4. Materials and Methods

### 4.1. Plant Material and Growth

Seeds of *Arabidopsis thaliana* Col-0 (ecotype Columbia) and the mutants *abi1-1* (AT4G26080) CS22, *ost1-2* (AT4G33950), *rboh-D* (AT5G47910) CS9555, *rboh-F* (AT1G64060) CS68522, *serat tko* (SALK_ 050213 x SALK_ 099019 x Kazusa_KG752, [45]), were sown on soil (Tonsubstrat from Ökohum, Herbertingen) supplemented with 10% (v/v) vermiculite and 2% quartz sand. Seeds were stratified at 4 °C for two days in the dark. The plants were grown under long-day conditions for five weeks before the experiment (16 h light, 100 µmol $m^{-2}$ $s^{-1}$ at 22 °C and eight h dark at 18 °C for day and night, respectively). Relative humidity was set at 50%.

### 4.2. Stomatal Aperture Bioassay

Epidermal peels were prepared from the abaxial side of Arabidopsis leaves as described in [16] and allowed to float on distilled water for 2 hours under constant light. The peels were then transferred to distilled water pH 5.5 supplemented without (control) or with effectors (15mM $MgSO_4$ and 50 µM ABA) for the indicated periods. Images of the stomata were captured after the treatment with a conventional wide-angle microscope (Leica DMIRB). The stomatal aperture was determined with ImageJ (http://fiji.sc/), using a µm ruler for calibration.

### 4.3. $H_2O_2$ Quantification in Guard Cells

ROS were determined in intact stomata of epidermal peels from the abaxial side of the leaf according to [58]. The peels were allowed to float on water without and with effectors for up to three hours. Subsequently, the epidermal peels were stained with 50 µM 2′,7′-dichlorodihydrofluorescein diacetate ($H_2$DCFA) for 10 min and transferred to a microscope slide. The ROS-specific fluorescence was detected at 525 nm after specific excitation at 488 nm using a confocal microscope (Nikon A1R) according to [59]. Images were captured from peels of five individual plants, and the fluorescent signal of intact stomata (50) was quantified using the open source software ImageJ (http://fiji.sc/).

### 4.4. Statistical Analysis

The Statistical analysis for the experimental data was done using SigmaPlot 12.5 (Systat Inc., San Jose, CA, USA). The analysis of all data sets was analyzed through One Way Repeated Measures Analysis of Variance (one-Way ANOVA) for all of the multiple pairwise comparisons using the Holm-Sidak method. The Shapiro-Wilk method (p to reject was $p > 0.05$) was used to test the normality distribution of data. In the figures, letters are used to indicate the significance difference ($p < 0.05$).

**Author Contributions:** H.R. performed the stomata closure experiments and ROS quantification. M.S.K., M.M., R.H. and M.W. supervised H.R. M.W. and R.H. designed the study. All authors participated in writing the manuscript.

**Funding:** This research was funded by the German Research Association, grant number (HE1848/14-1, -/15-1,-16/1 and WI3560/1-1, -/2-1. H.R. received funding from the Higher Education Commission (HEC)-Pakistan, and the German Academic Exchange Service (DAAD).

**Acknowledgments:** The authors want to thank Prof. Rainer Hedrich (University of Wuerzburg, Germany) and Dietmar Geiger (University of Wuerzburg, Germany) for the kind gift of *abi1-1*, *ost1-2*, *rboh-D* and *rboh-F* mutants. We are grateful for the Higher Education Commission (HEC)-Pakistan, and the German Academic Exchange Service DAAD for their financial support.

**Conflicts of Interest:** The authors declare no conflict of interest. The funders had no role in the design of the study; in the collection, analyses, or interpretation of data; in the writing of the manuscript, or in the decision to publish the results.

## References

1. Vlad, F.; Rubio, S.; Rodrigues, A.; Sirichandra, C.; Belin, C.; Robert, N.; Leung, J.; Rodriguez, P.L.; Lauriere, C.; Merlot, S. Protein phosphatases 2C regulate the activation of the Snf1-related kinase OST1 by abscisic acid in Arabidopsis. *Plant Cell* **2009**, *21*, 3170–3184. [CrossRef] [PubMed]
2. Geiger, D.; Scherzer, S.; Mumm, P.; Stange, A.; Marten, I.; Bauer, H.; Ache, P.; Matschi, S.; Liese, A.; Al-Rasheid, K.A.; et al. Activity of guard cell anion channel SLAC1 is controlled by drought-stress signaling kinase-phosphatase pair. *Proc. Natl. Acad. Sci. USA* **2009**, *106*, 21425–21430. [CrossRef]
3. Kuromori, T.; Seo, M.; Shinozaki, K. ABA Transport and Plant Water Stress Responses. *Trends Plant Sci.* **2018**, *23*, 513–522. [CrossRef] [PubMed]
4. Sirichandra, C.; Gu, D.; Hu, H.C.; Davanture, M.; Lee, S.; Djaoui, M.; Valot, B.; Zivy, M.; Leung, J.; Merlot, S.; et al. Phosphorylation of the Arabidopsis AtrbohF NADPH oxidase by OST1 protein kinase. *FEBS Lett.* **2009**, *583*, 2982–2986. [CrossRef]
5. Sierla, M.; Waszczak, C.; Vahisalu, T.; Kangasjärvi, J. Reactive Oxygen Species in the Regulation of Stomatal Movements. *Plant Physiol.* **2016**, *171*, 1569–1580. [CrossRef]
6. Chan, K.X.; Wirtz, M.; Phua, S.Y.; Estavillo, G.M.; Pogson, B.J. Balancing metabolites in drought: The sulfur assimilation conundrum. *Trends Plant Sci.* **2013**, *18*, 18–29. [CrossRef] [PubMed]
7. Ahmad, N.; Malagoli, M.; Wirtz, M.; Hell, R. Drought stress in maize causes differential acclimation responses of glutathione and sulfur metabolism in leaves and roots. *BMC Plant Biol.* **2016**, *16*, 247. [CrossRef] [PubMed]
8. Estavillo, G.M.; Crisp, P.A.; Pornsiriwong, W.; Wirtz, M.; Collinge, D.; Carrie, C.; Giraud, E.; Whelan, J.; David, P.; Javot, H.; et al. Evidence for a SAL1-PAP Chloroplast Retrograde Pathway That Functions in Drought and High Light Signaling in Arabidopsis. *Plant Cell* **2011**, *23*, 3992–4012. [CrossRef]
9. Marty, L.; Siala, W.; Schwarzlander, M.; Fricker, M.D.; Wirtz, M.; Sweetlove, L.J.; Meyer, Y.; Meyer, A.J.; Reichheld, J.-P.; Hell, R. The NADPH-dependent thioredoxin system constitutes a functional backup for cytosolic glutathione reductase in Arabidopsis. *Proc. Natl. Acad. Sci. USA* **2009**, *106*, 9109–9114. [CrossRef]
10. Meyer, A.J.; Brach, T.; Marty, L.; Kreye, S.; Rouhier, N.; Jacquot, J.P.; Hell, R. Redox-sensitive GFP in Arabidopsis thaliana is a quantitative biosensor for the redox potential of the cellular glutathione redox buffer. *Plant J.* **2007**, *52*, 973–986. [CrossRef]
11. Foyer, C.H.; Noctor, G. Ascorbate and Glutathione: The Heart of the Redox Hub. *Plant Physiol.* **2011**, *155*, 2–18. [CrossRef]
12. Noctor, G.; Mhamdi, A.; Chaouch, S.; Han, Y.; Neukermans, J.; Marquez-Garcia, B.; Queval, G.; Foyer, C.H. Glutathione in plants: An integrated overview. *Plant Cell Environ.* **2012**, *35*, 454–484. [CrossRef]
13. Chan, K.X.; Mabbitt, P.D.; Phua, S.Y.; Mueller, J.W.; Nisar, N.; Gigolashvili, T.; Stroeher, E.; Grassl, J.; Arlt, W.; Estavillo, G.M.; et al. Sensing and signaling of oxidative stress in chloroplasts by inactivation of the SAL1 phosphoadenosine phosphatase. *Proc. Natl. Acad. Sci. USA* **2016**, *113*, E4567–E4576. [CrossRef]
14. Klein, M.; Papenbrock, J. The multi-protein family of Arabidopsis sulphotransferases and their relatives in other plant species. *J. Exp. Bot.* **2004**, *55*, 1809–1820. [CrossRef]
15. Mugford, S.G.; Yoshimoto, N.; Reichelt, M.; Wirtz, M.; Hill, L.; Mugford, S.T.; Nakazato, Y.; Noji, M.; Takahashi, H.; Kramell, R.; et al. Disruption of Adenosine-5′-Phosphosulfate Kinase in Arabidopsis Reduces Levels of Sulfated Secondary Metabolites. *Plant Cell* **2009**, *21*, 910–927. [CrossRef]
16. Ernst, L.; Goodger, J.Q.D.; Alvarez, S.; Marsh, E.L.; Berla, B.; Lockhart, E.; Jung, J.; Li, P.; Bohnert, H.J.; Schachtman, D.P. Sulphate as a xylem-borne chemical signal precedes the expression of ABA biosynthetic genes in maize roots. *J. Exp. Bot.* **2010**, *61*, 3395–3405. [CrossRef]

17. Malcheska, F.; Ahmad, A.; Batool, S.; Müller, H.M.; Ludwig-Müller, J.; Kreuzwieser, J.; Randewig, D.; Hänsch, R.; Mendel, R.R.; Hell, R.; et al. Drought-Enhanced Xylem Sap Sulfate Closes Stomata by Affecting ALMT12 and Guard Cell ABA Synthesis. *Plant Physiol.* **2017**, *174*, 798–814. [CrossRef]
18. Rotte, C.; Leustek, T. Differential subcellular localization and expression of ATP sulfurylase and 5′-adenylylsulfate reductase during ontogenesis of Arabidopsis leaves indicates that cytosolic and plastid forms of ATP sulfurylase may have specialized functions. *Plant Physiol.* **2000**, *124*, 715–724. [CrossRef]
19. Mugford, S.G.; Lee, B.R.; Koprivova, A.; Matthewman, C.; Kopriva, S. Control of sulfur partitioning between primary and secondary metabolism. *Plant J.* **2011**, *65*, 96–105. [CrossRef]
20. Khan, M.S.; Haas, F.H.; Allboje Samami, A.; Moghaddas Gholami, A.; Bauer, A.; Fellenberg, K.; Reichelt, M.; Hansch, R.; Mendel, R.R.; Meyer, A.J.; et al. Sulfite Reductase Defines a Newly Discovered Bottleneck for Assimilatory Sulfate Reduction and Is Essential for Growth and Development in Arabidopsis thaliana. *Plant Cell* **2010**, *22*, 1216–1231. [CrossRef]
21. Vauclare, P.; Kopriva, S.; Fell, D.; Suter, M.; Sticher, L.; von Ballmoos, P.; Krahenbuhl, U.; den Camp, R.O.; Brunold, C. Flux control of sulphate assimilation in *Arabidopsis thaliana*: Adenosine 5′-phosphosulphate reductase is more susceptible than ATP sulphurylase to negative control by thiols. *Plant J.* **2002**, *31*, 729–740. [CrossRef]
22. Heeg, C.; Kruse, C.; Jost, R.; Gutensohn, M.; Ruppert, T.; Wirtz, M.; Hell, R. Analysis of the Arabidopsis *O*-acetylserine(thiol)lyase gene family demonstrates compartment-specific differences in the regulation of cysteine synthesis. *Plant Cell* **2008**, *20*, 168–185. [CrossRef]
23. Birke, H.; Heeg, C.; Wirtz, M.; Hell, R. Successful Fertilization Requires the Presence of at Least One Major *O*-Acetylserine(thiol)lyase for Cysteine Synthesis in Pollen of Arabidopsis. *Plant Physiol.* **2013**, *163*, 959–972. [CrossRef]
24. Watanabe, M.; Kusano, M.; Oikawa, A.; Fukushima, A.; Noji, M.; Saito, K. Physiological Roles of the b -Substituted Alanine Synthase Gene Family in Arabidopsis. *Plant Physiol.* **2008**, *146*, 310–320. [CrossRef]
25. Watanabe, M.; Mochida, K.; Kato, T.; Tabata, S.; Yoshimoto, N.; Noji, M.; Saito, K. Comparative genomics and reverse genetics analysis reveal indispensable functions of the serine acetyltransferase gene family in Arabidopsis. *Plant Cell* **2008**, *20*, 2484–2496. [CrossRef]
26. Birke, H.; Muller, S.J.; Rother, M.; Zimmer, A.D.; Hoernstein, S.N.; Wesenberg, D.; Wirtz, M.; Krauss, G.J.; Reski, R.; Hell, R. The relevance of compartmentation for cysteine synthesis in phototrophic organisms. *Protoplasma* **2012**, *249* (Suppl. 2), 147–155. [CrossRef]
27. Birke, H.; De Kok, L.J.; Wirtz, M.; Hell, R. The Role of Compartment-Specific Cysteine Synthesis for Sulfur Homeostasis During $H_2S$ Exposure in Arabidopsis. *Plant Cell Physiol.* **2015**, *56*, 358–367. [CrossRef]
28. Birke, H.; Haas, F.H.; De Kok, L.J.; Balk, J.; Wirtz, M.; Hell, R. Cysteine biosynthesis, in concert with a novel mechanism, contributes to sulfide detoxification in mitochondria of *Arabidopsis thaliana*. *Biochem J.* **2012**, *445*, 275–283. [CrossRef]
29. Kimura, H. Signaling molecules: Hydrogen sulfide and polysulfide. *Antioxid. Redox Signal.* **2015**, *22*, 362–376. [CrossRef]
30. Hancock, J.T.; Whiteman, M. Hydrogen sulfide and cell signaling: Team player or referee? *Plant Physiol. Biochem.* **2014**, *78*, 37–42. [CrossRef]
31. Wang, L.; Wan, R.; Shi, Y.; Xue, S. Hydrogen Sulfide Activates S-Type Anion Channel via OST1 and $Ca^{2+}$ Modules. *Mol. Plant* **2016**, *9*, 489–491. [CrossRef]
32. Laureano-Marín, A.M.; Moreno, I.; Romero, L.C.; Gotor, C. Negative Regulation of Autophagy by Sulfide Is Independent of Reactive Oxygen Species. *Plant Physiol.* **2016**, *171*, 1378–1391. [CrossRef]
33. Jin, Z.; Xue, S.; Luo, Y.; Tian, B.; Fang, H.; Li, H.; Pei, Y. Hydrogen sulfide interacting with abscisic acid in stomatal regulation responses to drought stress in Arabidopsis. *Plant Physiol. Biochem.* **2013**, *62*, 41–46. [CrossRef]
34. Scuffi, D.; Alvarez, C.; Laspina, N.; Gotor, C.; Lamattina, L.; Garcia-Mata, C. Hydrogen sulfide generated by L-cysteine desulfhydrase acts upstream of nitric oxide to modulate abscisic acid-dependent stomatal closure. *Plant Physiol.* **2014**, *166*, 2065–2076. [CrossRef]
35. Honda, K.; Yamada, N.; Yoshida, R.; Ihara, H.; Sawa, T.; Akaike, T.; Iwai, S. 8-Mercapto-Cyclic GMP Mediates Hydrogen Sulfide-Induced Stomatal Closure in Arabidopsis. *Plant Cell Physiol.* **2015**. [CrossRef]

36. Batool, S.; Uslu, V.V.; Rajab, H.; Ahmad, N.; Waadt, R.; Geiger, D.; Malagoli, M.; Xiang, C.-B.; Hedrich, R.; Rennenberg, H.; et al. Sulfate is Incorporated into Cysteine to Trigger ABA Production and Stomatal Closure. *Plant Cell* **2018**, in press. [CrossRef]
37. Speiser, A.; Haberland, S.; Watanabe, M.; Wirtz, M.; Dietz, K.J.; Saito, K.; Hell, R. The significance of cysteine synthesis for acclimation to high light conditions. *Front. Plant Sci.* **2015**, *5*, 776. [CrossRef]
38. Mueller, S.M.; Wang, S.; Telman, W.; Liebthal, M.; Schnitzer, H.; Viehhauser, A.; Sticht, C.; Delatorre, C.; Wirtz, M.; Hell, R.; et al. The redox-sensitive module of cyclophilin 20-3, 2-cysteine peroxiredoxin and cysteine synthase integrates sulfur metabolism and oxylipin signaling in the high light acclimation response. *Plant J.* **2017**, *91*, 995–1014. [CrossRef]
39. Koornneef, M.; Reuling, G.; Karssen, C.M. The isolation and characterization of abscisic acid-insensitive mutants of Arabidopsis thaliana. *Physiol. Plant.* **1984**, *61*, 377–383. [CrossRef]
40. Pei, Z.M.; Kuchitsu, K.; Ward, J.M.; Schwarz, M.; Schroeder, J.I. Differential abscisic acid regulation of guard cell slow anion channels in Arabidopsis wild-type and *abi1* and *abi2* mutants. *Plant Cell* **1997**, *9*, 409–423. [CrossRef]
41. Mustilli, A.C.; Merlot, S.; Vavasseur, A.; Fenzi, F.; Giraudat, J. Arabidopsis OST1 protein kinase mediates the regulation of stomatal aperture by abscisic acid and acts upstream of reactive oxygen species production. *Plant Cell* **2002**, *14*, 3089–3099. [CrossRef]
42. Kwak, J.; Mori, I.; Pei, Z.; Leonhardt, N.; Torres, M.; Dangl, J.; Bloom, R.; Bodde, S.; Jones, J.; Schroeder, J. NADPH oxidase AtrbohD and AtrbohF genes function in ROS-dependent ABA signaling in Arabidopsis. *EMBO J.* **2003**, *22*, 2623–2633. [CrossRef]
43. García-Mata, C.; Lamattina, L. Hydrogen sulphide, a novel gasotransmitter involved in guard cell signalling. *New Phytol.* **2010**, *188*, 977–984. [CrossRef] [PubMed]
44. Papanatsiou, M.; Scuffi, D.; Blatt, M.R.; Garia-Mata, C. Hydrogen sulphide regulates inward-rectifying K+ channels in conjunction with stomatal closure. *Plant Physiol.* **2015**. [CrossRef]
45. Dong, Y.; Silbermann, M.; Speiser, A.; Forieri, I.; Linster, E.; Poschet, G.; Allboje Samami, A.; Wanatabe, M.; Sticht, C.; Teleman, A.A.; et al. Sulfur availability regulates plant growth via glucose-TOR signaling. *Nat. Commun.* **2017**, *8*, 1174. [CrossRef] [PubMed]
46. Takahashi, F.; Suzuki, T.; Osakabe, Y.; Betsuyaku, S.; Kondo, Y.; Dohmae, N.; Fukuda, H.; Yamaguchi-Shinozaki, K.; Shinozaki, K. A small peptide modulates stomatal control via abscisic acid in long-distance signalling. *Nature* **2018**, *556*, 235–238. [CrossRef]
47. Cao, M.J.; Wang, Z.; Zhao, Q.; Mao, J.L.; Speiser, A.; Wirtz, M.; Hell, R.; Zhu, J.K.; Xiang, C.B. Sulfate availability affects ABA levels and germination response to ABA and salt stress in *Arabidopsis thaliana*. *Plant J.* **2014**, *77*, 604–615. [CrossRef] [PubMed]
48. Cao, M.J.; Wang, Z.; Wirtz, M.; Hell, R.; Oliver, D.J.; Xiang, C.B. SULTR3;1 is a chloroplast-localized sulfate transporter in *Arabidopsis thaliana*. *Plant J.* **2013**, *73*, 607–616. [CrossRef] [PubMed]
49. Goodger, J.Q.; Sharp, R.E.; Marsh, E.L.; Schachtman, D.P. Relationships between xylem sap constituents and leaf conductance of well-watered and water-stressed maize across three xylem sap sampling techniques. *J. Exp. Bot.* **2005**, *56*, 2389–2400. [CrossRef]
50. Korovetska, H.; Novák, O.; Jůza, O.; Gloser, V. Signalling mechanisms involved in the response of two varieties of *Humulus lupulus* L. to soil drying: I. changes in xylem sap pH and the concentrations of abscisic acid and anions. *Plant Soil* **2014**, *380*, 375–387. [CrossRef]
51. Imes, D.; Mumm, P.; Bohm, J.; Al-Rasheid, K.A.; Marten, I.; Geiger, D.; Hedrich, R. Open stomata 1 (OST1) kinase controls R-type anion channel QUAC1 in Arabidopsis guard cells. *Plant J.* **2013**, *74*, 372–382. [CrossRef]
52. Meyer, S.; Mumm, P.; Imes, D.; Endler, A.; Weder, B.; Al-Rasheid, K.A.; Geiger, D.; Marten, I.; Martinoia, E.; Hedrich, R. AtALMT12 represents an R-type anion channel required for stomatal movement in Arabidopsis guard cells. *Plant J.* **2010**, *63*, 1054–1062. [CrossRef] [PubMed]
53. Speiser, A.; Silbermann, M.; Dong, Y.; Haberland, S.; Uslu, V.V.; Wang, S.; Bangash, S.A.K.; Reichelt, M.; Meyer, A.J.; Wirtz, M.; et al. Sulfur Partitioning between Glutathione and Protein Synthesis Determines Plant Growth. *Plant Physiol* **2018**, *177*, 927–937. [CrossRef] [PubMed]
54. Mittler, R.; Blumwald, E. The Roles of ROS and ABA in Systemic Acquired Acclimation. *Plant Cell* **2015**, *27*, 64–70. [CrossRef]

55. Scuffi, D.; Nietzel, T.; Di Fino, L.M.; Meyer, A.J.; Lamattina, L.; Schwarzlander, M.; Laxalt, A.M.; Garcia-Mata, C. Hydrogen Sulfide Increases Production of NADPH Oxidase-Dependent Hydrogen Peroxide and Phospholipase D-Derived Phosphatidic Acid in Guard Cell Signaling. *Plant Physiol.* **2018**, *176*, 2532–2542. [CrossRef] [PubMed]
56. Murata, Y.; Pei, Z.-M.; Mori, I.C.; Schroeder, J. Abscisic Acid Activation of Plasma Membrane $Ca^{2+}$ Channels in Guard Cells Requires Cytosolic NAD(P)H and Is Differentially Disrupted Upstream and Downstream of Reactive Oxygen Species Production in abi1-1 and abi2-1 Protein Phosphatase 2C Mutants. *Plant Cell* **2001**, *13*, 2513–2523. [CrossRef]
57. Pornsiriwong, W.; Estavillo, G.M.; Chan, K.X.; Tee, E.E.; Ganguly, D.; Crisp, P.A.; Phua, S.Y.; Zhao, C.; Qiu, J.; Park, J.; et al. A chloroplast retrograde signal, 3′-phosphoadenosine 5′-phosphate, acts as a secondary messenger in abscisic acid signaling in stomatal closure and germination. *eLife* **2017**, *6*, e23361. [CrossRef] [PubMed]
58. Pei, Z.M.; Murata, Y.; Benning, G.; Thomine, S.; Klusener, B.; Allen, G.J.; Grill, E.; Schroeder, J.I. Calcium channels activated by hydrogen peroxide mediate abscisic acid signalling in guard cells. *Nature* **2000**, *406*, 731–734. [CrossRef]
59. LeBel, C.P.; Ischiropoulos, H.; Bondy, S.C. Evaluation of the probe 2′,7′-dichlorofluorescin as an indicator of reactive oxygen species formation and oxidative stress. *Chem. Res. Toxicol.* **1992**, *5*, 227–231. [CrossRef]

Article

# Assessment of Sulfur Deficiency under Field Conditions by Single Measurements of Sulfur, Chloride and Phosphorus in Mature Leaves

**Philippe Etienne [1,2,3,*,†], Elise Sorin [1,2,3,4,†], Anne Maillard [1,2,3], Karine Gallardo [5], Mustapha Arkoun [4], Jérôme Guerrand [6], Florence Cruz [4], Jean-Claude Yvin [4] and Alain Ourry [1,2,3]**

1 Normandie Université, 14032 Caen CEDEX 5, France; elisesorin@live.fr (E.S.); Anne.Maillard@roullier.com (A.M.); alain.ourry@unicaen.fr (A.O.)

2 Etienne Philippe, Normandie Université-INRA 950 Ecophysiologie Végétale, Agronomie et nutritions N, C, S, Esplanade de la Paix, UNICAEN, CS14032, 14032 Caen CEDEX 5, France

3 INRA, UMR 950 Ecophysiologie Végétale, Agronomie et nutritions N, C, S, Esplanade de la Paix, UNICAEN, CS14032, 14032 Caen CEDEX 5, France

4 Centre Mondial d'Innovation, CMI, Groupe Roullier, 27 Avenue Franklin Roosevelt, 35400 Saint-Malo, France; Mustapha.Arkoun@roullier.com (M.A.); fcruz@roullier.com (F.C.); jcyvin@roullier.com (J.-C.Y.)

5 UMR Agroécologie, AgroSup Dijon, INRA, Université Bourgogne Franche-Comté, F-21000 Dijon, France; karine.gallardo-guerrero@inra.fr

6 Vegenov BBV, Penn ar Prat, 29250 Saint Pol de Léon, France; jeromeguerrand@hotmail.com

* Correspondence: philippe.etienne@unicaen.fr; Tel.: +33-231-565-374; Fax: +33-231-565-360

† These authors contributed equally to this work.

Received: 6 April 2018; Accepted: 24 April 2018; Published: 28 April 2018

**Abstract:** Determination of S status is very important to detect S deficiency and prevent losses of yield and seed quality. The aim of this study was to investigate the possibility of using the ($[Cl^-]+[NO_3{}^-]+[PO_4{}^{3-}]$):$[SO_4{}^{2-}]$ ratio as an indicator of S nutrition under field conditions in *Brassica napus* and whether this could be applied to other species. Different S and nitrogen (N) fertilizations were applied on a S deficient field of oilseed rape to harvest mature leaves and analyze their anion and element contents in order to evaluate a new S nutrition indicator and useful threshold values. Large sets of commercial varieties were then used to test S deficiency scenarios. As main results, this study shown that, under field conditions, leaf ($[Cl^-]+[NO_3{}^-]+[PO_4{}^{3-}]$):$[SO_4{}^{2-}]$ ratio was increased by lowering S fertilization, indicating S deficiency. The usefulness of this ratio was also found for other species grown under controlled conditions and it could be simplified by using the elemental ([Cl]+[P]):[S] ratio. Threshold values were determined and used for the clustering of commercial varieties within three groups: S deficient, at risk of S deficiency and S sufficient. The ([Cl]+[P]):[S] ratio quantified under field conditions, can be used as an early and accurate diagnostic tool to manage S fertilization.

**Keywords:** *Brassica napus*; diagnostic tools; indicator of S nutrition; S fertilization; *Triticum aestivum*; *Zea mays*

## 1. Introduction

Sulfur deficiency in crops has been a major concern at the global scale for a number of years, and especially in crops that require higher levels of sulfur (S) than other cultivated species, such as oilseed rape. Sulfur limitation in oilseed rape crops provokes multiple changes in plant physiology leading to losses of yield and seed quality through modified lipid and protein compositions [1–5]. Sulfur deficiency in crops has increasingly been observed over the last 50 years [3,6–8]. The limited S availability in soil can be explained by several factors: significant reductions in S emission from industrial sources, use of mineral fertilizers without S, decreases in use of organic fertilizers, and changes in cropping systems including the use of high yielding commercial coupled with intensive management practices [3,9]. Seed yield from oilseed rape is usually improved by S fertilization, with doses depending on the multiple environment factors under which the crop is being grown [9]. Consequently, a precise recommendation for S fertilization is difficult, resulting in a wide variation in S doses applied under production conditions from 0 to 112 kg·S·ha$^{-1}$ [10]. Consequently, S indicators are needed for an appropriate use of fertilizers and to avoid cases of excessive fertilization having environmental and economic consequences, and particularly to avoid cases of S deficiency so as to prevent losses of yield and seed quality [6,7].

The visual symptoms of severe S deficiency are characterized by general yellowing and purpling of leaves, which can be quantified by reflectance measurements. However, these visual symptoms of S deficiency can be confused with other deficiencies like N deficiency [6,11] and can be also imperceptible during moderated S deficiency which can decrease seed quality [12]. In order to diagnose S status in crops, methods derived from soil testing, modeling or plant analysis have already been proposed. Different soil tests have been developed to evaluate the amounts of inorganic $SO_4^{2-}$ in soil and the potential mineralization capacity of soil organic S compounds [13]. However, chemical analyses of soil have not proved successful to evaluate the amount of S available for the crops. Indeed, from these analyses, it's difficult to predict the requirement of S fertilization because S availability in soils is highly variable due to fluctuations of mineralization, immobilization or leaching throughout the growing season [14]. A previous study has proposed a computer model to predict the risk of S deficiency in cereals in Great Britain using soil, atmospheric S deposition, and meteorological data [15]. This modeling approach identifies where S deficiency is likely to occur for a large scale assessment, but it is not designed to be used at the field scale [6]. In oilseed rape, a model based on the process of S allocation and partitioning might be relevant for determining the response to S nutrition under controlled conditions [16] but it needs to be validated under field conditions.

Plant analyses appear to be a better method for identifying S deficiency and evaluate soil fertility and crop management strategies [17]. Different indicators of S nutrition using plant analysis have been proposed, i.e., the N:S ratio [18], the $SO_4^{2-}$:total S ratio ([19], $SO_4^{2-}$[20]), total S (Pinkerton 1998 [21], the malate:$SO_4^{2-}$ ratio [22], the molybdenum:S ratio [23],glutathione [24], amides [24], or expression of sulfur deficiency responsive genes[25,26], each presenting the advantages and drawbacks. Gene expression has potential as a sensitive indicator of S nutritional status because of its very early response of plant to S deficiency. Indeed, Howarth et al. [25] have shown that the expression of the *sdi1* gene was induced especially in leaf and root tissues in response to S deficiency in wheat. In oilseed rape, the up-regulation of genes encoding $SO_4^{2-}$transporters (*BnaSultr1;1, BnaSultr1;2, BnaSultr4;1* and *BnaSultr4;2*) is one of the first responses to S deprivation [26]. However, none of these molecular indicators can be easily used under field conditions as they require laboratory analysis and the use of control plants. The use of biochemical indicators in response to S deficiency has been suggested, such as a rapid decrease in glutathione [22,24,27,28] and a large accumulation of asparagine and glutamine [24]. However, the concentrations of these molecules can be influenced by factors other than S nutrition, such as salinity, water or temperature stress [29,30]) and their diagnostic usefulness has been questioned. Use of $SO_4^{2-}$and total S concentrations have been proposed by Scaife and Burns [20] and Pinkerton [21] as the most satisfactory indices of plant S status because of the very wide range of values with clear changes from deficiency to sufficiency. These parameters present the added advantage of requiring only one measurement.

It has been shown that ratios of mineral (or organic) plant compounds can predict S nutrition more reliably than absolute values of one or more contents of mineral (or organic) compounds. Indeed, it has been suggested that ratios between compounds varied less with plant growth stages and growth rates than absolute values [7,17,22,31]. The $SO_4^{2-}$:total S ratio has been proposed by Spencer and Freney [19] as a useful guide to S fertilization because it is not strongly affected by the age of the plant or N supply. However, Scaife and Burns [20] criticized this use of the $SO_4^{2-}$:total S ratio instead of $SO_4^{2-}$or total S alone because the change in total S is partly due to the levels of $SO_4^{2-}$, thus the numerator ($SO_4^{2-}$) depends on the denominator (total S), which reduces the sensitivity of the indicator. Rasmussen et al. [18] proposed the N:S ratio for wheat at the vegetative stage. Compared to the total S concentration of vegetative tissues, the N:S ratio also decreases, but to a lesser extent, so that its use as an indicator requires sampling at a precise growth stage [2]. Moreover, this ratio should be interpreted with caution because it can be affected by oversupply of N without S deficiency. Recently, the N:S ratio in grain has been used as the most accurate diagnostic tool by many agronomists and the farming industry for S deficiency in crops. However, the grain N:S ratio is not a predictive test as it indicates that the harvested crop may have been S deficient, so that S fertilization can be increased for the following crop [32,33]. Alternately, Blake-Kalff et al. [22,32] proposed the malate: $SO_4^{2-}$ ratio as a diagnostic indicator of S deficiency in wheat and oilseed rape. This ratio is based on the inverse relationship between malate and $SO_4^{2-}$concentrations in leaves; when the $SO_4^{2-}$concentrations decrease, the malate concentrations increase, and vice versa. This ratio has the advantage of being stable throughout plant development when the S supply is sufficient. Moreover, it can be calculated from a single analytical run by using the peak areas of malate and $SO_4^{2-}$, avoiding the problems of accurate calibration. Studies on the reliability of the malate:$SO_4^{2-}$ ratio indicate that it overestimated diagnosis of S deficiency and suggested that threshold values need to be reconsidered [33,34]). More recently, Maillard et al. [23] have demonstrated that the [Mo]/[S] constitutes an accurate indicator to diagnose S deficiency. However, because the range of Mo leaf contents are usually low, measurements of this element need complex and senesitive laboratory methods (such as ICPMS-HR) which make difficult to use routinely this diagnostic tool. Finally, Sorin et al. [26] proposed that the ($[Cl^-]+[NO_3^-]+[PO_4^{3-}]$):$[SO_4^{2-}]$ratio could provide a relevant early indicator of S deficiency in oilseed rape. This new early indicator of S nutrition identified under controlled conditions in oilseed rape is based on early physiological regulation that occurs before significant metabolic perturbations. It was shown that vacuolar $SO_4^{2-}$, acting as an osmoticum, was efficiently remobilized during S deprivation to sustain plant growth. This was compensated osmotically by a vacuolar accumulation of $Cl^-$, $NO_3^-$ and $PO_4^{3-}$. This mechanism of osmotic compensation occurred as early as 3 days after S deprivation and long before metabolic disturbances and growth reduction in oilseed rape.

The main objectives of the current work were to validate the use of the ($[Cl^-]+[NO_3^-]+[PO_4^{3-}]$):$[SO_4^{2-}]$ ratio as an indicator of S nutrition under field conditions. The first objective was to test the ($[Cl^-]+[NO_3^-]+[PO_4^{3-}]$):$[SO_4^{2-}]$ ratio in leaves of oilseed rape grown under field conditions with different S and N fertilization rates on an S deficient field, and consequently if it constitutes a relevant indicator to predict the risk of S deficiency under field conditions. The second objective was to test the genericity of the ($[Cl^-]+[NO_3^-]+[PO_4^{3-}]$):$[SO_4^{2-}]$ ratio as an indicator of S deficiency in other species of different families such as *Brassica oleracea*, *Triticum aestivum*, *Zea mays*, *Medicago truncatula* and *Solanum lycopersicum* under controlled conditions. Finally, the last aim was to simplify the quantification of the ($[Cl^-]+[NO_3^-]+[PO_4^{3-}]$):$[SO_4^{2-}]$ ratio, by using elemental analysis of Cl, P and S in order to facilitate its quick measurement on samples harvested in field and allow an adjustment of S fertilization in short time delay after the diagnostic of S deficiency.

## 2. Materials and Methods

### 2.1. Field Experiments and Plant Sampling

#### 2.1.1. Field Experiment 1: Study of Different Fertilization Rates on an S-Deficient Field

The experimental site was selected from a previous study [10] showing an S deficiency in *Brassica napus* L. grown in 2009. This field of 11.3 ha was located at Ondefontaine, France (48°59′18.69″ N, 00°41′56.70″ W). A winter oilseed rape variety (*Brassica napus* L., 95% cv DK Exstorm and 5% cv Alicia) was sown with a density of 35–40 plants $m^{-2}$ on 27 august 2013. The previous crop was spring barley (*Hordeum vulgare* L.). Chemical analyses of the loam soil provided an S content of 220 mg $kg^{-1}$

and a pH value of 6.4. Before crop establishment, organic fertilizer (40 $m^3$ $ha^{-1}$ of bovine manure) was applied on the whole field corresponding to 122 kg N $ha^{-1}$ and 44 kg S $ha^{-1}$. The field was separated into seven randomized plots fertilized with different doses of N and S mineral fertilizer. One plot was unfertilized. Other plots were fertilized on 27 February 2014 with three different doses of S fertilizer: 0, 12, and 36 kg S $ha^{-1}$assulfuric anhydride (36% $SO_3$) and ammonium nitrate to obtain 65 kg N $ha^{-1}$on each plot. Then, on 27 March 2014, a second N fertilization with 60 kg N $ha^{-1}$ (liquid N containing 50% urea, 25% ammonium N and 25% nitrate N) was applied to three plots that thus received 125 kg N $ha^{-1}$ after two fertilizer applications.

Five harvests were performed between January and July 2014: before fertilization (at the rosette stage and at the start of stem elongation (GS30), 30 January 2014 and 12February 2014), 15 days after S fertilization (start of the visible bud stage (GS50), 14March 2014), 47 days after S fertilization (stage of pods formation (GS70), 15 April 2014) and finally after 60 days (stage of pods formation (GS70), 28 April 2014). Two batches of leaves, mature leaves (from the lower canopy) and young leaves (from the higher canopy), were harvested using 20 leaves randomly collected from each plot and with three replicates. Leaves were freeze-dried for dry weight (DW) determination and ground, using an oscillating grinder (MM400, Retsch, Haan, Germany) to fine powder for further analysis.

#### 2.1.2. Field Experiment 2: Study of 45 Commercial Varieties before Fertilization and Flowering

Forty-five commercial varieties of oilseed rape were selected in France according to different locations (Supplemental data SD1), different agricultural practices (dose of fertilizers, previous crop, tillage), and under contrasting soil and climate conditions that may affect $SO_4{}^{2-}$leaching and hence $SO_4{}^{2-}$availability. The farmers, identified with the help of DATAGRI (Lyon, France), collaborated in the present study. Plots have been numbered from 1′ to 45′ according to the increasing ($[Cl^-]$+$[NO_3{}^-]$+$[PO_4{}^{3-}]$):$[SO_4{}^{2-}]$ ratio. Two batches of leaves comprising mature leaves (from the lower canopy) and young leaves (from the higher canopy), each of them consisting of 20 leaves, were randomly collected from each of the 45 fields in February 2014, just at the end of winter corresponding to the start of stem elongation (GS30) and before fertilization. Leaves were freeze-dried before laboratory processing.

In addition, four commercial varieties were studied before and after fertilization. Agricultural practices were managed by the farmers and the authors did not intervene in the choice of fertilizer doses. These four plots were located in France at Indre (N 46°36′48″, E 02°06′41″)corresponding to plot number 8′, Loiret (N 48°06′59″, E 02°25′04″) corresponding to plot number 11′, Indre-et-Loire (N 47°06′05″, E 00°22′14″) corresponding to plot number 20′ and Seine-et-Marne (N 48°30′28″, E 02°49′56″) corresponding to plot number 36′. Two harvests occurred between January and May 2014: before fertilization (at the rosette stage and at the start of stem elongation (GS30), February 2014) and after N and S fertilizers were applied by the farmers (stage of pods formation (GS70), April 2014). Leaves were collected as for the 45 plots, with three replicates of leaves collected.

#### 2.1.3. Field Experiment 3: Study of 56 Commercial Varieties after Fertilization and Flowering

In order to maximize variability for S status at the late stage of oilseed rape development, 56 commercial varieties were selected in Calvados (Lower Normandy, France) during a previous study [10]. The numbering of plots was performed in accordance with Sarda et al. [10]. Fifty senescent leaves (i.e., yellowing, just before their abscission) were randomly collected after flowering (May 2009) from each of the 56 fields. Four biological replicates of plants were used per condition and harvesting date. Leaf samples were kept at 4 °C for a short period before drying at 60 °C for 72 h.

### *2.2. Multispecies Experiment under Controlled Conditions*

In order to assess if data obtained with *B. napus* could be extrapolated to other species, experiments using *B. oleracea*, *T. aestivum*, *Z. mays*, *S. lycopersicum*, *M, truncatula* and *B. napus* grown under controlled conditions were conducted under different culture conditions described in Supplemental data SD2 with optimal and sub-optimal S fertilization. Leaves were harvested and frozen immediately in liquid nitrogen and stored at −80 °C until freeze-drying for further analysis.

### 2.3. $SO_4^{2-}$, $PO_4^{3-}$, $Cl^-$ and $NO_3^-$ Analysis

Ions were extracted from 30 mg of freeze-dried plant material and were initially mixed with 1.5 mL of 50% ethanol solution. After incubation at 40 °C for 1 h, the extract was centrifuged at 12,000× *g* for 20 min and the supernatant was collected. This step was repeated on the pellet and the resulting supernatant obtained was pooled with the previous one. All these operations (i.e., incubation and centrifugation) were repeated twice, but with 1.5 mL of ultra-pure water and incubation at 95 °C. All the supernatants were pooled and evaporated under vacuum (Concentrator Evaporator RC 10.22, Jouan, Saint-Herblain, France). The dry residue was re-suspended in 1.5 mL of ultra-pure water and was filtered on 45 µm filters.

Thereafter, anion contents were determined using high performance liquid chromatography (HPLC) with a conductivity detector (ICS3000, Thermo Scientific-Dionex, Villebon-sur-Yvette, France). The eluent solution for anion analysis consisted of 4.05 mM $Na_2CO_3$ and 1.26 mM $NaHCO_3$ and was pumped isocratically over an analytical column (AS22 4 × 250 mm).

### 2.4. S, P, Cl and N Analysis

After drying and grinding, leaf samples were placed sample cups and S, P, and Cl were determined with X-ray fluorescence (XRF) analysis (Portable XRF S1 TITAN 800, Bruker, Kalkar, Germany). Quantification of each element was performed using calibration curves of samples quantified by High Resolution Inductively Coupled Plasma Mass Spectrometry (HR ICP-MS, Thermo Scientific, Element 2$^{TM}$, Bremen, Germany) [35]. For N analysis, an aliquot of each dried sample was placed into tin capsules using a microbalance and the total N content was determined with a continuous flow isotope ratio mass spectrometer (Horizon, NU Instruments, Wrexham, UK) linked to a C/N/S analyzer (EA3000, Euro Vector, Milan, Italy).

## 3. Statistical Analysis

For field experiment 1, three replicates corresponding to a pool of 20 leaves of 20 independent plants were collected. Data are reported as mean ± SE for $n$ = 3 and were analyzed by Student's test (Excel software, Microsoft Corporation, Redmond, Washington, DC, USA) and marked by different letters when they were significantly different between treatments at a given date, at $p < 0.05$. For the second part of field experiment 2, three replicates corresponding to a pool of 20 leaves of 20 independent plants were collected. These data are also reported as mean ± SE for $n$ = 3 and were analyzed by Student's test (Excel software) and marked by different letters when they were significantly different between plots at a given date, at $p < 0.05$, and marked by one or several asterisks when significantly different before and after fertilization. All experiments on different species in controlled conditions were conducted with four independent biological replicates. Data are reported as mean ± SE for $n$ = 4. All data of the multispecies experiment were analyzed by Student's test (Excel software) and marked by one or several asterisks when significantly different between controls and S-deprived plants.

## 4. Results

### 4.1. Under Field Conditions, a Decrease in the $SO_4^{2-}$ Content Was Compensated by an Increase in the ($Cl^-$+$NO_3^-$+$PO_4^{3-}$) Contents in Oilseed Rape Leaves Leading to an Increase in the ([$Cl^-$]+[$NO_3^-$]+[$PO_4^{3-}$]):[$SO_4^{2-}$]Ratio

For field experiment 1, $SO_4^{2-}$, $Cl^-$, $NO_3^-$and $PO_4^{3-}$ contents and the ([$Cl^-$]+[$NO_3^-$]+[$PO_4^{3-}$]):[$SO_4^{2-}$] ratio were quantified in leaves of oilseed rape at 15, 47, and 60 days after S fertilization. Similar observations were found at 47 and 60 days after S fertilization, and to simplify readings, only the results at 60 days after S fertilization are presented. The $SO_4^{2-}$content

(Figure 1a–c) was decreased significantly in mature leaves by a reduction in S fertilization (from 36, 12, to 0 kg S $ha^{-1}$), whatever the dose of N fertilization (65 or 125 kg N $ha^{-1}$). The decrease in $SO_4^{2-}$ content in mature leaves was compensated by an increase in $Cl^-$, $NO_3^-$, and $PO_4^{3-}$ contents (Figure 1d–f). The lowest ($Cl^-$+$NO_3^-$+$PO_4^{3-}$) content was found in mature leaves of plants fertilized with the highest dose of S (i.e., 36 kg S $ha^{-1}$). The ([$Cl^-$]+[$NO_3^-$]+[$PO_4^{3-}$]):[$SO_4^{2-}$]ratio (Figure 1g–i) differentiated the three plots according to S fertilization: 0, 12, 36 kg S $ha^{-1}$. Mature leaves of plants receiving N fertilization (65 or 125 kg N $ha^{-1}$) without S had the highest ([$Cl^-$]+[$NO_3^-$]+[$PO_4^{3-}$]):[$SO_4^{2-}$]ratio, whereas the mature leaves of plants with 36 kg S $ha^{-1}$ had the lowest, and mature leaves of plants with 12 kg S $ha^{-1}$ presented an intermediate ([$Cl^-$]+[$NO_3^-$]+[$PO_4^{3-}$]):[$SO_4^{2-}$]ratio.

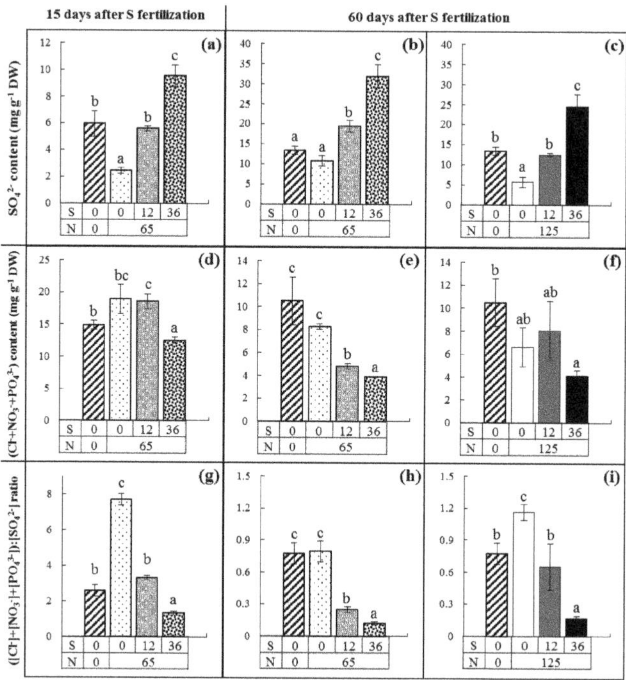


**Figure 1.** (**a–c**) $SO_4^{2-}$contents (mg $g^{-1}$ DW); (**d–f**) ($Cl^-$+$NO_3^-$+$PO_4^{3-}$) contents (mg $g^{-1}$ DW) and (**g–i**) the ([$Cl^-$]+[$NO_3^-$]+[$PO_4^{3-}$]):[$SO_4^{2-}$] ratio in mature leaves of oilseed rape grown under field conditions (field experiment 1), after (**a,d,g**) 15 and (**b,c,e,f,h,i**) 60 days of S fertilization. Plants received no mineral fertilization (hatched bars, 0 kg S $ha^{-1}$, 0 kg N $ha^{-1}$) or 0 (white bars), 12 (gray bars) or 36 (black bars) kg S $ha^{-1}$, with 65 (dashed bars) or 125 (full bars) kg N $ha^{-1}$. Within the same graph, letters when different indicate significant difference between fertilization treatments for $p < 0.05$.

Interactions between N and S fertilization were also found. It can be assumed that the growth rate was mostly reduced by low N availability. For example, 15 days after S fertilization, the biomass of mature leaves was 5.03 ± 0.38 g DW $leaf^{-1}$ in the unfertilized plot whereas in the plot with 65 kg N $ha^{-1}$ the biomass was 6.70 ± 0.01 g DW $leaf^{-1}$ ($p < 0.001$). This reduced growth of N unfertilized plants had a direct consequence on the $SO_4^{2-}$ content because $SO_4^{2-}$ requirements were lower than in N fertilized plants, which used $SO_4^{2-}$ to ensure growth. For example, for the same S fertilization rate, the $SO_4^{2-}$ content (Figure 1a–c) in mature leaves of plants without N fertilization was significantly higher than (Figure 1a,c) or similar to (Figure 1b) the content in leaves of plants with N fertilization. Moreover, compared to leaves from plants with N fertilization and without S fertilization, the $Cl^-$+$NO_3^-$+$PO_4^{3-}$ content (Figure 1d–f) in mature leaves of the N unfertilized plot was unchanged after 15 and 60 days. As a consequence, the ([$Cl^-$]+[$NO_3^-$]+[$PO_4^{3-}$]):[$SO_4^{2-}$] ratio (Figure 1g–i) decreased significantly (Figure 1g,i) or was similar (Figure 1h) in the absence of N mineral fertilization in mature leaves, reflecting lower S requirements. In the same way, plant growth was stimulated by a second N fertilization of 60 kg N $ha^{-1}$ (corresponding to plots with 125 kg N $ha^{-1}$) and consequently the S requirements were increased, reducing the $SO_4^{2-}$ content in mature leaves ($p < 0.05$, when comparing Figure 1b,c). The second N fertilization had no significant impact on the ($Cl^-$+$NO_3^-$+$PO_4^{3-}$) contents of mature leaves (Figure 1f) but these ions tended to increase for a given dose of S fertilization (0, 12 or 36 kg S $ha^{-1}$) compared to plants with 65 kg N $ha^{-1}$ (Figure 1e). So, when comparing Figure 1h and i, the ([$Cl^-$]+[$NO_3^-$]+[$PO_4^{3-}$]):[$SO_4^{2-}$] ratio was increased in mature leaves of plants receiving high N fertilization (Figure 1i), which reflected a higher requirement for S.

The same analyses were also performed with younger leaves (data not shown), which showed a similar trend to mature leaves but with a lower amplitude of response of the ([$Cl^-$]+[$NO_3^-$]+[$PO_4^{3-}$]):[$SO_4^{2-}$] ratio between plants receiving N or S fertilizations. Consequently, for the other experiments only data obtained from mature leaves will be given.

### *4.2. The ([$Cl^-$]+[$NO_3^-$]+[$PO_4^{3-}$]):[$SO_4^{2-}$]Ratio Can Be Used in Other Plant Species to Detect S Deficiency under Controlled Conditions*

The ([$Cl^-$]+[$NO_3^-$]+[$PO_4^{3-}$]):[$SO_4^{2-}$] ratio was also calculated in leaves of *B. oleracea*, *T. aestivum*, *Z. mays*, *M. truncatula*, *S. lycopersicum*, and *B. napus* grown under controlled conditions with or without S restriction (Table 1). In all species, S deficiency decreased the $SO_4^{2-}$ content in leaf tissue and significantly increased ($Cl^-$+$NO_3^-$+$PO_4^{3-}$) contents. Consequently, the ([$Cl^-$]+[$NO_3^-$]+[$PO_4^{3-}$]):[$SO_4^{2-}$] ratio was highly and significantly increased by S deficiency as soon as threedays after treatment in *B. napus*, eight days in *T. aestivum*, five days in *Z. mays*, and eight days in *M. truncatula*, and this increase was even amplified in the mid- to long-term. However, if no early harvest was performed with *B. oleracea* and *S. lycopersicum*, a significant increase in the ([$Cl^-$]+[$NO_3^-$]+[$PO_4^{3-}$]):[$SO_4^{2-}$] ratio during S deficiency was found in the longer term (i.e., after 135 days of S deficiency in *B. oleracea* and after 75 days in *S. lycopersicum*). Finally, it must be pointed out that the absolute value of this ratio was species specific, and probably a result of the intrinsic capacity of a plant species to store $SO_4^{2-}$ in leaves, as illustrated by the two *Brassica* species as well as tomato which accumulated higher levels of $SO_4^{2-}$ (Table 1).

### *4.3. The Simplified ([Cl]+[P]):[S] Ratio Could Be Used for an Easier and Faster Determination of S Deficiency under Field Conditions*

The resulting correlations between ion contents measured by HPLC and element contents measured by handheld X-Ray Fluorescence spectrometry (XRF) are given in Figure 2. In oilseed rape leaves, the ions $SO_4^{2-}$, $PO_4^{3-}$, and $Cl^-$were highly linearly correlated to their respective elements S, P, and Cl (Figure 2a–c) with correlation coefficients of 0.91, 0.70, and 0.97, respectively. Such strong linear relationships can be easily explained by the fact that these ions represented the main proportion of the corresponding element in mature leaves. A maximum (top of the curve) of 56%, 70%, and 95% of P, S and Cl were in the form of $PO_4^{3-}$, $SO_4^{2-}$, and $Cl^-$, respectively. On the other hand, no significant correlation was found between $NO_3^-$ and N contents in leaves of oilseed rape (Figure 2d), which was the result of very low $NO_3^-$ contents in leaves of plants grown under field conditions (data not shown). Consequently, for plants from field experiment 1, the anionic ratio ([$Cl^-$]+[$NO_3^-$]+[$PO_4^{3-}$]):[$SO_4^{2-}$] was significantly correlated to the ([Cl]+[P]):[S] ratio (Figure 3a) by a second order polynomial equation with a correlation coefficient of 0.94 showing a saturation plateau for higher values of the ([Cl]+[P]):[S] ratio. This previous graph was split in order to separate data from plants harvested before flowering (Figure 3b, plants harvested before fertilization and 15 days after fertilization when plants were at the start of the visible bud stage) and after flowering (Figure 3c, plants harvested 47 and 60 days after S fertilization for which plants were at the stage of pod formation) to assess whether this ratio was affected by plant development and then to determine potential threshold values. Before flowering (Figure 3b), both ratios were higher than after flowering (Figure 3c) with the values being between 1.69 and 3.83 for the ([Cl]+[P]):[S] ratio in plots with high S and low S fertilization, respectively. After flowering the ([Cl]+[P]):[S] ratio values (Figure 3d) were lower, between 0.19 (high S fertilization) and 2.14 (no S fertilization). Before any mineral fertilization, the ([Cl]+[P]):[S] ratios of leaves were between 2.02 and 2.68 (Figure 3a,b purple diamonds) and without S fertilization (i.e., 0 kg S ha$^{-1}$, orange squares) the ratio increased and to a lesser extent for the unfertilized plot (i.e., 0 kg S ha$^{-1}$ and 0 kg N ha$^{-1}$, red squares), whereas it decreased at 36 kg S ha$^{-1}$ (green squares).

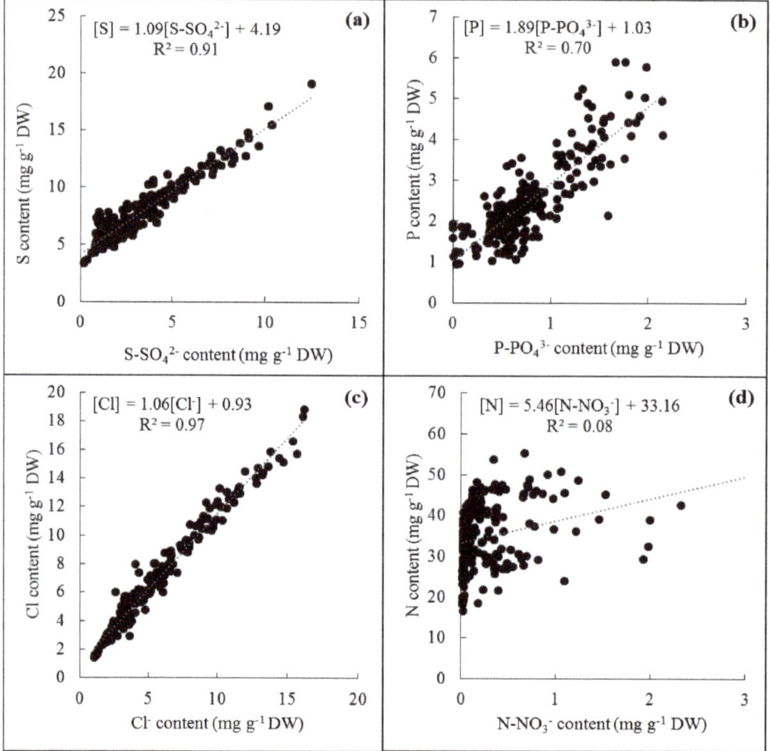


**Figure 2.** Correlation between (**a**) S-$SO_4^{2-}$ and S content (mg g$^{-1}$ DW); (**b**) P-$PO_4^{3-}$ and P content (mg g$^{-1}$ DW); (**c**) $Cl^-$ and Cl content (mg g$^{-1}$ DW) and (**d**) N-$NO_3^-$ and N content (mg g$^{-1}$ DW). Ion contents were quantified by HPLC and element contents by X-ray fluorescence (XRF), except N which was quantified by IRMS, in leaves of oilseed rape grown under field conditions (field experiment 1 and the second part of field experiment 2). Data used were obtained from plants of all fertilization treatments.

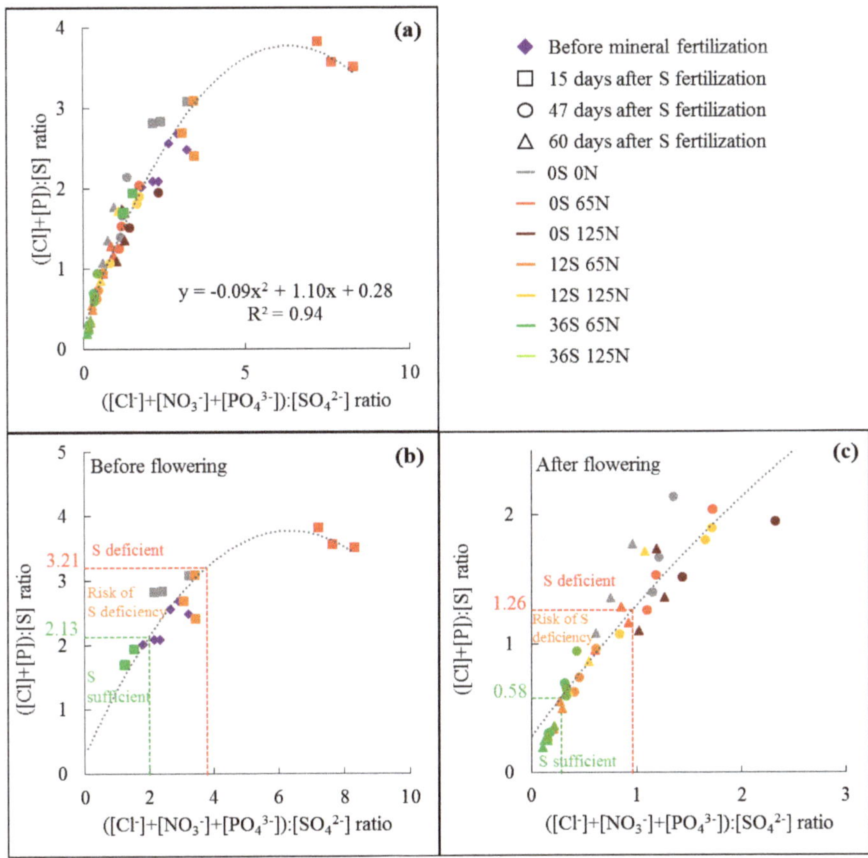


**Figure 3.** Correlation between the ($[Cl^-]+[NO_3^-]+[PO_4^{3-}]$):$[SO_4^{2-}]$ ratio and the ([Cl]+[P]):[S] ratio in mature leaves of oilseed rape submitted to different N and S fertilization rates (field experiment 1), (**a**) for all data points (all fertilization treatments, leaves harvested before mineral fertilization, or 15, 47 or 60 days after S fertilization), or for mature leaf samples harvested (**b**) before or (**c**) after flowering. Thresholds of the ([Cl]+[P]):[S] ratio were determined as the difference or sum between the mean and the 95% confidence interval of plants receiving 0 or 36 kg S ha$^{-1}$, defining three groups of plants: S deficient, at risk of S deficiency and S sufficient plants.

### 4.4. Determination of Threshold Values of the ([Cl]+[P]):[S]Ratio and Their Use on Independent Fields

Two sets of threshold values of the ([Cl]+[P]):[S]ratio were determined: before flowering with a potential use to determine whether or not S fertilization was required and after flowering to assess whether plants were S deficient. Threshold values were then calculated from data given in Figure 3b and c as the sum or difference between the ([Cl]+[P]):[S]ratio mean and the 95% confidence interval of S sufficient plots (36 kg S ha$^{-1}$) and S deficient plots (0 kg S ha$^{-1}$), respectively. The threshold values were then 2.13 and 3.21 before flowering, whereas they were 0.58 and 1.26 after flowering. Such thresholds would allow clustering of the plots within three groups of plots: S sufficient (ratio $<2.13$ or 0.58 after flowering), at risk of S deficiency (ratio between 2.13 and 3.21 or between 0.58 and 1.26 after flowering), and S deficient (ratio $>3.21$ or 1.26 after flowering) (Figure 3b,c).

This classification of oilseed rape plots into three groups, according to their S nutrition status evaluated by the leaf ([Cl]+[P]):[S]ratio, has been tested on two sets of independent commercial varieties(Figure 4). The first set corresponded to 45 commercial varieties from different locations in France and for which leaves were harvested before flowering and S fertilization (Figure 4a, field experiment 2). The second set corresponded to 56 other commercial varieties located in Calvados (Lower Normandy, France) and for which leaves were harvested after flowering and S fertilization (Figure 4b, field experiment 3). In the first case (Figure 4a), the use of the threshold values suggested that 27%, 33%, and 40% of the plots analyzed before flowering would be classified as S sufficient, at risk of S deficiency or as S deficient, respectively. After flowering (Figure 4b) 9%, 36%, and 55% of the analyzed plots fell into the previously cited groups.

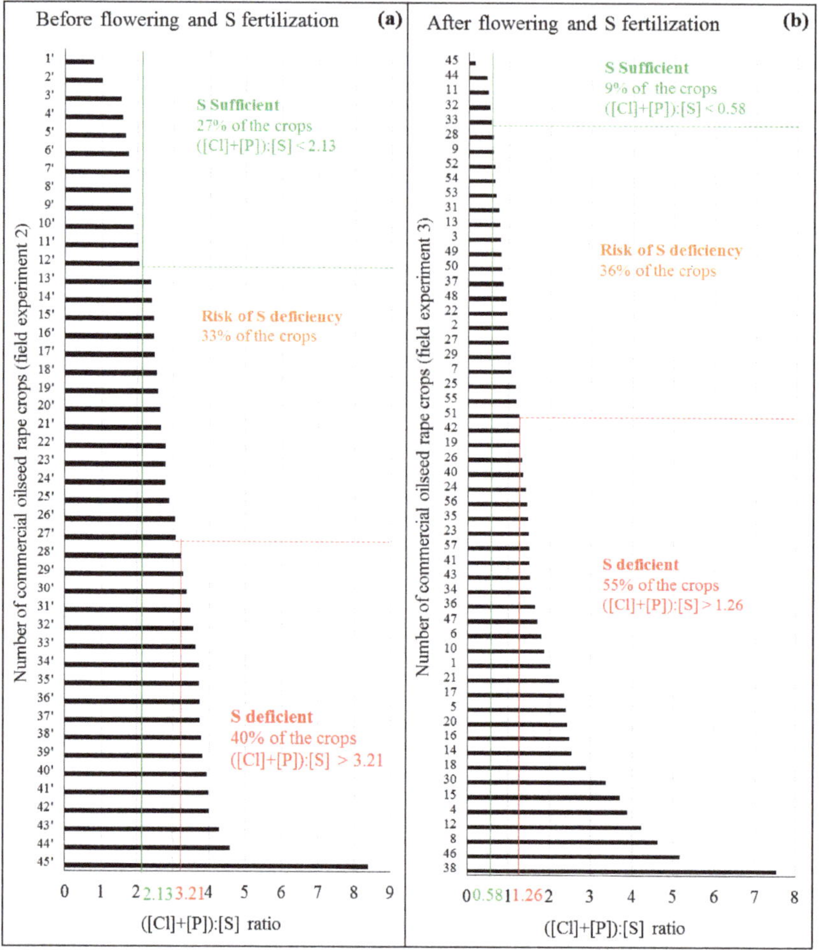


**Figure 4.** Commercial oilseed rape varieties (field experiments 2 and 3) classified according to decreasing values of the ([Cl]+[P]):[S] ratio in mature leaves quantified (**a**) before flowering and S fertilization (Field experiment 2 using 45 commercial varieties from different locations in France, see SD1) or (**b**) after flowering and S fertilization (Field experiment 3 using 56 commercial varietieslocalized in Calvados, Lower Normandy, France). Threshold values of the ([Cl]+[P]):[S] ratio classified these oilseed rape plots into three S status groups: S deficient, at risk of S deficiency and S sufficient plants.

**Table 1.** $SO_4^{2-}$contents (mg $g^{-1}$ DW), ($Cl^-$+$NO_3^-$+$PO_4^{3-}$) contents (mg $g^{-1}$ DW) and ([$Cl^-$]+[$NO_3^-$]+[$PO_4^{3-}$]):[$SO_4^{2-}$] ratios in leaves of *B. napus*, *B. oleracea*, *T. aestivum*, *Z. mays*, *S. lycopersicum*,and *M. truncatula* following different S treatments (+S: control treatment, white columns, -S: S deprivation treatment, grey columns) of plants grown under controlled conditions (given in Supplemental Data SD2). Data are given as the mean ± SE ($n$ = 4). *, ** and *** indicate significant difference between control and S deprived plants for $p < 0.05$, $p < 0.01$ and $p < 0.001$, respectively.

| Species | Day of Treatment | $SO_4^{2-}$ Content (mg $g^{-1}$ DW) | | ($Cl^-$+$NO_3^-$+$PO_4^{2-}$) Content (mg $g^{-1}$ DW) | | ([$Cl^-$]+[$NO_3^-$]+[$PO_4^{2-}$])/[$SO_4^{2-}$] Ratio | |
|---|---|---|---|---|---|---|---|
| | | +S | −S | +S | −S | +S | −S |
| *B. napus* | 0 | 24.38 ± 1.83 | | 59.79 ± 7.37 | | 2.32 ± 0.28 | |
| | 3 | 26.20 ± 1.23 | 15.90 ± 1.44*** | 63.65 ± 0.96 | 70.86 ± 5.68 | 2.43 ± 0.11 | 4.46 ± 0.39** |
| | 13 | 29.85 ± 0.50 | 8.05 ± 0.48*** | 62.83 ± 5.86 | 86.42 ± 4.66** | 2.10 ± 0.20 | 10.74 ± 0.17*** |
| *B. oleracea* | 135 | 7.05 ± 1.22 | 0.35 ± 0.04*** | 44.9 ± 6.85 | 61.87 ± 3.27* | 6.36 ± 1.06 | 176.77 ± 40.89** |
| *T. aestivum* | 0 | 3.03 ± 0.09 | | 63.58 ± 0.80 | | 20.98 ± 0.76 | |
| | 8 | 1.94 ± 0.05 | 1.15 ± 0.03 | 59.34 ± 2.10 | 68.04 ± 0.84** | 30.58 ± 0.53 | 59.16 ± 2.36*** |
| | 16 | 2.17 ± 0.06 | 0.50 ± 0.05*** | 64.79 ± 1.29 | 70.19 ± 1.80* | 29.85 ± 0.53 | 140.38 ± 15.81*** |
| *Z. mays* | 0 | 2.02 ± 0.49 | | 74.74 ± 1.55 | | 37.00 ± 7.16 | |
| | 5 | 3.44 ± 0.07 | 1.39 ± 0.08*** | 78.31 ± 2.82 | 91.71 ± 1.28** | 22.76 ± 1.09 | 56.33 ± 3.45*** |
| | 18 | 1.38 ± 0.05 | 0.27 ± 0.02*** | 70.38 ± 3.46 | 103.28 ± 0.79*** | 51.00 ± 2.19 | 382.51 ± 23.44*** |
| *S. lycopersicum* | 75 | 63.75 ± 4.79 | 50.39 ± 2.42* | 68.84 ± 4.61 | 87.77 ± 2.44* | 1.07 ± 0.15 | 1.74 ± 0.10** |
| *M. truncatula* | 0 | 3.31 ± 0.16 | | 25.32 ± 0.60 | | 7.55 ± 0.21 | |
| | 8 | 3.60 ± 0.12 | 0.15 ± 0.01*** | 25.40 ± 1.54 | 34.27 ± 3.44* | 7.05 ± 0.48 | 228.47 ± 25.21*** |
| | 21 | 4.75 ± 0.18 | 0.26 ± 0.01*** | 25.92 ± 1.94 | 34.90 ± 2.46* | 5.49 ± 0.50 | 134.23 ± 12.48*** |

In order to test if the ([Cl]+[P]):[S]ratio and associated thresholds take into account the availability of soil S and its interaction with the rate of S fertilization, the mature leaves, and the soil of four plots derived from field experiment 2 (i.e., plots # 8′, 11′, 20′, and 36′ represented in Figure 4a and supplemental data SD1) were harvested and analyzed before and after S fertilization (Table 2). These four oilseed rape plots were then classified within three groups (from S deficient to S sufficient) according to the ([Cl]+[P]):[S]ratio, determined before and after S fertilization. It was found that before flowering and S fertilization, the lower the S content in the soil, the higher the ([Cl]+[P]):[S]ratio and vice versa, which validated the possibility of using the([Cl]+[P]):[S]ratio as an indicator of sulfur status (Table 2). After S fertilization (from 0 to 100kg S $ha^{-1}$), the ([Cl]+[P]):[S]ratio was decreased partly due to plant development (see plot # 11′ without S fertilization in Table 2 as well as data of experiment 1 Figure 3a,c) but mostly as a result of the S fertilization rate. For example, plot # 20′, classified at risk of S deficiency before S fertilization, received 100 kg S $ha^{-1}$ which decreased its ([Cl]+[P]):[S]ratio by 78% (from 2.64 to 0.59) and thus ascribed it to the sufficient S availability group. Overall, the simple use of the leaf ([Cl]+[P]):[S]ratio before fertilization on these four plots indicated that plots # 8′ and 11′ would not have required any S fertilization. Nevertheless, the decrease in this ratio after flowering suggested that these plots were S deficient with little or no S fertilization. While plots # 36′ and 20′ seemed to require S fertilization, plot # 36′ would have required a higher dose of S, and plot # 20′ would have required a dose of S probably below those that were actually applied.

**Table 2.** Soil type, soil S content (mg S $kg^{-1}$), S and N fertilization managed by the farmers (kg S or N $ha^{-1}$) and the value of the ([Cl]+[P]):[S] ratio in mature leaves harvested before or after fertilization in four oilseed rape commercial varieties from field experiment 2 (crops # 36′, 20′, 11′, and 8′). Status of plots have been determined within three groups (S deficient, at risk of S deficiency and S sufficient) according to threshold values of the ([Cl]+[P]):[S] ratio. Data are given as mean ± SE ($n$ = 3). *, ** and *** indicate significant difference before and after fertilization for $p < 0.05$, $p < 0.01$ and $p < 0.001$, respectively. Letters when different indicate significant differences between crops at a given date, at $p < 0.05$.

| | | | Before Flowering and Fertilization | | S-(N) Fertilization (kg S-(N) $ha^{-1}$ | After Flowering and Fertilization | |
|---|---|---|---|---|---|---|---|
| Plots number | Soil type | Soil S content (mg $kg^{-1}$) | ([Cl]+[P])/[S] ratio | Status of plots | | ([Cl]+[P])/[S] ratio | Status of plots |
| 36′ | Compact silt | 136 | 3.72 ± 0.43 b | S deficient | 10-(166) | 1.54 ± 0.21 b** | S deficient |
| 20′ | Superficial clay-limestone | 239 | 2.64 ± 0.13 c | Risk of S deficiency | 100-(146) | 0.59 ± 0.10 a*** | S sufficient |
| 11′ | Superficial clay-limestone | 350 | 1.99 ± 0.14 a | S sufficient | 0-(110) | 1.42 ± 0.05 b** | S deficient |
| 8′ | deep clay-limestone | 668 | 1.80 ± 0.06 a | S sufficient | 15-(151) | 1.41 ± 0.24 b | S deficient |

## 5. Discussion

As previously reported, oilseed rape requires fairly high levels of S in order to maintain yield and seed quality [2,5,14]. Since a general S deficiency in cultivated soils has been described [3,6–8], S fertilization is usually provided to oilseed rape, but less frequently on other crops. Moreover, the dose used or recommended does not take into account the real needs of the plants and growth potential. This can lead to excessive or insufficient S fertilization [10]. Hence, the development of new diagnostic tools of S deficiency, usable under field conditions, is required. The aims of this work were to identify and validate a diagnostic tool derived from early physiological processes occurring during oilseed rape adaptations to S deficiency.

### 5.1. The ($[Cl^-]$+$[NO_3^-]$+$[PO_4^{3-}]$):$[SO_4^{2-}]$ Ratio as an Indicator of S Nutrition under Field Conditions

Under field conditions, the highest $Cl^-$, $NO_3^-$, $PO_4^{3-}$contents were found in mature leaves of plants with the lowest S fertilizations having the lowest $SO_4^{2-}$content, and hence the highest ($[Cl^-]$+$[NO_3^-]$+$[PO_4^{3-}]$):$[SO_4^{2-}]$ ratio (Figure 1a–i). These higher contents of $Cl^-$, $NO_3^-$ and $PO_4^{3-}$ might help to maintain the osmotic potential when $SO_4^{2-}$is mobilized from the vacuole of mature leaves (Figure 1d–f). It was also found that leaves of plants grown under field conditions have accumulated $Cl^-$ with the same order of magnitude that was previously found under hydroponic conditions. Indeed, oilseed rape leaves of field grown plants accumulated anions in the same range of concentrations found under hydroponic conditions: 2–16 mg $Cl^-$ $g^{-1}$ DW (4–20 mg $Cl^-$ $g^{-1}$ DW in hydroponic culture), 0.1–3 mg $PO_4^{3-}$ $g^{-1}$ DW (2–7 mg $PO_4^{3-}$ $g^{-1}$ DW in hydroponic culture) and 1–37 mg $SO_4^{2-}$ $g^{-1}$ DW (2–42 mg $SO_4^{2-}$ $g^{-1}$ DW in hydroponic culture). Only $NO_3^-$was found to accumulate in leaves at a much lower level under field conditions (0.1–5 mg *versus* 40–70 $NO_3^-$ $g^{-1}$ DW under hydroponic culture). The same range of $NO_3^-$ contents in mature leaves has been previously reported by Blake-Kalff et al. [22] and Sarda et al. [10] in oilseed rape grown under field conditions. Nevertheless, the ($[Cl^-]$+$[NO_3^-]$+$[PO_4^{3-}]$):$[SO_4^{2-}]$ ratio was able to significantly differentiate plants receiving different doses of S fertilization under field conditions as soon as 15 days after S supply (Figure 1g–i). This ratio was also found to be sensitive to N fertilization. Janzen and Bettany [1] and Zhao et al. [36] reported that higher N fertilization increased plant growth rates and consequently S requirements and hence may aggravate the S deficiency of oilseed rape plants.

Whatever the plant species, the leaf ($[Cl^-]$+$[NO_3^-]$+$[PO_4^{3-}]$):$[SO_4^{2-}]$ ratio was significantly and precociously increased by reduced $SO_4^{2-}$ availability in nutrient solutions, resulting from a disappearance in $SO_4^{2-}$in leaves that was accompanied by an increase in ($Cl^-$+$NO_3^-$+$PO_4^{3-}$)content.

These results strongly suggest that this ratio could be used as a diagnostic tool for a wide range of cultivated plant species.

### 5.2. Using the ([Cl]+[P]):[S] Ratio Instead of the ($[Cl^-]$+$[NO_3^-]$+$[PO_4^{3-}]$):$[SO_4^{2-}]$ Ratio

Because $Cl^-$, $NO_3^-$, $PO_4^{3-}$, and $SO_4^{2-}$ cannot be quantified with enough accuracy with simple methods such as colorimetric test strips, another aim of this study was to find a way to simplify the analysis of this anionic ratio. Due to their consequent accumulation in their mineral form and the resulting strong linear correlations between anions and elements (Figure 2a–c), quantification of elements such as Cl, P, or S can provide an easy way to determine accurately the $Cl^-$, $PO_4^{3-}$, and $SO_4^{2}$ contents in leaf tissues. As $NO_3^-$was not massively accumulated in leaves of plants grown under field conditions [10,22], we found that the ($[Cl^-]$+$[NO_3^-]$+$[PO_4^{3-}]$):$[SO_4^{2-}]$ and ([Cl]+[P]):[S]ratios were highly correlated by a second-degree polynomial equation (Figure 3a). This ([Cl]+[P]):[S]ratio was measured accurately by handheld XRF spectrometry, directly from ground dry leaf samples, which could be done from leaf samples harvested in the field. Another advantage of quantifying elements instead of anions relies on the fact that the plant samples can be stored for several hours or days after harvest without affecting elementary quantification, which will not be the case with anions.

### 5.3. Potential Thresholds of the ([Cl]+[P]):[S] Ratio

A successful diagnostic indicator as defined by Blake-Kalff et al. [22] needs to show a large response to different S fertilizations, remain stable during development, and be relatively easy to measure with little effort and with great accuracy. Most of these criteria are respected by the ([Cl]+[P]):[S] indicator. The precaution to be taken with this indicator is the use of threshold values depending on the developmental stage. According to previous study [17], threshold values should never be considered as absolute, but as a representation of a possible range resulting from uncontrolled or unknown variables. Having an index of nutrition with constant threshold values at different growing stages has already been questioned [6,20]. In this study, two threshold values for the two groups of development stages have been determined, allowing classification of plots into three S status groups: S deficient, at risk of S deficiency, and S sufficient, with the additional intermediate class taking into account the area of uncertainty. The survey of the S status of 45 commercial oilseed rape varieties in France, field experiment 2, has enabled mapping of their locations according to S requirements (Supplemental data SD1). The plots located in the north-east of France, which corresponds to the cereal plain, was mostly S sufficient. Other areas were more subject to S deficiency. Previous studies have shown that in response to a moderate S deficiency or a decrease in S availability during late stage of reproductive development, oilseed rape is able to maintain its growth and its yield components by increasing remobilization of previously stored S (especially $SO_4^{2-}$) in leaves [37,38]. This was probably the case for plots # 8′ and # 11′ (Table 2), which were classified as S sufficient before flowering and with little or no S fertilization, whereas they were classified as S deficient after flowering. The example of these two plots also suggests that the determination of soil S content is not reliable on its own to predict the plant S status, and which highlights the need for indicators at the plant level.

## 6. Conclusions

This study firstly showed that the accumulation of anions such as $Cl^-$, $NO_3^-$, and $PO_4^{3-}$ in response to $SO_4^{2-}$remobilization triggered by S deficiency also occurred in oilseed rape grown under field conditions as previously shown in hydroponically grown plants. The same process was also found in species of different families. The derived leaf elementary ratio, ([Cl]+[P]):[S], seems to be a reliable means to detect S deficiency under field conditions. Measurement of this ratio by XRF presents the advantage of requiring only one measurement to be performed on grounded dry matter. Its use as a diagnostic tool of plant S nutrition requires accurate threshold values to adjust S fertilization accordingly. A determination method for these threshold values was suggested but will need to be tested at a larger scale (larger number of plots, adaptation to other cultivated species). Finally, the S fertilizer recommendation derived from this indicator will have to be verified a posteriori on yield measurement and/or seed quality.

**Supplementary Materials:** The following are available online at http://www.mdpi.com/2223-7747/7/2/37/s1, Location of 45 commercial plots in France of field experiment 2; Multispecies experiment under controlled conditions; Composition of the two nutrient solutions used for control and S deprivation during the treatment period of *B. napus and Z. mays*.

**Author Contributions:** P.E., E.S., A.M. and A.O. participated in experiment design on *B. napus*, data analysis, and interpretation and wrote the manuscript. K.G. and J.G. provided data for other plant species. M.A., F.C. and J.-C.Y. participated to design of experiments.

**Acknowledgments:** The authors thank M. Chédeville for providing the experimental site at Ondefontaine and for technical help to fertilize plots, and all farmers identified by DATAGRI who collaborated in this study. We are also grateful to Vegenov for their participation in the multispecies experiment. Acknowledgements go to the PLATIN' (Plateau d'Isotopie de Normandie) core facility for the element analysis used in this study. We thank Laurence Cantrill for improving the English in the manuscript. This work, conducted through the SERAPIS project, was supported by the Regional Council of Lower Normandy (grant number 12P03057), the Regional Council of Brittany (grant number 12008011), the European Regional Development Fund (grant number 33525) and the Centre Mondial de lInnovation (CMI)

**Conflicts of Interest:** The authors declare no conflict of interest.

## Abbreviations

| | |
|---|---|
| DW | dry weight |
| HPLC | high performance liquid chromatography |
| HR ICP-MS | high resolution inductively coupled plasma mass spectrometry |
| XRF | X-ray fluorescence |

## References

1. Janzen, H.H.; Bettany, J.R. Sulfur nutrition of rapeseed: I. Influence of fertilizer nitrogen and sulfur rates. *Soil Sci. Soc. Am. J.* **1984**, *48*, 100–107. [CrossRef]
2. McGrath, S.P.; Zhao, F.J. Sulphur uptake, yield responses and the interactions between nitrogen and sulphur in winter oilseed rape (*Brassica napus*). *J. Agric. Sci.* **1996**, *126*, 53–62. [CrossRef]
3. Scherer, H.W. Sulphur in crop production. *Eur. J. Agron.* **2001**, *14*, 81–111. [CrossRef]
4. Malhi, S.S.; Gan, Y.; Raney, J.P. Yield, seed quality, and sulfur uptake of oilseed crops in response to sulfur fertilization. *Agron. J.* **2007**, *99*, 570–577. [CrossRef]
5. D'Hooghe, P.; Dubousset, L.; Gallardo, K.; Kopriva, S.; Avice, J.C.; Trouverie, J. Evidence for proteomic and metabolic adaptations associated with alterations of seed yield and quality in Sulfur-limited *Brassica napus* L. *Mol. Cell. Proteom.* **2014**, *13*, 1165–1183. [CrossRef] [PubMed]
6. Zhao, F.J.; Hawkesford, M.J.; McGrath, S.P. Sulphur assimilation and effects on yield and quality of wheat. *J. Cereal Sci.* **1999**, *30*, 1–17. [CrossRef]
7. Blake-Kalff, M.M.A.; Zhao, F.J.; Hawkesford, M.J.; McGrath, S.P. Using plant analysis to predict yield losses caused by sulphur deficiency. *Ann. Appl. Biol.* **2001**, *138*, 123–127. [CrossRef]
8. McNeill, A.M.; Eriksen, J.; Bergström, L.; Bergström, L.; Smith, K.A.; Marstorp, H.; Kirchmann, H.; Nilsson, I. Nitrogen and sulphur management: Challenges for organic sources in temperate agricultural systems. *Soil Use Manag.* **2005**, *21*, 82–93. [CrossRef]
9. Grant, C.A.; Mahli, S.S.; Karamanos, R.E. Sulfur management for rapeseed. *Field Crops Res.* **2012**, *128*, 119–128. [CrossRef]
10. Sarda, X.; Diquelou, S.; Abdallah, M.; Nési, N.; Cantat, O.; Le Gouee, P.; Avice, J.-C.; Ourry, A. Assessment of sulphur deficiency in commercial oilseed rape crops from plant analysis. *J. Agric. Sci.* **2014**, *152*, 616–633. [CrossRef]
11. Ayala-Silva, T.; Beyl, C.A. Changes in spectral reflectance of wheat leaves in response to specific macronutrient deficiency. *Adv. Space Res.* **2005**, *35*, 305–317. [CrossRef] [PubMed]
12. Zuber, H.; Poignavent, G.; Le Signor, C.; Aimé, D.; Vieren, E.; Tadla, C.; Lugan, R.; Belghazi, M.; Labas, V.; Santoni, A.L.; et al. Legume adaptation to sulfur deficiency revealed by comparing nutrient allocation and seed traits in *Medicago truncatula*. *Plant J.* **2013**, *76*, 982–996. [CrossRef] [PubMed]

13. Scherer, H.W. Sulfur in soils. *J. Plant Nutr. Soil Sci.* **2009**, *172*, 326–335. [CrossRef]
14. Malhi, S.S.; Schoenau, J.J.; Grant, C.A. A review of sulphur fertilizer management for optimum yield and quality of canola in the Canadian Great Plains. *Can. J. Plant Sci.* **2005**, *85*, 297–307. [CrossRef]
15. McGrath, S.P.; Zhao, F.J. A risk assessment of sulphur deficiency in cereals using soil and atmospheric deposition data. *Soil Use Manag.* **1995**, *11*, 110–114. [CrossRef]
16. Brunel-Muguet, S.; Mollier, A.; Kauffmann, F.; Avice, J.C.; Sénécal, E.; Goudier, D.; Bataillé, M.P.; Etienne, P. SuMoToRI, an ecophysiological model to predict growth and sulfur allocation and partitioning in oilseed rape (*Brassica napus* L.). *Front. Plant Sci.* **2015**, *6*, 1–14. [CrossRef] [PubMed]
17. Melsted, S.W.; Motto, H.L.; Peck, T.R. Critical plant nutrient composition values useful in *interpreting plant analysis data. Agron. J.* **1969**, *61*, 17–20. [CrossRef]
18. Rasmussen, P.E.; Ramig, R.E.; Rohde, C.R. Tissue analyses guidelines for diagnosing sulfur deficiency in white wheat. *Plant Soil* **1977**, *46*, 153–163. [CrossRef]
19. Spencer, K.; Freney, J.R. Assessing the sulfur status of field-grown wheat by plant analysis. *Agron. J.* **1980**, *72*, 469–472. [CrossRef]
20. Scaife, A.; Burns, I.G. The sulphate-S/total S ratio in plants as an index of their sulphur status. *Plant Soil* **1986**, *91*, 61–71. [CrossRef]
21. Pinkerton, A. Critical sulfur concentrations in oilseed rape (*Brassica napus*) in relation to nitrogen supply and to plant age. *Aust. J. Exp. Agric.* **1998**, *38*, 511–522. [CrossRef]
22. Blake-Kalff, M.M.A.; Hawkesford, M.J.; Zhao, F.J.; McGrath, S.P. Diagnosing sulfur deficiency in field-grown oilseed rape (*Brassica napus* L.) and wheat (*Triticum aestivum* L.). *Plant Soil* **2000**, *225*, 95–107. [CrossRef]
23. Maillard, A.; Sorin, E.; Etienne, P.; Diquélou, S.; Kopriva, A.; Kopriva, S.; Arkoun, M.; Gallardo, K.; Turner, M.; Cruz, F.; et al. Non-specific root transport of nutrient gives access to an early nutritional indicator: The case of sulfate and molybdate. *PLoS ONE* **2016**, *11*, e0166910. [CrossRef] [PubMed]
24. Zhao, F.J.; Hawkesford, M.J.; Warrilow, A.G.S.; McGrath, S.P.; Clarkson, D.T. Responses of two wheat commercial to sulphur addition and diagnosis of sulphur deficiency. *Plant Soil* **1996**, *181*, 317–327. [CrossRef]
25. Howarth, J.R.; Parmar, S.; Barraclough, P.B.; Hawkesford, M.J. A sulphur deficiency-induced gene, *sdi1*, involved in the utilization of stored sulphate pools under sulphur-limiting conditions has potential as a diagnostic indicator of sulphur nutritional status. *Plant Biotechnol. J.* **2009**, *7*, 200–209. [CrossRef] [PubMed]
26. Sorin, E.; Etienne, P.; Maillard, A.; Zamarreno, A.M.; Garcia-Mina, J.M.; Arkoun, M.; Jamois, F.; Cruz, F.; Yvin, J.C.; Ourry, A. Effect of sulphur deprivation on osmotic potential components and nitrogen metabolism in oilseed rape leaves: Identification of a new early indicator. *J. Exp. Bot.* **2015**, *66*, 6175–6189. [CrossRef] [PubMed]
27. Blake-Kalff, M.M.A.; Harrison, K.R.; Hawkesford, M.J.; Zhao, F.J.; McGrath, S.P. Distribution of sulfur within oilseed rape leaves in response to sulfur deficiency during vegetative growth. *Plant Physiol.* **1998**, *118*, 1337–1344. [CrossRef] [PubMed]
28. Honsel, A.; Kojima, M.; Haas, R.; Frank, W.; Sakakabara, H.; Herschbach, C.; Heinz, R. Sulphur limitation and early sulphur deficiency responses in poplar: Significance of gene expression, metabolites, and plant hormones. *J. Exp. Bot.* **2012**, *63*, 1873–1893. [CrossRef] [PubMed]
29. Rabe, E. Altered nitrogen metabolism under environmental stress conditions. In *Handbook of Plant and Crop Stress*, 2nd ed.; Pessarakli, M., Ed.; Marcel Dekker: New York, NY, USA, 1999; pp. 349–363, ISBN 978-1439813966.
30. Tausz, M. The role of glutathione in plant response and adaptation to natural stress. In *Significance of Glutathione to Plant Adaptation to the Environment*; Grill, D., Tausz, M., Kok, L.J.D., Eds.; Springer: Houten, The Netherlands, 2001; pp. 101–122, ISBN 978-1402001789.
31. Maynard, D.; Stewart, J.; Bettany, J. Use of plant analysis to predict sulfur deficiency in rapeseed (*Brassica napus* and *B. campestris*). *Can. J. Soil Sci.* **1983**, *63*, 387–396. [CrossRef]
32. Blake-Kalff, M.M.A.; Zhao, F.J.; McGrath, S.P.; Withers, P.J.A. *Development of the Malate:Sulphate Ratio Test for Sulphur Deficiency in Winter Wheat and Oilseed Rape*; HGCA Project Report; No. 327; Home-Grown Cereals Authority: London, UK, 2004.
33. Carver, M.F.F. *Monitoring Winter Barley, Wheat, Oilseed Rape and Spring Barley for Sulphur in England and Wales to Predict Fertiliser Need*; HGCA Project Report; Rep, No. 374; Home-Grown Cereals Authority: London, UK, 2005.
34. Reussi, N.; Echeverría, H.; Rozas, H.S. Diagnosing sulfur deficiency in spring red wheat; plant analysis. *J. Plant Nutr.* **2011**, *34*, 573–589. [CrossRef]

35. Maillard, A.; Diquélou, S.; Billard, V.; Laîné, P.; Garnica, M.; Prudent, M.; Garcia-Mina, J.-M.; Yvin, J.C.; Ourry, A. Leaf mineral nutrient remobilization during leaf senescence and modulation by nutrient deficiency. *Front. Plant Sci.* **2015**, *6*, 1–15. [CrossRef] [PubMed]
36. Zhao, F.; Evans, E.J.; Bilsborrow, P.E.; Syers, J.K. Influence of sulphur and nitrogen on seed yield and quality of low glucosinolate oilseed rape (*Brassica napus* L.). *J. Sci. Food Agric.* **1993**, *63*, 29–37. [CrossRef]
37. Dubousset, L.; Etienne, P.; Avice, J.C. Is the remobilization of S and N reserves for seed filling of winter oilseed rape modulated by sulphate restrictions occurring at different growth stages? *J. Exp. Bot.* **2010**, *61*, 4313–4324. [CrossRef] [PubMed]
38. Abdallah, M.; Etienne, P.; Ourry, A.; Meuriot, F. Do initial S reserves and mineral S availability alter leaf S-N mobilization and leaf senescence in oilseed rape? *Plant Sci.* **2011**, *180*, 511–520. [CrossRef] [PubMed]

*Article*

# The Effect of Granular Commercial Fertilizers Containing Elemental Sulfur on Wheat Yield under Mediterranean Conditions

**Dimitris L. Bouranis [1,*], Dionisios Gasparatos [2], Bernd Zechmann [3], Lampros D. Bouranis [1] and Styliani N. Chorianopoulou [1]**

1 Plant Physiology and Morphology Laboratory, Crop Science Department, Agricultural University of Athens, 11855 Athens, Greece; lampros.bouranis@ucdconnect.ie (L.D.B.); s.chorianopoulou@aua.gr (S.N.C.)

2 Soil Science Laboratory, Department of Hydraulics, Soil Science and Agricultural Engineering, School of Agriculture, Faculty of Agriculture, Forestry and Natural Environment, Aristotle University of Thessaloniki, 54124 Thessaloniki, Greece; gasparatos@agro.auth.gr

3 Center for Microscopy and Imaging, Baylor University, One Bear Place 97046, Waco, TX 76798-7046, USA; Bernd_Zechmann@baylor.edu

* Correspondence:bouranis@aua.gr

Received: 1 November 2018; Accepted: 17 December 2018; Published: 20 December 2018

**Abstract:** The demand to develop fertilizers with higher sulfur use efficiency has intensified over the last decade, since sulfur deficiency in crops has become more widespread. The aim of this study was to investigate whether fertilizers enriched with 2% elemental sulfur (ES) via a binding material of organic nature improve yield when compared to the corresponding conventional ones. Under the scanning electron microscope, the granules of the ES-containing fertilizer were found to be covered by a layer of crystal-like particles, the width of which was found to be up to 60 μm. Such a layer could not be found on the corresponding conventional fertilizer granules. Several fertilization schemes with or without incorporated ES were tested in various durum wheat varieties, cultivated in commercial fields. The P-Olsen content of each commercial field was found to be correlated with the corresponding relative change in the yields (YF/YFBES) with a strong positive relationship. The content of 8 ppm of available soil phosphorus was a turning point. At higher values the incorporation of ES in the fertilization scheme resulted in higher yield, while at lower values it resulted in lower yield, compared with the conventional one. The experimental field trials that established following a randomized block design, were separated in two groups: One with P-Olsen ranging between 18–22 ppm and the other between 12–15 ppm, the results of which corroborated the aforementioned finding. The use of ES in all portions of fertilization schemes provided higher relative yields. The coexistence of ES with sulfate in the granule was more efficient in terms of yield, when compared to the granule enriched with ES alone under the same fertilization scheme and agronomic practice. The application of fertilizer mixtures containing the urease inhibitor N-(n-butyl) thiophosphoric triamide (NBPT), ES and ammonium sulfate resulted in even higher relative yields. Yield followed a positive linear relationship with the number of heads per square meter. In this correlation, the P-Olsen content separated the results of the two groups of blocks, where the applied linear trend line in each group presented the same slope.

**Keywords:** elemental sulfur; granular S containing fertilizers; durum wheat; yield; NBPT

---

## 1. Introduction

Sulfur (S) deficiency in crops has become a common agronomic issue in many countries, mainly due to reduced atmospheric inputs and the use of sulfur free high-analysis fertilizers (i.e., fertilizer products that contain a high percentage of nutrients, usually more than 30 per cent). The development of S deficiency in plants during the last decades has been discussed in detail [1,2]. Therefore, the demand for and interest in S fertilizers has been intensified. Wheat (*Triticum aestivum*) is the most widely cultivated crop in the world, a crop that has a relatively low requirement for S. However, S deficiency in wheat has become increasingly widespread [3].

A detailed discussion of S fertilizers has been provided by Grant and Hawkesford [4]. The most important inorganic S fertilizers are ammonium sulfate (AS, 24% S), gypsum (18% S), single superphosphate (12% S) and elemental sulfur (ES, 100% S) [5]. The granular fertilizers that provide S when applied in crops are mainly based on sulfate salts, because sulfate-S (i.e., S in the form of sulfate) is readily available to plants, however highly mobile in the soil profile.

On the other hand, ES, a soil amendment with a long history in agricultural practice, is insoluble in water and works as a slow release S fertilizer, which becomes available to plants after oxidation to sulfate by microorganisms [6]. According to Boswell and Friesen [7], 80% of ES (particles of 75–150 μm) was found to be oxidized after six months of application in a warm temperate region of New Zealand. The incorporation of ES into the granules of commercial fertilizers provide an efficient use of ES in agriculture, as S component in the fertilization management practices [8].

In a recent industrial approach, the granules of conventional fertilizers (F) were enriched with 2% ES ($w/w$) via a binding material of organic nature (B, i.e., a mixture of molasses and glycerol) (FBES technology, [9]). The aforementioned approach for the production of granular fertilizers of this type can be applied to various granular fertilizers, excluding nitrates, thus providing a family of ES-enriched fertilizers.

While several studies have been conducted to assess the efficiency of ES as fertilizer, there are scarce results from agronomic field trials on the granular ES-enriched macronutrient fertilizers, in order to estimate if these products may have a positive effect on crop production [10]. ES-enriched urea (U) produced by various technological approaches is used as fertilizer over the last 25 years. On the other hand, NBPT (i.e., N-(n-butyl)-thiophosphoric triamide) is a commonly used urease inhibitor for retarding U hydrolysis in the past 10 years [11]. U or fertilizer materials derived primarily from U, enriched with other additional useful additives and plant nutrient materials, can be treated with a solution of NBPT dissolved in one or more amino alcohols to reduce nitrogen volatilization. The solution may be applied as a coating for granular U fertilizers [12].

The initial field experiment showed that the commercial ES-fertilizers enriched with the FBES technology increased yield of durum wheat (*Triticum durum* or *Triticum turgidum* subsp. *durum*) crop by 27.3% compared with the corresponding conventional ones [13]. In order to verify whether the fertilizer products of the aforementioned FBES technology were able to provide a higher yield under the common fertilization schemes and agricultural practices, commercial field trials with durum wheat crops were established and monitored in central and northern Greece. In these schemes, U was the common N fertilizer. Furthermore, U in mixtures with AS (UAS) was also of common use. Both U and UAS can be enriched with ES by the FBES technology (U/ES, UAS/ES). UAS/ES can release readily available sulfate-S for immediate use by the crop, while ES can provide available S after S oxidation to sulfate-S. In order to further explain the findings obtained from the commercial field trials, experimental field trials were established and monitored. U or UAS along with their stabilized form with NBPT are also of common use in the commercial fertilization schemes (U/NBPT). No product was available containing both of the aforementioned chemicals, i.e., ES and NBPT. Was the use of U/NBPT or UAS/NBPT compatible with the use of U/ES or UAS/ES? In order to verify to what degree the enriched fertilizer schemes with ES and NBPT were influencing the yield, the following question was addressed: Could a mixture of both fertilizer types be an effective one?

## 2. Materials and Methods

### 2.1. Commercial Field Trials

In the wheat growing season of 2015–2016, several crops with durum wheat were established in different areas of central and northern Greece (Groups A1 and A2). Sowing and harvesting took place mid November 2015 and end June 2016 respectively. In 2016–2017, more crops were established in the same or other fields (Group A3). Sowing and harvesting took place mid November 2016 and end June 2017 respectively. All fields were of commercial use. Each production field was divided into two equal areas; one of them was subject to conventional fertilization treatment according to the local agricultural practices (F-trial), while the other one received the corresponding FBES-trial. The fertilization scheme and the agronomic details per trial (i.e., area in ha, sowing rate, cultivar, fertilizer type and rate) are provided in Table 1, along with the percentage relative yield. Soil fertility of each field was determined prior to crop establishment. In order to ensure comparable soil conditions between the corresponding F- and FBES-trials, each area was arbitrarily divided into plots 15 m × 7 m (105 square meters) each [14]. All perimetric plots were excluded, while the internal plots were grouped into five successive groups. Within each group of plots, one composite sample per group (0.5 kg) was collected at the depth of 0–20 cm, these samples were pooled together and a final composite sample per trial (2.5 kg) was formed and analyzed. Soil analysis data and yields are provided in Table 2. Only the fields with comparable F- and FBES-trials with regard to soil fertility are further analyzed and discussed in this study.

**Table 1.** Fertilization schemes per trial. FT, field trial; C, field trial with the conventional fertilization (F-)treatment, S, the corresponding field trial with incorporated elemental sulfur in the fertilization scheme (FBES-treatment); F1, initial fertilization along with sowing; F2a, first additional fertilization; F2b, second additional fertilization; Δx/x, percentage relative change of grain yield; 46-0-0, urea; 40-0-0, urea plus ammonium sulfate; 34.5-0-0, ammonium nitrate; /ES, fertilizer granules with incorporated 2% elemental sulfur.

| Group | Field Trial | Area ha | Seed kg ha$^{-1}$ | Cultivar | F1 kg ha$^{-1}$ | Fertilizer Type | F2a kg ha$^{-1}$ | Fertilizer Type | F2b kg ha$^{-1}$ | Fertilizer Type | Δx/x (%) |
|---|---|---|---|---|---|---|---|---|---|---|---|
| | | | | | | 2015–2016 | | | | | |
| | FT_13C | 1.4 | 280 | Meridiano | 250 | 16-20-0 | 120 | 46-0-0 | | | |
| | FT_13S | 1.4 | 280 | Meridiano | 250 | 16-20-0/ES | 120 | 46-0-0 | | | 28.5 |
| | FT_08C | 1.4 | 250 | Quadrato | 300 | 20-10-0 | 200 | 40-0-0 | | | |
| | FT_08S | 1.4 | 250 | Quadrato | 300 | 20-10-0/ES | 200 | 40-0-0 | | | 26.9 |
| A1 | FT_03C | 1.5 | 250 | Quadrato | 300 | 16-20-0 | 250 | 40-0-0 | | | |
| | FT_03S | 1.5 | 250 | Quadrato | 300 | 16-20-0/ES | 250 | 40-0-0/ES | | | 18.5 |
| | FT_12C | 1.35 | 250 | Simeto | 250 | 10-20-0 | 100 | 46-0-0 | 100 | 34.5-0-0 | |
| | FT_12S | 1.35 | 250 | Simeto | 250 | 10-20-0/ES | 100 | 46-0-0 | 100 | 34.5-0-0 | 8.1 |
| | FT_07C | 1.4 | 220 | Normano | 280 | 20-10-0 | 250 | 45-0-0 | | | |
| | FT_07S | 1.4 | 220 | Normano | 280 | 20-10-0/ES | 250 | 45-0-0 | | | −10.9 |
| | FT_02C | 1.0 | 250 | Quadrato | 300 | 16-20-0 | 100 | 40-0-0 | 150 | 34.5-0-0 | |
| | FT_02S | 1.0 | 250 | Quadrato | 300 | 16-20-0/ES | 100 | 40-0-0 | 150 | 34.5-0-0 | 7.8 |
| | FT_10C | 1.1 | 280 | Meridiano | 300 | 16-20-0 | 220 | 40-0-0 | | | |
| | FT_10S | 1.1 | 280 | Meridiano | 300 | 16-20-0/ES | 220 | 40-0-0 | | | 4.2 |
| A2 | FT_11C | 1.5 | 280 | Normano | 300 | 16-20-0 | 300 | 40-0-0 | | | |
| | FT_11S | 1.5 | 280 | Normano | 300 | 16-20-0/ES | 300 | 40-0-0 | | | 2.7 |
| | FT_01C | 0.9 | 250 | Simeto | 275 | 16-20-0 | 180 | 40-0-0 | 180 | 40-0-0 | |
| | FT_01S | 0.9 | 250 | Simeto | 275 | 16-20-0/ES | 180 | 40-0-0/ES | 180 | 40-0-0/ES | −3.4 |
| | FT_09 | 1.5 | 250 | Meridiano | 300 | 16-20-0 | 200 | 40-0-0 | | | |
| | FT_09 | 1.5 | 250 | Meridiano | 300 | 16-20-0/ES | 200 | 40-0-0/ES | | | −26.8 |
| | | | | | | 2016–2017 | | | | | |
| | FT_6C | 1.3 | 280 | Meridiano | 300 | 16-20-0 | 200 | 40-0-0 | | | |
| | FT_6S | 1.3 | 280 | Meridiano | 300 | 16-20-0/ES | 200 | 40-0-0/ES | | | 9.5 |
| | FT_7C | 1.5 | 280 | Meridiano | 300 | 16-20-0 | 200 | 40-0-0 | | | |
| | FT_7S | 1.5 | 280 | Meridiano | 300 | 16-20-0/ES | 200 | 40-0-0/ES | | | 3.2 |
| | FT_3C | 1.5 | 280 | Meridiano | 300 | 16-20-0 | 200 | 40-0-0 | | | |
| A3 | FT_3S | 1.5 | 280 | Meridiano | 300 | 16-20-0/ES | 200 | 40-0-0/ES | | | −13.0 |
| | FT_4C | 0.7 | 250 | Simeto | 300 | 20-10-0 | 230 | 45-0-0 | | | |
| | FT_4S | 0.7 | 250 | Simeto | 300 | 20-10-0/ES | 230 | 45-0-0 | | | −5.9 |
| | FT_5C | 0.9 | 250 | Simeto | 300 | 20-10-0 | 220 | 40-0-0 | | | |
| | FT_5S | 0.9 | 250 | Simeto | 300 | 20-10-0/ES | 220 | 40-0-0 | | | −18.0 |

**Table 2.** Soil analysis data and yields. OM, soil organic matter; HS, soil humic substances; YF, yield of the conventional crop; YFBES, yield of the elemental sulfur treated crop; Δx/x, YFBES percentage change relative to YF.

| Group | Field Trial | Sand (%) | Silt (%) | Clay (%) | Type | pH | ECe μS/cm | $NO_3^-N$ mg/kg | YF Mg ha$^{-1}$ | YFBS | Δx/x (%) |
|---|---|---|---|---|---|---|---|---|---|---|---|
| | | | | | | 2015–2016 | | | | | |
| A1 | FT_13 | 34 | 30 | 36 | CL | 6.20 | 479 | 20.2 | 5.19 | 6.67 | 28.5 |
| | FT_08 | 24 | 30 | 46 | C | 7.53 | 505 | 16.0 | 3.60 | 4.57 | 26.9 |
| | FT_03 | 26 | 24 | 50 | C | 6.95 | 352 | 11.5 | 4.65 | 5.51 | 18.5 |
| | FT_12 | 22 | 44 | 34 | CL | 7.17 | 603 | 25.7 | 4.30 | 4.65 | 8.1 |
| | FT_07 | 22 | 44 | 34 | CL | 6.65 | 386 | 17.1 | 4.60 | 4.10 | −10.9 |
| A2 | FT_02 | 28 | 40 | 32 | CL | 8.02 | 572 | 15.6 | 6.16 | 6.64 | 7.8 |
| | FT_10 | 28 | 34 | 38 | CL | 8.10 | 453 | 9.1 | 4.30 | 4.48 | 4.2 |
| | FT_11 | 34 | 38 | 28 | CL | 7.96 | 675 | 16.6 | 4.40 | 4.52 | 2.7 |
| | FT_01 | 28 | 30 | 42 | C | 8.01 | 613 | 24.6 | 5.06 | 4.89 | −3.4 |
| | FT_09 | 10 | 42 | 48 | SiC | 8.00 | 592 | 23.2 | 3.32 | 2.43 | −26.8 |
| | | | | | | 2016–2017 | | | | | |
| A3 | FT_6 | 30 | 34 | 36 | CL | 7.81 | 667 | 23.1 | 1.68 | 1.84 | 9.5 |
| | FT_7 | 14 | 40 | 46 | SiC/C | 7.88 | 610 | 25.5 | 4.73 | 4.88 | 3.2 |
| | FT_3 | 16 | 38 | 46 | C | 7.74 | 491 | 10.5 | 2.62 | 2.28 | −13.0 |
| | FT_4 | 26 | 28 | 46 | C | 7.80 | 684 | 19.6 | 1.53 | 1.44 | −5.9 |
| | FT_5 | 30 | 28 | 42 | C | 7.74 | 571 | 19.9 | 1.78 | 1.46 | −18.0 |
| B1 | | 23 | 25 | 52 | C | 8.57 | 543 | 19.2 | | | |
| B2 | | 23 | 24 | 53 | C | 8.13 | 527 | 17.1 | | | |
| | **Field Trial** | **$CaCO_3$ (%)** | **P-Olsen mg/kg** | **Kexch mg/kg** | **Fe-DTPA mg/kg** | **Mn-DTPA mg/kg** | **Cu-DTPA mg/kg** | **Zn-DTPA mg/kg** | **SOM (%)** | **HS (%)** | **HS/SOM (%)** |
| | | | | | | 2015–2016 | | | | | |
| A1 | FT_13 | 0 | 19.4 | 230 | 22.7 | 13.4 | 1.5 | 0.8 | 1.5 | 0.2 | 11.7 |
| | FT_08 | 0 | 16.8 | 300 | 17.7 | 12.1 | 1.5 | 0.7 | 2.3 | 0.5 | 21.1 |
| | FT_03 | 0 | 26.0 | 230 | 26.3 | 13.4 | 4.7 | 1.8 | 1.8 | 0.2 | 12.7 |
| | FT_12 | 0 | 17.4 | 270 | 25.8 | 12.2 | 2.2 | 1.8 | 1.9 | 0.2 | 8.1 |
| | FT_07 | 0 | 6.4 | 180 | 27.3 | 13.4 | 2.2 | 0.7 | 2.3 | 0.3 | 14.3 |
| A2 | FT_02 | 5.2 | 65.4 | 240 | 13.0 | 8.4 | 1.5 | 1.1 | 2.1 | 0.2 | 9.9 |
| | FT_10 | 49.1 | 12.2 | 390 | 15.9 | 9.3 | 1.8 | 0.7 | 1.8 | 0.1 | 2.8 |
| | FT_11 | 6.6 | 12.4 | 270 | 11.2 | 7.3 | 1.1 | 0.7 | 1.8 | 0.1 | 5.0 |
| | FT_01 | 24.5 | 7.8 | 270 | 5.8 | 6.3 | 1.4 | 0.8 | 1.8 | 0.1 | 6.7 |
| | FT_09 | 26.6 | 3.2 | 290 | 8.0 | 6.2 | 1.3 | 0.5 | 2.9 | 0.1 | 2.4 |
| | | | | | | 2016–2017 | | | | | |
| A3 | FT_6 | 41.0 | 15.6 | 450 | 5.8 | 9.6 | 1.7 | 0.8 | 2.4 | 0.1 | 4.2 |
| | FT_7 | 1.4 | 9.8 | 350 | 14.8 | 10.1 | 1.5 | 0.7 | 3.7 | 0.4 | 10.8 |
| | FT_3 | 13.9 | 4.1 | 420 | 12.6 | 7.5 | 1.2 | 0.5 | 3.0 | 0.2 | 7.8 |
| | FT_4 | 11.1 | 2.8 | 240 | 10.1 | 7.3 | 0.9 | 0.9 | 1.9 | 0.1 | 7.2 |
| | FT_5 | 8.9 | 1.6 | 240 | 8.0 | 6.4 | 0.7 | 0.9 | 2.1 | 0.2 | 9.4 |
| B1 | | 18.1 | 20.2 | 230 | 24.3 | 12.1 | 1.8 | 0.9 | 2.5 | 0.2 | 9.6 |
| B2 | | 21.7 | 13.1 | 224 | 21.3 | 11.7 | 2.0 | 0.9 | 2.3 | 0.2 | 9.9 |

### *2.2. Experimental Field Trials*

Based on the experience gained during the seasons 2014–2015 and 2015–2016, two experiments with durum wheat crops (cv. Simeto) were established in the experimental fields of the Agricultural University of Athens in Aliartos at Viotia county during the season 2016–2017, based on a randomized block design. Each plot was 18 square meters and 1 m apart from the next ones. In each plot of the provided area for experimentation, the P-Olsen content was determined and two areas of almost 0.2 ha each were located, one with P-Olsen values ranging between 18–22 ppm (Group B1) and the other between 12–15 ppm (Group B2). Soil samples were collected at depth of 0–20 cm prior to the crop establishment and fertilizer application. The plot that received ammonium phosphate (16-20-0-13S) as initial fertilizer (starter) and urea (U; 46-0-0) as additional fertilizer (topdressing) served as the reference crop. The compatibility of urea-FBES (U + 2% ES; U/ES) or urea ammonium sulfate (UAS + 2% ES; UAS/ES) along with the corresponding ones containing the urease inhibitor NBPT (U/NBPT) or urea ammonium sulfate (UAS/NBPT) was also studied in this experiment. Each combination was repeated three times.

### *2.3. The Nature of the FBES Granules*

The FBES granules tested in this research (16-20-0/ES, 20-10-0/ES, 10-20-0/ES, U/ES, UAS/ES) have been prepared by Sulphur Hellas S.A. (Sulfogrow®) as it follows: The fertilizer granules were mixed initially with ES in the form of dust at a percentage of 2% to 4% (*w*/*w*, ES to fertilizer granules). The mixture passed through a shower of fine droplets consisting of a 1:1 mixture of molasses and glycerol, which acted as the binding system, in such a way that the whole surface was exposed. The binder was added at a percentage of 0.4% to 1.2% (*w*/*w*, binder to fertilizer granules). Then, the mixture was led to a mixer, where the sticky ES dust was evenly attached onto the sticky fertilizer granules, thus forming the final FBES product [9].

### 2.4. Soil Analysis

After the soil samples were collected, were air-dried and ground to 2 mm prior to analysis. The particle size analyses were conducted using the hydrometer method, with a 2-h reading for the clay content [15]. The $CaCO_3$ equivalent percentage was estimated using a digital calcimeter [16]. Soil pH was measured in a 1:1 soil: Distilled water (w-v) suspension [17] and electrical conductivity (ECe) in saturation extract. Soil organic matter (SOM) was determined using the Walkley-Black wet digestion method [18] and the available P (P-Olsen) according to the Olsen method [19]. Exchangeable K ($K_{exch}$) was determined using the ammonium acetate extraction method [20]. Extractable Fe, Mn, Cu and Zn were determined with diethylene-triamine-pentaaceticacid (DTPA) [21]. Many studies have shown that DTPA is the most widely used extractant for the determination of metal availability in soils [22,23]. The soil humic substances content (HS) was determined according to Jones (1999) [24]. The main soil properties of each field trial are presented in Table 2.

### 2.5. Head Measurements

In the experimental field trials harvesting took place on 1–3 July 2017. Per plot, all stems were collected and were placed into bags. In the laboratory, each stem was separated from its head, and the heads were weighed. Three sets of a hundred heads each were weighed, and the seeds per set were separated, collected and weighed.

### 2.6. Scanning Electron Microscopy

Cross sections of granules were made using razor blades and mounted on SEM pedestal using conductive silver epoxy paint (EMS, Hatfield, USA). Samples were then sputter coated with gold (Denton Vacuum Desk II Sputter coater, Denton Vacuum, Moorestown, USA) and observed with a JSM-5410 scanning electron microscope (JEOL, Tokyo, Japan) at a working distance of 40 mm at 10 kV.

### 2.7. Statistical Analysis

Yield levels were handled as continuous variables. A linear mixed effects (LME) model [25] was used to assess the statistical significance of the covariates on the response variable, taking into account the within-group variability that exists for the repeated measurements setting. Analysis of variance (ANOVA) was performed on the LME model to determine which of the model covariates (i.e., Phosphorus level, Initial Fertilization and Additional Fertilization) had a significant effect on the average value of yield. All higher-order interactions were tried. Tests were declared significant when $p < 0.05$. All analyses were performed using the R statistical software (R Development Core Team, 2018) [26].

## 3. Results

### 3.1. Characteristics of ES Enriched Fertilizer Granules

The first product of this family of fertilizers was of the 20-10-10 type, where nitrogen was provided as ammonium sulfate, phosphorus as triple superphosphate, while potassium was provided as potassium sulfate. The diameter of ES-containing fertilizer granules (20-10-10+2% ES) was 2.463 $\pm$ 0.579 mm (mean value $\pm$ standard deviation; n = 100), while the corresponding one of the conventional fertilizer granules was 2.447 $\pm$ 0.617 mm (n = 100). The two-tailed *P*-value was 0.8502 thus by conventional criteria (t = 0.1891, df = 198, standard error of difference = 0.085) there was no statistically significant difference between the means. The granules were then categorized according to their size. The ES-fertilizer granules presented the following size distribution: <2 mm = 20%, 2–2.49 mm = 33%, 2.5–2.99 mm = 19%, 3–3.49 mm = 25%, >3 mm = 3%. The corresponding distribution of the conventional fertilizer was: <2 mm = 18%, 2–2.49 mm = 42%, 2.5–2.99 mm = 11%, 3–3.49 mm = 24%, >3 mm = 5%.

Under the SEM, granules of the ES–containing fertilizer were found to be covered with a layer of loose material containing fine crystal-like particles, obviously representing the ES powder layer attached via the binder (Figure 1A,B). The width of the layer was found to be up to 60 μm. Such a layer could not be found on granules of the conventional fertilizer (Figure 1C).

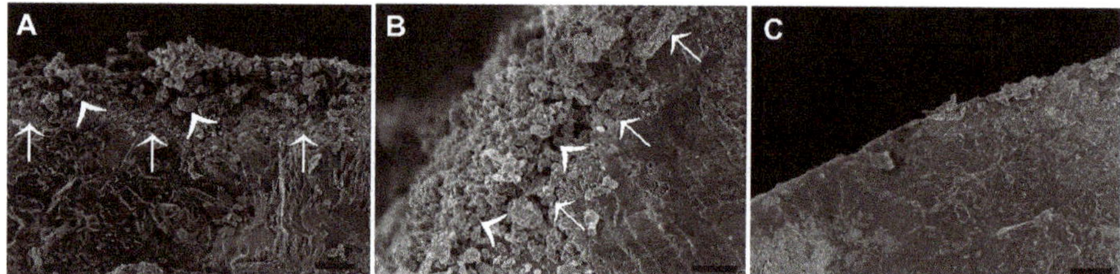


**Figure 1.** Scanning electron micrographs showing cross sections of fertilizer granules with (**A**,**B**) and without (**C**) an outer layer of sulphur crystals. (**A**) Cross-section of ES coated fertilizer granules containing sulphur crystals of different size (arrowheads) in the outer layer, attached to granule with a binder layer (arrows). (**B**) A bending view of the ES coated fertilizer bead, providing in addition an extended view of its surface covered by ES powder. (**C**) Control granules are characterized by a sharply defined edge without an outer layer. Bars = 20 μm.

### 3.2. The Effect of the Fertilization Schemes on Yield

Five fields of Group A1 contained no calcium carbonate or traces of it, with pH ranging from 6.20 to 7.74. Four of them characterized by adequate fertility levels, and the FBES-trials presented a relative yield increase of 8.1%, 18.5%, 26.9%, 28.5%, compared to F-trials. In the field FT_07 of Group A1 that was characterized by very low soil initial phosphorus content coupled with low potassium content, the yield of the FBES-trial presented a relative yield decrease (−10.9%) (Tables 1 and 2).

Five fields of Group A2 contained moderate or high calcium carbonate, with pH ranging from 7.96 to 8.20. In three of them, those with adequate fertility levels, the FBES-trials relative yield increased by 2.7%, 4.2%, 7.8%, respectively. The other two cases of this category presented relative decreases in the yield (−3.4%, −26.8%). The rhizosoil of these cases was characterized by very low initial phosphorus content coupled with marginal iron content, while the rhizosoil of the latter one (−26.8%) additionally presented low concentration of humic substances along with very low sand content (Tables 1 and 2).

Most of the five fields of Group A3 contained significant amounts of calcium carbonate, with pH ranging from 7.74 to 7.88. In two of them, characterized by adequate fertility levels, the yields of the FBES-trials presented relative yield increases (3.2%, 9.2%). In the fields that characterized by very low soil initial phosphorus content (FT_4, FT_3, FT_5), the yield of the FBES-trial presented a relative yield decrease (−5.9%, −13.0%, −18.0%) (Table 2).

### 3.3. P-Olsen as a Yield-Limiting Factor

The P-Olsen content of each field was correlated with the corresponding relative change in the yields ($Y_F/Y_{FBES}$) and a strong relationship was revealed. This held true for both wheat growing seasons (Figure 2A–C). The season 2015–2016 provided good yields, while the season 2016–2017 provided poor ones. Despite this fact, the relative change due to the application of the elemental sulfur seemed not to be influenced by the $Y_F$ level. As a next step the data of both years were combined and fitted by a power function ($R_1^2$ = 0.76, Figure 2D), suggesting that the relationship might not be influenced by the wheat growing season. The level of 8 ppm of available phosphorus was a turning point. At higher values the relative change was higher, i.e., the incorporation of elemental sulfur in the fertilization scheme of the FBES-trial resulted in higher yield compared with the corresponding conventional F-trial, while at lower values the relative change was lower. These results suggested an important role for the soil available phosphorous as a yield-limiting factor.

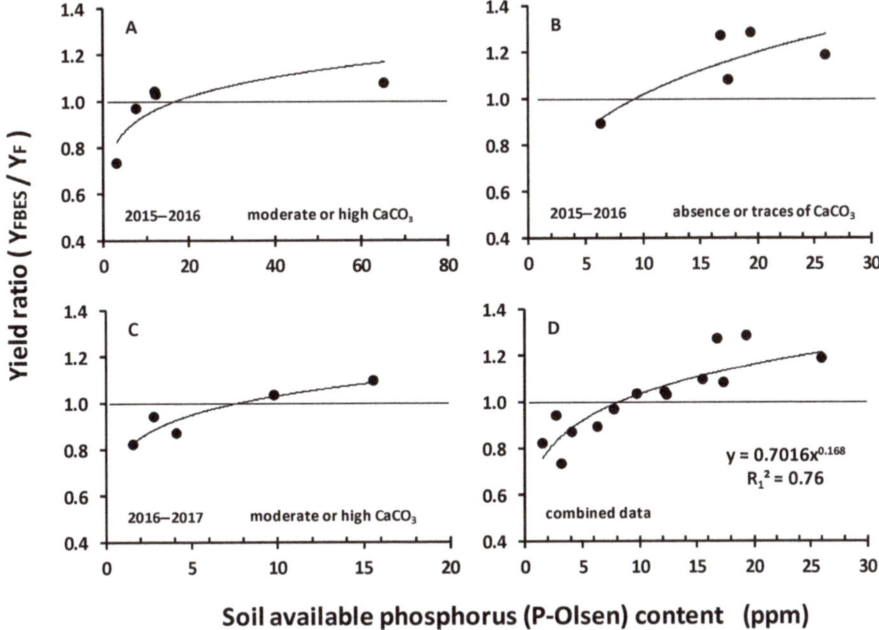


**Figure 2.** The relationship between the yield ratio (YFBES/YF) and available phosphorus content in the soil. YF, yield of the conventional crop; YFBES, yield of the elemental sulfur treated crop. (**A**) Wheat growing season in 2015–2016: Field trials with moderate of high calcium carbonate content (Group A2). (**B**) Wheat growing season in 2015–2016: Field trials lacking calcium carbonate content (Group A1). (**C**) Wheat growing season in 2016–2017: Field trials with moderate of high calcium carbonate content (Group A3). (**D**) The combination of data.

### *3.4. The effect of ES and NBPT Containing Fertilizers on Yield*

The results of experimental field trials supported the aforementioned finding. The corresponding treatments within Group B1 with P-Olsen values ranging between 18–22 ppm (Table 3 and Figure 3) provided higher yields compared with those of Group B2 with P-Olsen values ranging between 12–15 ppm, i.e., less phosphorus by −47% to −50%; (Table 3 and Figure 3). Within each group, the same rank was obtained, i.e., U < UAS < U/ES < UAS/ES. The use of incorporated ES along with the initial fertilization provided statistically higher yields in most cases. This held true within both groups B1 and B2.

**Table 3.** Results of the randomized plot experiments of 2016–2017 growing season (Group B). F (16-20-0) or FBES (16+20-0+2%ES): Initial fertilization (F1) along with sowing, U, urea 46-0-0; UAS, urea plus ammonium sulfate 40-0-0. "/NBPT" denotes fertilizer with NBPT (N-(n-butyl) thiophosphoric triamide; urease inhibitor). "/ES" denotes fertilizer with 2% elemental sulfur. "/NBPT/ES" denotes application of a 50:50 mixture of fertilizer with NBPT and the same fertilizer with 2% elemental sulfur. The F; U trial served as a reference, with rates of F1 and F2 (additional fertilizer application) 250 kg ha$^{-1}$ respectively. Simeto cultivar was used.

| Treatment | Group B1 | | | Group B2 | | | Group B1 | | Group B2 | |
|---|---|---|---|---|---|---|---|---|---|---|
| | P-Olsen (ppm) 18–22 | | | 12–15 | | | P-Olsen (ppm) 18–22 | | 12–15 | |
| | Grain yield (Mg ha$^{-1}$) | | | | | | Yield relative to the combination F; U (%) | | | |
| no fertilization | 1.75 | | | 1.30 | | | −25 | | −10 | |
| | F | FBES | RC (%) | F | FBES | RC (%) | F | FBES | F | FBS$^{0}$ |
| U | 2.32 | 2.51 | 8.2 | 1.44 | 1.52 | 5.6 | 0 | 8 | 0 | 6 |
| U/ES | 2.90 | 3.19 | 10.0 | 1.53 | 1.71 | 11.8 | 25 | 38 | 6 | 19 |
| U/NBPT | 2.78 | 2.98 | 7.2 | 1.64 | 1.76 | 7.3 | 20 | 28 | 14 | 22 |
| U/NBPT/ES | 3.04 | 3.54 | 16.4 | 1.74 | 1.93 | 10.9 | 31 | 53 | 21 | 34 |
| UAS | 2.68 | 3.18 | 18.7 | 1.50 | 1.58 | 5.3 | 16 | 37 | 4 | 10 |
| UAS/ES | 3.00 | 3.90 | 30.0 | 1.64 | 1.78 | 8.5 | 29 | 68 | 14 | 24 |
| UAS/NBPT | 2.95 | 3.45 | 16.9 | 2.09 | 2.25 | 7.7 | 27 | 49 | 45 | 56 |
| UAS/NBPT/ES | 3.52 | 4.22 | 19.9 | 2.27 | 2.49 | 9.7 | 52 | 82 | 58 | 73 |

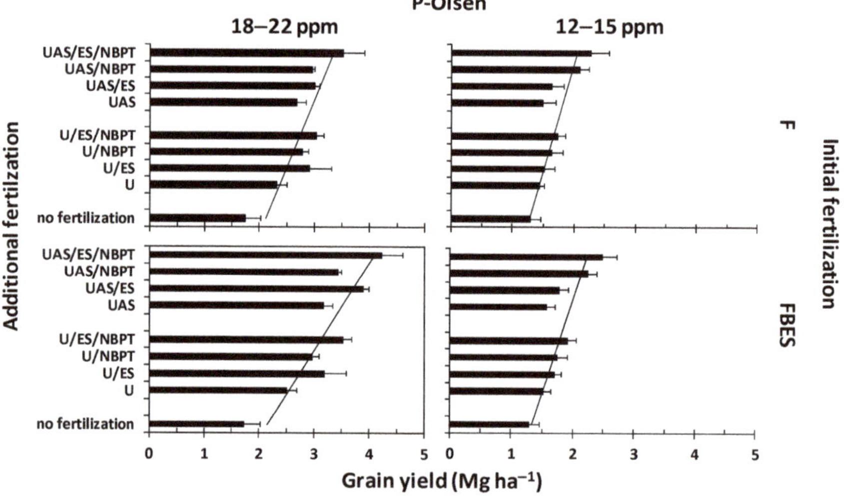


**Figure 3.** Visualization of the influence of the experiments' covariates (Phosphorus level, Initial Fertilization and Additional Fertilization) on yield.

The applied fertilizer mixtures that contained both ES and NBPT were more effective in both Groups B1 and B2 compared with their components applied alone (Table 3 and Figure 3). Two types of mixtures were studied: U/ES/NBPT and UAS/ES/NBPT. In Group B1, i.e., when P-Olsen was adequate (18–22 ppm), the rank was U < UAS < U/NBPT < U/ES < UAS/NBPT < UAS/ES < U/ES/NBPT < UAS/ES/NBPT. In any combination of this group, UAS provided higher yields compared to U alone. Furthermore, the incorporation of ES provided higher yields compared to NBPT. The incorporation of elemental sulfur in the initial fertilization provided even higher yields.

In Group B2, i.e., when P-Olsen was marginal (12–15 ppm), the situation altered. In this case the rank was U < UAS < U/ES < UAS/ES < U/NBPT < U/ES/NBPT < UAS/NBPT < UAS/ES/NBPT. Again, UAS provided higher yields compared to U alone. Incorporation of NBPT provided higher yields compared with those of incorporated ES, while UAS/ES/NBPT provided the highest possible yields under the circumstances. Again, the incorporation of elemental sulfur in the initial fertilization provided even higher yields.

In all cases the incorporation of ES in the fertilizer granule provided higher yields than the fertilizer granule alone. Moreover, the 50:50 combination of UAS/ES and UAS/NBPT (UAS/ES/NBPT) provided the highest yields. It is noteworthy that in such a combination actually 1% ES and half the rate of the urease inhibitor were applied to the plot with the additional fertilization.

ANOVA resulted in a model which shows that the linear effects of Phosphorus level (P-Olsen), Initial fertilization (IF) and Additional Fertilization (AF), as well as the pairwise interaction of Phosphorus level with Initial fertilization have a statistically significant effect on the yields' levels (Tables 4 and 5).

**Table 4.** Analysis of Variance with Satterthwaite's method. P-Olsen: IF denotes the interaction between P-Olsen and IF.

| Variable | Sum Sq | Mean Sq | NumDF | DenDF | *P*-Value |
|---|---|---|---|---|---|
| P-Olsen | 11.6159 | 11.6159 | 1 | 87 | <0.001 |
| IF | 3.0111 | 1.5056 | 2 | 87 | <0.001 |
| AF | 10.5053 | 1.5008 | 7 | 87 | <0.001 |
| P-Olsen:IF | 1.7335 | 0.8667 | 2 | 87 | 0.0013 |

**Table 5.** Regression coefficients (95% confidence intervals) and *p*-values for the selected model. (IF, initial fertilization; AF, additional fertilization; NF, no fertilization)

| Variable | Beta | 95% CI | | *P*-Value |
|---|---|---|---|---|
| Intercept | 2.38 | 2.14 | 2.61 | <0.001 |
| P-Olsen; Low | −1.17 | −1.35 | −0.98 | <0.001 |
| AF; UAS/ES/NBPT | 1.18 | 0.91 | 1.44 | <0.001 |
| AF; UAS/NBPT | 0.74 | 0.47 | 1.00 | <0.001 |
| P-Olsen; Low: IF; NF | 0.72 | 0.16 | 1.28 | 0.019 |
| AF; UAS/ES | 0.63 | 0.37 | 0.90 | <0.001 |
| IF; FBES | 0.47 | 0.29 | 0.66 | <0.001 |
| IF; NF | −0.63 | −1.06 | −0.19 | 0.009 |
| AF; U/ES/NBPT | 0.61 | 0.35 | 0.88 | <0.001 |
| AF; U/ES | 0.38 | 0.12 | 0.65 | 0.008 |
| AF; U/NBPT | 0.34 | 0.08 | 0.61 | 0.018 |
| AF; UAS | 0.29 | 0.02 | 0.55 | 0.046 |
| P-Olsen; Low: IF; FBES | −0.33 | −0.59 | −0.06 | 0.024 |

According to Model results, the overall model predicting value has a total explanatory power (conditional ${R_2}^2$) of 83.55%, in which the fixed effects explain 82.41% of the variance (marginal ${R_2}^2$). The model's intercept was at 2.38 (SE = 0.13, 95% CI [2.14, 2.61]). The effects of low P-Olsen values and especially with no fertilization, UAS/NBPT, and UAS/ES/NBPT were significant and can be considered as large. The effects of UAS/ES, U/ES/NBPT, along with FBES as initial fertilizer were significant and can be considered as medium, while the effects of UAS, U/ES, U/NBPT, as well as the combination low *P*-values along with FBES as initial fertilizer were significant and can be considered as small. The F; U trial served as the reference.

### *3.5. Yield and Head Characteristics*

Yield presented a positive linear relationship with head mass per square meter (Figure 4A), with the same slope for both Groups B1 and B2. This held true for the correlation of yield with the number of heads per square meter (Figure 4B), however in this correlation Group B1 was separated from Group B2. The mass per head presented a negative linear relationship with the number of heads per square meter (Figure 4C), and again Group B1 was separated from Group B2. Such a negative linear relationship was also revealed when mass per head was correlated with the ratio of seed mass per head mass (Figure 4D). Again, the two groups were separated, however the slope was not steep.

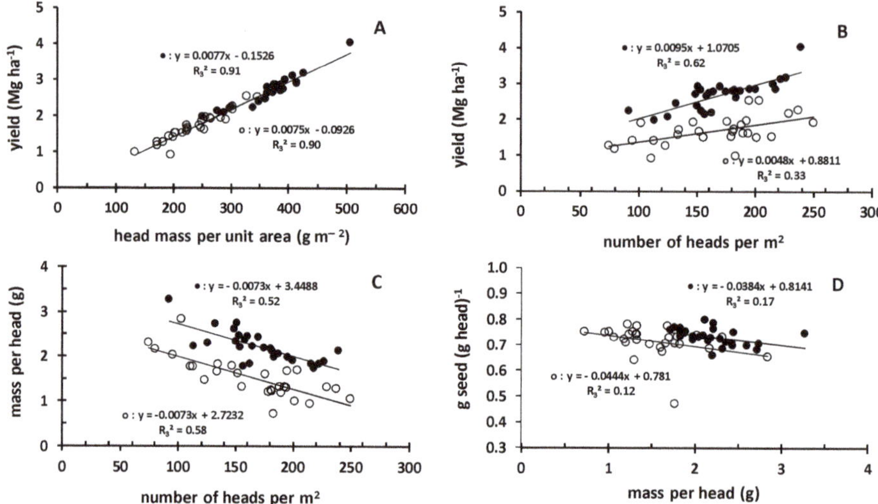


**Figure 4.** The relationships of the data provided by the treatments of the randomized plot experiments of the 2016–2017 growing season (Group B). Full circles: Data of Group B1 (P-Olsen: 18–22 ppm). Empty circles: Data of Group B2 (P-Olsen: 12–15 ppm). In each set of data, the tendency is described by including a trend line of linear type.

## 4. Discussion

### *4.1. The Role of Soil Available P on the Relative Yield*

The results of the relative yields were contradictory, as some FBES-trials provided lower yields compared to the corresponding conventional ones. It was further shown that soil available phosphorus was a strong limiting factor and the P-Olsen value of 8 ppm was established as the turning point. Above this threshold the relative yield was increased.

Sulfate-S is the available S source to the plants, while ES as S source is not readily available to plants. The conversion of ES to sulfate requires the action of microorganisms [27]. Recent studies showed that S-oxidizing bacteria activity depends on the variations in soil organic C and nutrient availability [28]. In a recent study, it was concluded that in soils with pH > 6.65, along with high S and organic matter content, the ES oxidation rate is higher compared to the other soils [29]. According to our results, it is speculated that the low organic matter content and the low available phosphorus are limiting factors for the activity of S-oxidizing rhizosoil microbes, the growth of which is boosted by the added ES with the fertilizer granule. We have shown that the application of FBES fertilizers increases the total numbers of microbial populations in the rhizosphere [30]. In the initial experiment [13], the 20-10-10/ES that was applied at sowing contained both ES and sulfate, i.e., both readily available and non-available S to plants, thus supporting crop growth with sulfate for an extended period. The role of soil microbes in the biogeochemical S cycle and S supply of plants has been reviewed; fungi and bacteria release S from sulfate-esters using sulfatases [31,32], while the interactions of roots with soil microorganisms, especially non-symbiotic plant growth promoting rhizobacteria, in relation to nutrient availability, as well as the mechanisms that are associated with plant growth promotion have been discussed [33]. We have found that a portion of the existing microorganisms in the rhizosphere presented arylsulfatase activity [30], thus releasing sulfate bound in the soil organic matter, the percentage of which increased with the addition of ES-enriched fertilizer. Moreover, the fertilization with FBES at sowing affected the iron fractions of the rhizosoil towards iron mobilization, thus providing more iron to the crop, which apart from the iron nutrition fortified the crop's sulfur nutrition, too [14].

On the other hand, phosphorus (P) is one of the most limiting essential nutrients for crop production. The dynamic processes determining P availability in the soil and in the rhizosphere, P mobilization, uptake, and utilization by plants have been reviewed [34,35]. Especially, the response of wheat grain yield to phosphorus has been reported by a number of researchers, as wheat crops have high phosphorus requirement, mostly during the early growth stages [36–39]. Although yield limiting factors are complex, the effect of soil available P was studied further in experimental plots under a randomized block design. Already existing spatial information of the experimental site [40] was used to guide sampling to representative plots, referred to in the literature as directed, smart or targeted sampling [41]. In accordance with Friesen (1996) [42], our results suggested that the presence of adequate P for the needs of the S-oxidizing bacteria may be a critical factor for the ES use efficiency. It is speculated that under low soil available P, the increased numbers of microbial populations, due to the addition of ES in the fertilization scheme were stronger P consumers compared to the wheat crop during the growing season, leaving less than adequate P for the crop's needs, thus possibly explaining the depressed yields. More research is needed to fully explain the effect of soil available P on yield under the circumstances.

### 4.2. The Efficacy of ES Fertilization Schemes

ES is used either as granules, or in the form of fine powder, the micronized ES. Micronized ES is preferable compared with ES granules, because the transformation to sulfate-S by the microorganisms is quick. On the other hand, the application of fine powder in large scale agriculture is a serious problem, because the handling of dust is involved. To overcome the problem, several technologies towards successful co-formulation of micronized ES with different materials have been arisen through the years, in order to produce granules applicable to large scale agriculture, including monoammonium phosphate, diammonium phosphate, and triple superphosphate, containing micronized ES without or with AS [43,44]. Scanning electron migrographs of 20-10-10/ES (2%) granules clearly show that elemental sulfur forms a dense coating layer on the surface, which is achieved via the sticky mixture of molasses plus glycerol. The product 20-10-10/ES produced dust. This held true for 16-20-0/ES(2%), too. The dust was estimated to be in the order of 4%, a minor fraction of which was ES (data not shown). On the other hand, U/ES(2%) and UAS/ES(2%) were indeed dustless.

The factors that influence the oxidation of elemental sulfur in soils have been discussed [27], while a "concept of the negative locality effect on ES oxidation" has been developed [8,44]. Briefly, when the fertilizer granule disintegrates it releases micronized ES particles to the soil. Due to the hydrophobic nature of the ES, the very fine ES particles create clusters and localize around the applied granule site. Clustering decreases contact between localized ES particles and soil, which in turn decreases the colonization of soil S-oxidizing bacteria on the surface of the ES particles. Moreover, the released micronized ES particles can coalesce to form larger aggregates, which in turn can further decrease ES oxidation [42,45]. It seems that the FBES technology diminishes the negative locality effect on ES oxidation. The binder utilized by the FBES technology is also a positive factor, because the binder around the ES and the granule particles can be consumed by the microorganisms as well.

The fertilizer type 16-20-0 was used and tested against 16-20-0/ES as NP starter fertilizer, which was combined with U or UAS, applied in spring as topdressing added N (and S). U and UAS are of common use and both water soluble. N from urea is available to the plant only after hydrolysis by the urease enzyme to ammonium carbonate, which can significantly increase soil pH around the applied urea granule sites (can be as high as 8–9), due to hydroxyl anion production. AS provides N and S, both critical plant nutrients. Compared with urea, AS may have some potential agronomic and environmental benefits [44]. One approach to enhance the N efficiency of urea is to partially substitute AS for urea in the mixture. The dilution effect of AS-N should be considered because ammonia volatilization is greater with increased urea concentration and/or rate of application [46]. An increase in crop yield should be expected by mixing AS and urea compared with urea alone. Another reason to mix AS and urea is to supply S along with N. Moreover, AS reacts with calcium carbonate to precipitate calcium sulfate. Sulfur-coated urea (SCU) as a source of S was used in wheat crop and it was found that 5% of SCU supplied 50% of the S wheat requirements and at the same time increased N recovery efficiency by 60.3% over prilled urea [47]. Our results showed that the incorporation of ES improved the performance of U or UAS, suggesting that the granules enriched with ES were fortified, while their performance was influenced by the soil available phosphorus.

### *4.3. The Efficacy of ES and NBPT Fertilization Schemes*

The stability of urea in the soil is affected by the enzyme urease, which is released by the soil microbial population or is derived from the decomposition of organic matter [11]. The incorporation of urease inhibitors into urea granules delays urea hydrolysis and urea is available for plant acquisition. N-(n-butyl) thiophosphoric triamide (NBPT) is the urease inhibitor that is often formulated into granular urea fertilizers. This compound is a structural analog of urea and inhibits urease activity by forming stable complexes with urease [11,48,49]. The use of granules co-enriched with ES and NBPT onto the same granule in a field experiment has been reported [50]. According to the authors, the inhibitor and ES were coated onto the fertilizers by Summit-Quinphos Ltd. (New Zealand) and applied directly onto the pasture of the respective plots by hand. No information was provided on the nature of the coating. In the present study, we tested the efficacy of a 50:50 mixture of NBPT containing granules and ES containing granules, using commercially available fertilizers produced by the same company. The 50:50 ratio was arbitrarily chosen. Such combination provided even higher yields compared to the yields provided by the application of the corresponding fertilizers alone and the soil available phosphorus influenced the performance of the mixture. According to the FBES technology, the core fertilizer can be enriched up to 4% with ES. To our knowledge, the least percentage of ES in NP fertilizer granules reported in the literature is 5% [10]. In this research, granules enriched with 2% ES were used, i.e., the minimum ES enrichment. Therefore, in the 50:50 mixture, more or less 1% of ES was applied, i.e. half of the amount, and the same holds true for NBPT.

In summary, UAS provided sulfate-S for immediate use, ES as a coating in its majority provided sulfate-S after oxidation, NBPT positively contributed to ammonium use efficiency provided by both U and AS constituents of the fertilizer granule, while the minor amount of the binding materials probably fed the microorganisms around the granule. A portion of these microorganisms were arylsulfatase producing ones, which implies that organic S was released as sulfate-S, thus comprising a third pool of sulfate-S. The more the available P the more effective the UAS/ES/NBPT mixture was. The efficacy of this activity was enhanced when FBES (16-20-0/ES) was used as starter fertilizer at sowing, instead of the corresponding 16-20-0 conventional fertilizer.

## 5. Conclusions

The use of UAS in spring applied fertilization scheme enriched with 2% ES and the urease inhibitor NBPT in the granules, significantly increased the yield of durum wheat crops, and in combination with the use of 2% ES as an ingredient of the starter fertilizer, it boosted the yield to even higher scores. The soil available phosphorus at P-Olsen value of 8 ppm seemed to be the threshold, below which the performance of the aforementioned combination was strongly depressed.

**Author Contributions:** D.L.B. and S.N.C. conceived and designed the experiment, elaborated the research questions, analyzed the data, and wrote the article; D.G carried out the soil chemical analyses, analyzed the soil data and contributed to the presentation and interpretation of soil fertility results; B.Z. carried out the SEM images and the FBES granule characteristics, and wrote the corresponding part of the paper; L.D.B. performed the statically analysis and wrote the corresponding part of the paper; D.L.B. supervised the field work, while S.N.C. supervised the analyses for head characteristics of the experiments of groups B1 and B2. All authors have read and approved the manuscript.

**Funding:** This research received no external funding.

**Acknowledgments:** The authors are grateful to the Greek fertilizer company Sulphur Hellas S.A. (Christos Papachatzopoulos, Dimitris Benardos, Dafni Papachatzopoulos, Dimitris Konstantelos) for providing the fertilizers and the consumables for chemical analyses; the farmers Pavlos Issaris, Christos Petropoulos, Nikos Theodosiou, Yiorgos Mitas, Elias Dasteridis, Vaios Tsiopoulos, Theodoros Karanikas, Anestis Ioannidis, Konstantinos Kontos, Apostolos Argyros, Stergios Girgiris, and Yiorgos Sirkilidis for applying the FBES products in their commercial crops; the agronomist Haris Mavrogiannis for his contribution in the coordination of the field trials, the agronomists Dimitris Argiros, Giorgos Georgoulas, Dimitris Petrakos, and Tassos Varnas, for their help during soil sampling and supervising various field trials; the agronomists of AUA Sratos Kanaros and Roikos Thanopoulos, the staff of AUA Alekos Dimitriou and Dimitris Athanasakis, the students Miltos Margetis, Alexandros Tataridas, Theodoros Andreou, Konstantinos Kaninis, Tassos Tsoungos and the two Pakistani workers for their help on the experiments of groups B1 and B2, and harvesting; as well as the anonymous reviewers for their suggestions. The soil samples of this study were collected by consent of the land owners.

**Conflicts of Interest:** The authors declare no conflict of interest. Sulphur Hellas S.A. had no role in the design of the study; in the collection, analyses, or interpretation of data; in the writing of the manuscript, nor in the decision to publish the results.

## Abbreviations

| | |
|---|---|
| RS | rhizosoil |
| OM | organic matter |
| HS | humic substances |
| F | granular conventional fertilizer |
| ES | elemental sulfur |
| B | binder |
| U | urea |
| UAS | urea ammonium sulfate |
| NBPT | N-(n-butyl) thiophosphoric triamide |

## References

1. Zhao, F.J.; Hawkesford, M.J.; McGrath, S.P. Sulphur assimilation and effects on yield and quality of wheat. *J. Cer. Sci.* **1999**, *30*, 1–17. [CrossRef]
2. Eriksen, J. Soil sulfur cycling in temperate agricultural systems. *Adv. Agron.* **2009**, *102*, 55–89.
3. Zhao, F.J.; McGrath, S.P.; Blake-Kalff, M.M.A.; Link, A.; Tucker, M. Crop responses to sulphur fertilisation in Europe. In *Proceedings No.504, International Fertiliser Society*; International Fertiliser Society: York, UK, 2002.
4. Grant, C.; Hawkesford, M.J. Sulfur. In *Handbook of Plant Nutrition*, 2nd ed.; Barker, A.V., Pilbeam, D.J., Eds.; CRC Press: Boca Raton, FL, USA, 2015; pp. 261–302.
5. Matamwa, W.; Blair, G.; Guppy, C.; Yunusa, I. Plant Availability of Sulfur Added to Finished Fertilizers. *Commun. Soil Sci. Plant Anal.* **2018**, *49*, 433–443. [CrossRef]
6. Somani, L.L.; Totawat, K.L. Mined and industrial waste products capable of generating gypsum in soil. In *Handbook of Soil Conditioners*; Wallace, A., Terry, R.E., Eds.; Marcel Dekker, Inc.: New York, NY, USA; Basel, Switzerland; Hong Kong, China, 1998; pp. 257–291.
7. Boswell, C.C.; Friesen, D.K. Elemental sulfur fertilizers and their use on crops and pastures. *Fert. Res.* **1993**, *35*, 127–149. [CrossRef]
8. Chien, S.H.; Prochnow, L.I.; Cantarella, H. Recent developments of fertilizer production and use to increase nutrient efficiency and minimize environmental impacts. *Adv. Agron.* **2009**, *102*, 261–316.
9. Benardos, D. Method for Coating Fertilizer Beads with Elemental Sulfur. U.S. Patent WO2017077350A, 5 November 2017.
10. Chien, S.H.; Teixeira, L.A.; Cantarella, H.; Rehm, G.W.; Grant, C.A.; Gearhart, M.M. Agronomic Effectiveness of Granular Nitrogen/Phosphorus Fertilizers Containing Elemental Sulfur with and without Ammonium Sulfate: A Review. *Agron. J.* **2016**, *108*, 1203–1213. [CrossRef]
11. Cantarella, H.; Otto, R.; Rodrigues Soares, J.; Gomes de Brito Silva, A. Agronomic efficiency of NBPT as a urease inhibitor: A review. *J. Adv. Res.* **2018**, *13*, 19–27. [CrossRef] [PubMed]
12. Whitehurst, G.B.; Whitehurst, B.M. Aminoalcohol Solutions of N-(N-butyl)thio-Phosphoric Triamine (NBPT) and Urea Fertilizers Using such Solutions as Urease Inhibitors. U.S. Patent 2010/0206031A1, 19 August 2010.

13. Chorianopoulou, S.N.; Saridis, G.I.; Sigalas, P.; Margetis, M.; Benardos, D.; Mavrogiannis, H.; Bouranis, D.L. The application of $S^0$-coated fertilizer to durum wheat crop. In *Plant Sulfur Proceedings*; Sulfur Metabolism in Higher Plants—Fundamental, Environmental and Agricultural Aspects; De Kok, L.J., Haneklaus, S.H., Hawkesford, M.J., Schnug, E., Eds.; Springer: Berlin, Germany, 2017; Volume 3, pp. 115–122.
14. Bouranis, D.L.; Chorianopoulou, S.N.; Margetis, M.; Saridis, G.I.; Sigalas, P.P. Effect of Elemental Sulfur as Fertilizer Ingredient on the Mobilization of Iron from the Iron Pools of a Calcareous Soil Cultivated with Durum Wheat and the Crop's Iron and Sulfur Nutrition. *Agriculture* **2018**, *8*, 20. [CrossRef]
15. Gee, G.W.; Bauder, J.W. Particle-size analysis. In *Methods of Soil Analysis, Part 1. Physical and Mineralogical Methods*; Klute, A., Ed.; ASA, SSSA: Madison, WI, USA, 1986; pp. 383–411.
16. Bilias, F.; Barbayiannis, N. Evaluation of sodium tetraphenylboron ($NaBPh_4$) as a soil test of potassium availability. *Arch. Agron. Soil Sci.* **2017**, *63*, 468–476.
17. McLean, E.O. Soil pH and lime requirement. In *Methods of Soil Analysis, Part 2. Chemical and Microbiological Properties*; Page, A.L., Miller, R.H., Keeney, D.R., Eds.; ASA, SSSA: Madison, WI, USA, 1982; pp. 199–224.
18. Nelson, D.W.; Sommers, L.E. Total carbon, organic carbon and organic matter. In *Methods of Soil Analysis, Part 2. Chemical and Microbiological Properties*; Page, A.L., Miller, R.H., Keeney, D.R., Eds.; ASA, SSSA: Madison, WI, USA, 1982; pp. 539–547.
19. Olsen, S.R.; Sommers, L.E. Phosphorus. In *Methods of Soil Analysis, Part 2*, 2nd ed.; Page, A.L., Ed.; Agron Monogr. ASA and ASSA: Madison, WI, USA, 1982; pp. 403–430.
20. Thomas, G.W. Exchangeable cations. In *Methods of Soil Analysis, Part 2. Chemical and Microbiological Properties*; Page, A.L., Miller, R.H., Keeney, D.R., Eds.; ASA, SSSA: Madison, WI, USA, 1982; pp. 159–166.
21. Lindsay, W.L.; Norvell, W.A. Development of DTPA soil test for zinc, iron, manganese and copper. *Soil Sci. Soc. Am. J.* **1978**, *42*, 421–428.
22. Roussos, P.A.; Gasparatos, D.; Kechrologou, K.; Katsenos, P.; Bouchagier, P. Impact of organic fertilization on soil properties, physiology and yield in two newly planted Greek olive (*Olea europaea* L.) cultivars under Mediterranean conditions. *Sci. Hortic.* **2017**, *220*, 11–19. [CrossRef]
23. Chatzistathis, T.; Papaioannou, A.; Gasparatos, D.; Molassiotis, A. From which soil metal fractions Fe, Mn, Zn and Cu are taken up by olive trees (Olea europaea L., 'Chondrolia Chalkidikis') in organic groves? *J. Environ. Manag.* **2017**, *203*, 489–499. [CrossRef] [PubMed]
24. Jones, J.B., Jr. *Soil Analysis Handbook of Reference Methods*; CRC Press: Boca Raton, FL, USA; London, UK; New York, NY, USA; Washington, DC, USA, 1999.
25. Stroup, W.W. *Generalized Linear Mixed Models*; CRC Press: Boca Raton, FL, USA, 2012.
26. R Core Team. *R: A Language and Environment for Statistical Computing*; R Foundation for Statistical Computing: Vienna, Austria, 2018. Available online: https://www.R-project.org/ (accessed on 2 December 2018).
27. Germida, J.J.; Jansen, H.H. Factors affecting the oxidation of elemental sulfur in soils. *Fertil. Res.* **1993**, *35*, 101–114. [CrossRef]
28. Zhao, C.; Gupta, V.V.S.R.; Degryse, F.; McLaughlin, M.J. Abundance and diversity of sulphur-oxidising bacteria and their role in oxidizing elemental sulphur in cropping soils. *Biol. Fertil. Soils* **2017**, *53*, 157–169. [CrossRef]
29. Zhao, C.; Degryse, F.; Gupta, V.; McLaughlin, M.J. Elemental Sulfur oxidation in Australian cropping soils. *Soil Sci. Soc. Am. J.* **2015**, *75*, 89–96. [CrossRef]
30. Chorianopoulou, S.N.; Venieraki, A.; Maniou, F.; Mavrogiannis, H.; Benardos, D.; Katinakis, P.; Bouranis, D.L. Sulfogrow®: A new type of plant biostimulant? In Proceedings of the 13th International Conference on Protection and Restoration of the Environment, Mykonos, Greece, 3–8 July 2016; pp. 733–738.
31. Kertesz, M.A.; Mirleau, P. The role of soil microbes in plant sulphur nutrition. *J. Exp. Bot.* **2004**, *55*, 1939–1945. [CrossRef] [PubMed]
32. Gahan, J.; Schmalenberger, A. The role of bacteria and mycorrhiza in plant sulfur supply. *Front. Plant Sci.* **2014**, *5*, 723. [CrossRef] [PubMed]
33. Richardson, A.E.; Barea, J.-M.; McNeill, A.M.; Prigent-Combaret, C. Acquisition of phosphorus and nitrogen in the rhizosphere and plant growth promotion by microorganisms. *Plant Soil* **2009**, *321*, 305–339. [CrossRef]
34. Shen, J.; Yuan, L.; Zhang, J.; Li, H.; Bai, Z.; Chen, X.; Zhang, W.; Zhang, F. Phosphorus Dynamics: From Soil to Plant. *Plant Physiol.* **2011**, *156*, 997–1005. [CrossRef]
35. Ziadi, N.; Whalen, J.K.; Messiga, A.J.; Morel, C. Assessment and modeling of soil available phosphorus in sustainable cropping systems. *Adv. Agron.* **2013**, *122*, 85–126.

36. Rodriguez, D.; Andrade, F.H.; Goudriaan, J. Effects of phosphorus nutrition on tiller emergence in wheat. *Plant Soil* **1999**, *209*, 283–295. [CrossRef]
37. Otto, W.M.; Kilian, W.H. Response of Soil Phosphorus Content, Growth and Yield of Wheat to Long-Term Phosphorus Fertilization in A Conventional Cropping System. *Nutr. Cycl. Agroecos.* **2001**, *61*, 283–292. [CrossRef]
38. Alam, S.M.; Azam, S.; Ali, S.; Iqbal, M. Wheat yield and P fertilizer efficiency as influenced by rate and integrated use of chemical and organic fertilizers. *Pak. J. Soil Sci.* **2003**, *22*, 72–76.
39. Gill, H.S.; Singh, H.B.A.; Sethia, S.K.; Behla, R.K. Phosphorus uptake and efficiency in different varieties of bread wheat (*Triticum aestivum* L.). *Arch. Agron. Soil Sci.* **2004**, *50*, 563–572. [CrossRef]
40. Kosmas, C.; Vassiliou, P.; Dolopoulou, C.; Gouvela, T. *Spatial Analysis of Soil Characteristics of Aliartos Field*; Agricultural University of Athens: Athens, Greece, 2015. (In Greek)
41. Haneklaus, S.; Schnug, E. An agronomic, ecological and economic assessment of site-specific fertilisation. *Landbauforsch. Völkenrode* **2002**, *52*, 123–133.
42. Friesen, D.K. Influence of co-granulated nutrients and granule size on plant response to elemental sulfur in compound fertilizers. *Nutr. Cycl. Agroecosyst.* **1996**, *46*, 81–90. [CrossRef]
43. Degryse, F.; da Silva, R.C.; Baird, R.; Beyrer, T.; Below, F.; McLaughlina, M.J. Uptake of elemental or sulfate-S from fall- or spring-applied co-granulated fertilizer by corn—A stable isotope and modeling study. *Field Crops Res.* **2018**, *221*, 322–332. [CrossRef]
44. Chien, S.H.; Prochnow, L.I.; Tu, S.; Synder, C.S. The agronomic and environmental aspects of phosphate fertilizers varying in source and solubility: An update review. *Nutr. Cycl. Agroecosyst.* **2011**, *89*, 229–255. [CrossRef]
45. Janzen, H.H.; Bettany, J.R. Measurement of sulfur oxidation in soils. *Soil Sci.* **1987**, *143*, 444–452. [CrossRef]
46. Cantarella, H.; Mattos Jr, D.; Quaggio, J.A.; Rigolin, A.T. Fruit yield of Valencia sweet orange fertilized with different N sources and the loss of applied N. *Nutr. Cycl. Agroecosyst.* **2003**, *67*, 215–223. [CrossRef]
47. Shivay, Y.S.; Pooniya, V.; Prasad, R.; Pal, M.; Bansal, R. Sulphur-coated urea as a source of sulphur and an enhanced efficiency of N fertilizer for spring wheat. *Cereal Res. Commun.* **2016**, *44*, 513–523. [CrossRef]
48. Ding, W.X.; Chen, Z.M.; Yu, H.Y.; Luo, J.F.; Yoo, G.Y.; Xiang, J.; Zhang, H.J.; Yuan, J.J. Nitrous oxide emission and nitrogen use efficiency in response to nitrophosphate, N-(n-butyl) thiophosphoric triamide and dicyandiamide of a wheat cultivated soil under sub-humid monsoon conditions. *Biogeosciences* **2015**, *12*, 803–815. [CrossRef]
49. Zanin, L.; Venuti, S.; Tomasi, N.; Zamboni, A.; De Brito Francisco, R.M.; Varanini, Z.; Pinton, R. Short-term treatment with the uease inhibitor N-(n-Butyl) Thiophosphoric Triamide (NBPT) alters urea assimilation and modulates transcriptional profiles of genes involved in primary and secondary metabolism in maize seedlings. *Front. Plant Sci.* **2016**, *8*, 845.
50. Zaman, M.; Nguyen, M.L.; Blennerhassett, J.D.; Quin, B.F. Reducing $NH_3$, $N_2O$ and $NO_3^-$–N losses from a pasture soil with urease or nitrification inhibitors and elemental S-amended nitrogenous fertilizers. *Biol. Fertil. Soils* **2008**, *44*, 693–705. [CrossRef]

*Article*

# Impact of Elemental Sulfur on the Rhizospheric Bacteria of Durum Wheat Crop Cultivated on a Calcareous Soil

**Dimitris L. Bouranis [1,*], Anastasia Venieraki [2], Styliani N. Chorianopoulou [1] and Panagiotis Katinakis [2,*]**

1 Plant Physiology and Morphology Laboratory, Crop Science Department, Agricultural University of Athens, Iera Odos 75, 118 55 Athens, Greece; s.chorianopoulou@aua.gr

2 General and Agricultural Microbiology Laboratory, Crop Science Department, Agricultural University of Athens, Iera Odos 75, 118 55 Athens, Greece; venieraki@aua.gr

* Correspondence: bouranis@aua.gr (D.L.B.); katp@aua.gr (P.K.)

Received: 23 August 2019; Accepted: 25 September 2019; Published: 27 September 2019

**Abstract:** Previous experiments have shown that the application of fertilizer granules containing elemental sulfur ($S^0$) as an ingredient ($FBS^0$) in durum wheat crops produced a higher yield than that produced by conventional ones (F), provided that the soils of the experimental fields (F vs. $FBS^0$) were of comparable quality and with the Olsen P content of the field's soil above 8 mg $kg^{-1}$. In this experiment the $FBS^0$ treatment took place in soil with Olsen P at 7.8 mg $kg^{-1}$, compared with the F treatment's soil with Olsen P of 16.8 mg $kg^{-1}$, aiming at reducing the imbalance in soil quality. To assess and evaluate the effect of $FBS^0$ on the dynamics of the rhizospheric bacteria in relation to F, rhizospheric soil at various developmental stages of the crops was collected. The agronomic profile of the rhizospheric cultivable bacteria was characterized and monitored, in connection with the dynamics of phosphorus, iron, organic sulfur, and organic nitrogen, in both the rhizosoil and the aerial part of the plant during development. Both crops were characterized by a comparable dry mass accumulation per plant throughout development, while the yield of the $FBS^0$ crop was 3.4% less compared to the F crop's one. The $FBS^0$ crop's aerial part showed a transient higher P and Fe concentration, while its organic N and S concentrations followed the pattern of the F crop. The incorporation of $S^0$ into the conventional fertilizer increased the percentage of arylsulfatase (ARS)-producing bacteria in the total bacterial population, suggesting an enhanced release of sulfate from the soil's organic S pool, which the plant could readily utilize. The proportion of identified ARS-producing bacteria possessing these traits exhibited a maximum value before and after topdressing. Phylogenetic analysis of the 68 isolated ARS-producing bacterial strains revealed that the majority of the isolates belonged to the *Pseudomonas* genus. A large fraction also possessed phosphate solubilization, and/or siderophore production, and/or ureolytic traits, thus improving the crop's P, Fe, S, and N balance. The aforementioned findings imply that the used $FBS^0$ substantially improved the quality of the rhizosoil at the available phosphorus limiting level by modulating the abundance of the bacterial communities in the rhizosphere and effectively enhancing the microbially mediated nutrient mobilization towards improved plant nutritional dynamics.

**Keywords:** elemental sulfur; calcareous soil; arylsulfatase-producing bacteria; P mobilization; Fe mobilization; nutritional fortification

## 1. Introduction

Plant nutrient acquisition from the soil via the root system is influenced by a wide range of factors. Plant species and genotype, the soil type and chemical-physical characteristics, the soil microorganism communities, the fertilization regime, and environmental conditions are all factors of central agronomic importance. In this context, the biological activities of microorganisms can play an important role [1]. Regardless of the plant species, a fraction of the soil microbiome will eventually colonize the soil bordering the root surface (i.e., the rhizosphere), because plant roots release a wide range of compounds involved in attracting microorganisms, which may be beneficial, neutral, or detrimental to plants [2]. Therefore, a zone of intense microbial activity is generated that can affect plant nutrient acquisition processes by influencing nutrient availability in the rhizosphere and/or the plant biochemical mechanisms underlying the nutritional process [3], thus profoundly influencing crop fortification and production [4]. The beneficial bacteria residing in the rhizosphere, referred to as plant growth-promoting rhizobacteria (PGPR), are in general capable of enhancing the growth of plants and/or protecting them from biotic or abiotic stresses [5]. The PGPR that are involved in soil nutrients' mobilization, thus enhancing the availability of the mobilized nutrients to the plant, are usually endowed with traits that stimulate the mobilization of unavailable plant nutrients through various mechanisms, including phosphate solubilization and production of siderophores, among several others.

Phosphate mineralization by microorganisms involves either acidification of the extracellular environment by production of organic acid anions, such as gluconic acid, (acid or alkaline) phosphatase activity, or phytase activity. The use of P-solubilizing microorganisms has received much attention over the last few decades as a means of mobilizing P in different types of soils and fertilization regimes. Given that micronutrient deficiency generally occurs in calcareous soils because of high pH, acidification of the rhizosphere can increase the solubility of micronutrients [1,6–12].

A large fraction of the agricultural soils in the Mediterranean area show an alkaline pH, thus the bioavailable Fe fraction is lower than that required for optimal plant growth. PGPR might increase the bioavailability of insoluble Fe by the production and secretion of siderophores, rendering it available for both bacteria and plants. There is a considerable body of evidence showing that several siderophore-producing microorganisms are able to enhance Fe uptake in monocot and dicot plants species [4,13–16].

The role of bacteria capable of mobilizing S from the soil organic matter, thus providing S available to plants, has attracted little attention so far, despite the fact that most of the S in soils is bound to organic molecules, making up more than 90% of the S present in soils [17–20]. Evidence has been provided that plant S fortification of *Miscanthus giganthus* is correlated with the population density of rhizobacteria that transform organic sulfur into bioavailable forms [21]. Furthermore, it has been demonstrated that S-demanding plants, such as rape, appear to recruit higher numbers of arylsulfatase (ARS)-producing bacteria as compared to barley [22]. Arylsulfatase hydrolyzes sulfate esters to sulfate and is a key enzyme of soil organic S mineralization, so is often utilized as a soil fertility indicator [23–25].

The application of fertilizer granules with incorporated elemental sulfur ($FBS^0$) in durum wheat crop produced a higher yield and higher S and Fe contents compared with the application of the conventional (F) ones, provided that the experimental fields (F and $FBS^0$) were of comparable quality. The fertilization with $FBS^0$ at sowing mobilized iron from the rhizosoil, thus providing more iron to the crop; in addition, it boosted the crop's sulfur content [26]. On the other hand, the level of available phosphorus was found to be correlated with the corresponding relative change in the yields ($YFBS^0/YF$), presenting a strong positive relationship, and the content of 8 mg $kg^{-1}$ was a turning point; lower values resulted in a lower yield compared with the conventional crop [27].

In this experimental approach, the $FBS^0$ treatment took place in soil of inferior quality and with an Olsen *p* value of 7.8 mg $kg^{-1}$, compared to the F treatment's soil with adequate quality and an Olsen P of 16.8 mg $kg^{-1}$. The working hypothesis was that the $FBS^0$ treatment could possibly alleviate the anticipated yield loss produced by the control treatment at an Olsen P of 7.8 mg $kg^{-1}$. Moreover, it was hypothesized that the small amount (2% w/w) of $S^0$ added in with the fertilizer application at sowing stimulated the recruitment of rhizobacteria that are able to mobilize unavailable P, Fe, and S into bioavailable forms. Thus, the aims of this study were (i) to monitor the effect of elemental sulfur as an ingredient of the applied fertilization scheme on phosphorus and iron dynamics in the soil explored by the root system mass, i.e., the rhizosoil (RS), as well as in the aerial part of the plant during development under the specific regimes, i.e., when the $FBS^0$ treatment takes place in soil with available phosphorus at a turning point level of 8 mg $kg^{-1}$, compared to the level of 16 mg $kg^{-1}$ of the F treatment's soil; and, in parallel, (ii) to monitor the dynamics of rhizospheric bacteria capable of transforming unavailable P, Fe, and S into bioavailable forms, under the circumstances, in order to elaborate on the effect of the small added amount of $S^0$ on the crop's rhizospheric microbiome, if any, under the two aforementioned levels (marginal vs. adequate) of available phosphorus.

## 2. Materials and Methods

### 2.1. Experimental Setup

Durum wheat (*Triticum turgidum* L. *subsp. durum* (Desf.) Husn.), commercial cultivar SIMETO, was sown at Arma (latitude 38.35° N, longitude 23.49° E, 256 m asl) in Viotia county, central Greece, in a production area of 1.8 ha with calcareous soil. This area was chosen due to its heterogeneity in Olsen P content and was divided into two experimental fields of 0.9 ha each. The field with an Olsen *p* value of 16.8 mg $kg^{-1}$ was subject to conventional F treatment according to the local agricultural practices (F crop), while the other one, with an Olsen *p* value of 7.8 mg $kg^{-1}$, received the corresponding $FBS^0$ treatment ($FBS^0$ crop, Figure 1). Sowing and fertilizer application took place on 24 November 2015 (day 0). The control crop was fertilized with a commercial 16-20-0 fertilizer (nitrogen was provided as ammonium sulfate, phosphorus as triple superphosphate) at a rate of 275 kg $ha^{-1}$. The $FBS^0$-treated crop received 280.5 kg $ha^{-1}$ by applying the corresponding $FBS^0$ 16-20-0 commercial fertilizer carrying 2% $S^0$ (F: 275; $S^0$: 5.5). At 100 and 110 days after sowing (DAS), additional fertilization (topdressing) with a commercial 40-0-0 fertilizer (urea plus ammonium sulfate; UAS) took place at the rate of 180 kg $ha^{-1}$, according to the local practice and the same for both treatments. At 122 DAS, herbicide application took place at the rate of 500 mL $ha^{-1}$ (Pasifica Bayer CropScience Ltd, Cambridge, UK).

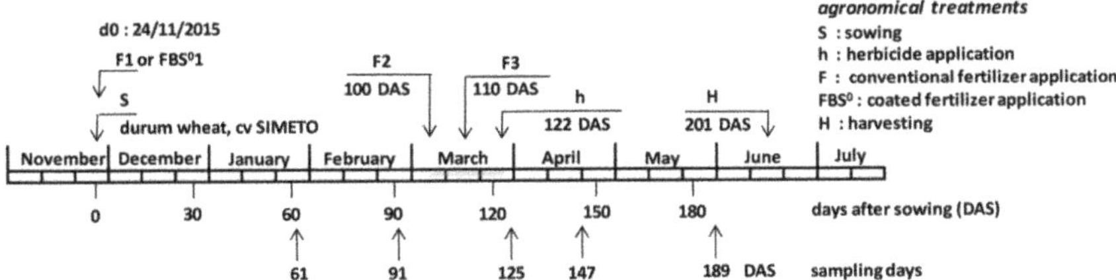


**Figure 1.** Overview of the experiment. The gray line in March shows the extent of the agronomical treatments. Phenological phases at each sampling day: 61 DAS, tillering; 91 and 125 DAS, stem elongation; 188 DAS, grain filling. Heading was apparent at 133 DAS.

Within each experimental field, the soil quality was tested prior to sowing by analyzing a representative composite sample collected at a depth of 20 cm. Then, an internal perimeter 7 m from the border was established and excluded from sampling and the internal area was divided into five groups. On each sampling day, at least five plants were collected from each group, with their root system and the surrounding soil by means of a shovel. The excess of soil was removed by hand and the soil explored by the root system mass, i.e., the rhizosoil (RS), was collected and analyzed. Afterward, the rhizosphere (Rhs) was collected from the root system by means of a brash. RS was monitored for phosphorus and iron concentration, while Rhs was used for microbiological procedures. In the dry mass (DM) of the aboveground plant part, the fluctuation dynamics of organic nitrogen, organic sulfur, phosphorus, and iron were monitored during the crops' development. Thus, on each sampling day one composite sample of DM, RS, and Rhs per group was constructed and analyzed.

### 2.2. Determinations of Soil Parameters

Determinations of pH, organic matter, $CaCO_3$, $NO_3^-$, Olsen P, exchangeable potassium, Fe-DTPA (DTPA: diethylenetriaminepentaacetic acid), Mn-DTPA, Cu-DTPA, and Zn-DTPA in the RS were performed according to the procedures described by Jones (1999) [28].

### 2.3. Determinations of Dry Matter, P, Fe, Organic N, and Organic S Concentrations in the Aerial Part of the Plant

Samples of the aerial part of the plants were oven-dried at 80 °C and the dry mass was weighed and recorded. Then, it was ground to pass through a 40-mesh screen using an analytical mill (IKA, model A10, IKA®-Werke GmbH & Co. KG, Staufen, Germany). Prior to Fe and P analysis, samples were digested with hot $H_2SO_4$ and repeated additions of 30% $H_2O_2$ until the digestion was complete [28]. In the diluted dry mass (DM) digests, Fe content was determined by atomic absorption spectrophotometry (GBC, Model Avanta spectrophotometer, GBC Scientific Equipment PTY Ltd., Dandenong, Australia), while the P content was determined using the ammonium molybdate and stannous chloride procedure. Organic nitrogen ($N_{org}$) was determined by micro-Kjeldahl digestion followed by distillation. The sulfate concentration was determined by extracting the oven-dried samples with 2% (*v*/*v*) acetic acid aqueous solution and by analyzing by a turbidimetric method [29,30]. Total sulfur content was determined after dry ashing at 600 °C [31]. The ash was dissolved in a 2% (*v*/*v*) acetic acid aqueous solution and filtered through Whatman No. 42 paper, and the total sulfate was determined turbidimetrically [29,30]. The content of organic sulfur ($S_{org}$) was calculated by subtracting the total sulfate content from the total sulfur.

### 2.4. Microorganism Sampling, Isolation, Culture Conditions, and Arylsulfatase Activity

Immediately after collection, Rhs was placed on ice for further analysis. Rhs (3 g) was suspended in 30 mL of sterile phosphate-buffered saline (PBS), pH 7.2 on an orbital shaker (120 rev $min^{-1}$) for 45 min. Cultivable bacteria were quantified by plating a serial dilution (three replicates, $10^{-4}$, $10^{-5}$, and $10^{-6}$) of soil suspension dilutions on nutrient agar medium, M9 mineral medium, and modified M9 mineral medium. The chromogenic arylsulfatase substrate 5-bromo-4-chloro-3-indolyl sulfate (X-Sulf, Sigma) was incorporated (100 mg $L^{-1}$) into a modified M9 medium, as the sole sulfur source [24]. Plates were incubated at 28 °C for 48–72 h and colony-forming units were counted and expressed as cfu $g^{-1}$ of soil. Microbial colonies possessing ARS activity were detected by their blue color on Petri dishes. Individual X-Sulf-utilizing strains were identified for further study from the highest dilutions showing growth in the modified M9 mineral medium. Single colonies were picked, re-cultured in modified M9 mineral medium, and maintained as glycerol stock (20%) at −80°C for further use.

### 2.5. Assessment of Plant Growth-Promoting (PGP) Traits

Apart from the arylsulfatase activity of the selected strains, we investigated other possible PGP traits, such as phosphate solubilization, siderophore production, urea solubilization, their swarming motility, and biocontrol activity. The phosphate-solubilizing activity of the isolates was studied on Pikovskaya agar [32]. Phosphate-solubilizing bacterial strains were identified by clear halo zones around their colonies. Sterile medium served as a control. Siderophore production was examined on chrome azurol-S agar medium [33]. CAS plates were spot-inoculated with a liquid culture of bacterial strains and observed for the development of an orange halo against a dark blue background around the colonies after 48 h of incubation at 28 °C. A change in color from blue to orange (hydroxamate-type siderophore) was considered a positive reaction. Urease production ability was tested by inoculating Urea Base Christensen ISO 6579, ISO 19250 (Conda, Madrid, Spain) medium Petri dishes. Phenol red indicator detects the alkalinity generated by the visible color change from orange to pink. Fluorescent pigment production was tested on King medium B and inspected under UV light. Bacterial strain swarming motility was tested by the inoculation of fresh swarm plates (nutrient agar, 0.5%). Plates were spotted with 1–5 μL of overnight broth culture of each strain, followed by overnight incubation (≤20 h) at 28 °C. Control plates were prepared in parallel with swarming ones with nutrient agar, 1.5% [34]. The bacterial isolates were also evaluated for their antagonistic potential against the soil-borne phytopathogenic fungi *Rhizoctonia solani* and *Fusarium oxysporum* [34]. Briefly, 5 μL drops of each bacterial culture ($10^8$ cfu $mL^{-f}$) were streaked equidistantly on the margins of NA plates. A mycelial agar plug of 5 mm diameter from a seven-day-old culture of *R. solani* and *F. oxysporum* grown on a PDA plate was placed in the center of the plate, between the two parallel streaks of the test bacterial isolate. Control plates were also prepared without the mycelial fragment. Two independent experiments were performed with each bacterial isolate. Plates were incubated at 25 °C for 7–10 days. All assays were done in triplicate.

### 2.6. PCR Amplification 16S rRNA Gene Sequence and Phylogenetic Analysis

Genomic DNA from bacterial cultures of the isolates was extracted using the NucleoSpin® Microbial DNA Kit (Macherey-Nagel GmbH & Co KG, Düren, Germany) according to the manufacturer's instructions. The DNA concentration and purity were assessed by a Nanodrop ND-1000 spectrophotometer (Thermo Fisher Scientific, Waltham, MA, USA). The primer set fD1 (5′-AGAGTTTGATCCTGGCTCAG-3′) and rP2 (5′-ACGGCTACCTTGTTACGACTT-3′) [35], targeting ribosomal DNA of approximately 1500 bp, was used for PCR amplification of the 16S ribosomal DNA. The polymerase chain reaction (PCR) amplification mixture contained 0.5 U KAPA2G Robust DNA polymerase (Kapa Biosystems Inc., Wilmington, MA, USA); 5 × KAPA2G GC buffer; 50 pmol of each oligonucleotide; and 50 ng of DNA template. A final volume of 50 μL was adjusted with distilled water. PCR reactions were cycled in a BIORAD thermocycler (MJ mini) with a hot start step at 94 °C for 5 min, followed by 35 cycles of 94 °C for 1 min, annealing at the appropriate temperature for 1.5 min, and 72 °C for 1 min, with a final extension step at 72 °C for 10 min. The 16S rRNA gene sequence was determined by direct sequencing of the PCR product and was performed by CeMIA SA, Larissa, Greece. Analysis of the sequence was performed with the basic sequence alignment BLAST program (National Center for Biotechnology Information, U.S. National Library of Medicine, Bethesda MD, USA) run against the database provided on the website of the National Center for Biotechnology Information (http://www.ncbi.nlm.nih.gov/BLAST, National Center for Biotechnology Information, U.S. National Library of Medicine, Bethesda MD, USA). Phylogenetic trees based on the nucleotide sequences of the 16S rRNA gene fragments were constructed with the Molecular Evolutionary Genetics Analysis software version 7.0 (Pennsylvania State University, State College, PA, USA) [36] using the neighbor-joining algorithm (1000 bootstrap replication).

### 2.7. Nucleotide Sequence Accession Numbers

The nucleotide sequence data have been submitted to the European Nucleotide Archive GenBank database under accession numbers LR027392 to LR027460 (16S rRNA sequences), BioProject: PRJEB28499.

### 2.8. Statistical Analysis

The comparisons between the corresponding $FBS^0$ crop and F crop values in each case were performed using one-way ANOVA and Tukey's honest significant difference post hoc test at $p < 0.05$.

## 3. Results

### 3.1. Dry Mass Accumulation per Plant

Despite the fact that the $FBS^0$ crop was grown in soil containing 53% less phosphorus (Table 1), the dry mass accumulation per plant was statistically the same as the F crop's one (Figure 2). Moreover, a tendency for higher accumulation (though not statistically significant) was observed at 189 DAS. The $FBS^0$ crop had a 3.4% lower yield (4.89 t $ha^{-1}$) compared to the F crop (5.06 t $ha^{-1}$).

**Table 1.** Soil quality of the experimental fields at sowing.

| Crop | Clay | Silt | Sand | Class | pH | EC | $CaCO_3$ | SOM |
|---|---|---|---|---|---|---|---|---|
| | % | % | % | | | μS $cm^{-1}$ | % | % |
| F | 38 | 34 | 28 | Silty Clay | 7.86 | 685 | 31.5 | 2.55 |
| $FBS^0$ | 42 | 30 | 28 | Clay | 8.01 | 613 | 24.5 | 1.79 |
| Δx/x (%) | | | | | 1.9 | −10.5 | −22.2 | −29.8 |
| | **Olsen P** | **$NO_3$-N** | **Kexc** | **Fe-DTPA** | **Mn-DTPA** | **Cu-DTPA** | **Zn-DTPA** | |
| | mg $kg^{-1}$ | mg $kg^{-1}$ | mg $kg^{-1}$ | mg $kg^{-1}$ | mg $kg^{-1}$ | mg $kg^{-1}$ | mg $kg^{-1}$ | |
| F | 16.8 | 28.48 | 410 | 7.08 | 8.97 | 1.76 | 1.26 | |
| $FBS^0$ | 7.8 | 24.57 | 270 | 5.78 | 6.25 | 1.43 | 0.75 | |
| Δx/x (%) | −53.6 | −13.7 | −34.1 | −18.4 | −30.3 | −18.8 | −40.5 | |

**F crop**: the crop that was subject to conventional fertilization (F) treatment according to the local agricultural practices; **$FBS^0$ crop**: the crop that received the corresponding $FBS^0$ treatment; **EC**: electric conductivity; **Kexch**: soil exchangeable potassium. **SOM**: soil organic matter; **DTPA**: diethylenetriaminepentaacetic acid.

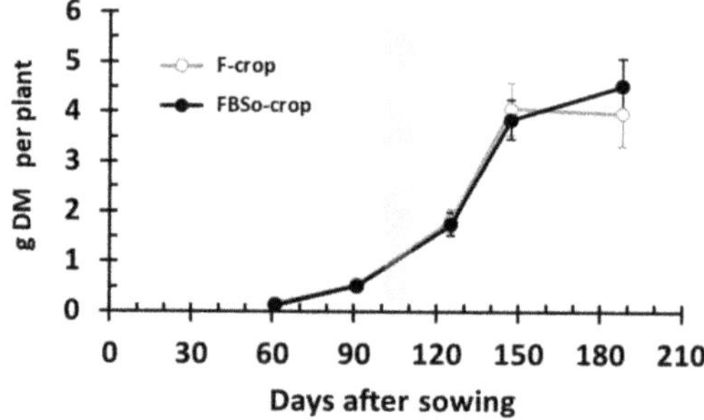


**Figure 2.** Dry mass accumulation per plant. The gray histogram indicates the topdressing application. Bars indicate standard deviation. An asterisk indicates a statistically significant difference ($p < 0.05$).

### *3.2. Developmental Dynamics of Organic Nitrogen versus Organic Sulfur in the Crop's Aerial Part*

Dynamics of both $N_{org}$ and $S_{org}$ concentrations followed the same pattern in both crops. Patterns of both nutrients in the $FBS^0$ crop had a tendency to be lower, but the difference was not statistically significant through development. As a result, both $N_{org}$ and $S_{org}$ accumulations per plant in the aerial part followed a sigmoid curve, with a tendency to be lower in the $FBS^0$ crop, especially after topdressing, although this difference was not statistically significant (Figure 3).

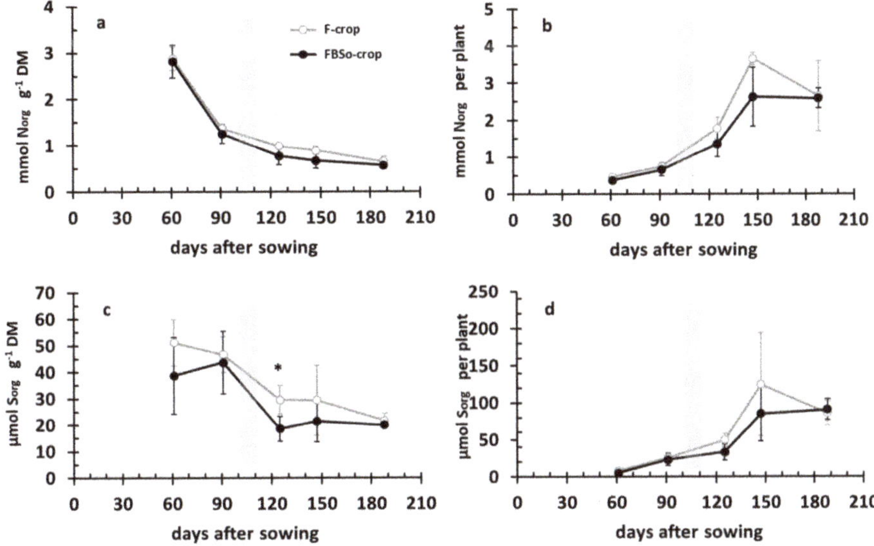


**Figure 3.** Time course of organic nitrogen ($N_{org}$; (**a**)) and organic sulfur ($S_{org}$; (**c**)) concentration along with the corresponding accumulations per plant in the crop's aerial part (**b**,**d**). The gray histogram indicates the application of topdressing. Bars indicate standard deviation. An asterisk indicates a statistically significant difference ($p < 0.05$).

### 3.3. Quantification of Cultivable ARS-Producing Bacterial Communities in the Rhizosphere of F- and FBS$^0$-Treated Wheat

Rhizospheric soil (Rhs) samples collected from the FBS$^0$ and F crops from 61 to 188 DAS were evaluated for total cultured bacterial population. Estimations of CFU per gram of dry Rhs were carried out in full media (nutrient broth agar) and minimal media (M9 mineral medium agar). The average total population of bacterial heterotrophs was similar at 61, 91, and 125 DAS and reached a higher density at 147 DAS in both the FBS$^0$ and F crops (Figure 4). However, the heterotrophic bacterial population at 147 DAS was notably higher in FBS$^0$ as compared to the F crop (Figure 4), suggesting that S$^0$ may have a positive effect on the population density of rhizospheric bacteria. Examination of the ARS-producing bacterial communities (blue colonies in M9 X-sulf medium) in all Rhs samples revealed that the relative abundance of ARS-possessing bacteria was radically higher in FBS$^0$- than in F-treated wheat. The densities of ARS-producing bacteria changed during the course of development of both FBS$^0$- and F-treated wheat (Figure 5). In the FBS$^0$ crop at 61 DAS, the ARS-possessing bacteria community represented 40% of the heterotrophs. At 125 and 147 DAS, almost 50% of the heterotrophs were identified as ARS-possessing bacteria, while at 188 DAS the ARS-possessing bacteria constituted the majority (84%) of the heterotrophs. In the F crop at 61 DAS, the ARS-producing bacteria represented 30% of the heterotrophs; this percentage was reduced to 4% at 188 DAS. Interestingly, at 61 DAS in the FBS$^0$ crop, the ARS-possessing bacteria represented about 40% and reached at 188 DAS 85% of the heterotrophs (Figure 5). Taken together, these results suggest that mixing conventional fertilizers with S$^0$ induces a noteworthy shift in the population density of ARS-possessing bacteria (i.e., the percentage of ARS-possessing bacteria in the total bacterial population), thus enhancing the levels of bioavailable sulfur supply to the plant.

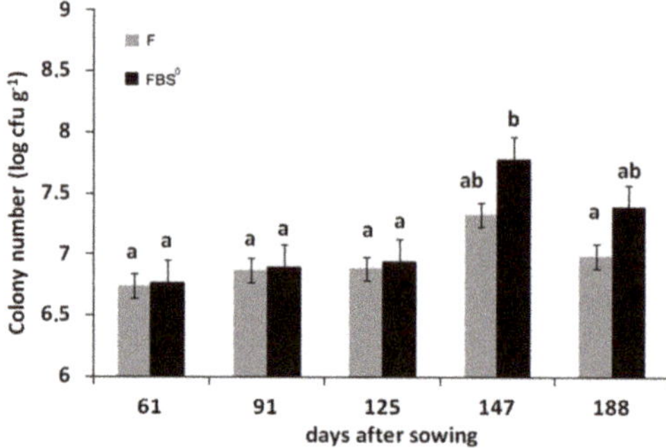


**Figure 4.** Abundance (cfu per $g^{-1}$ dry mass) of rhizospheric soil aerobic bacteria growing in M9-Xsulf minimal media at 61, 91, 125, 147, and 188 DAS for both (F and $FBS^0$) fertilization regimes.

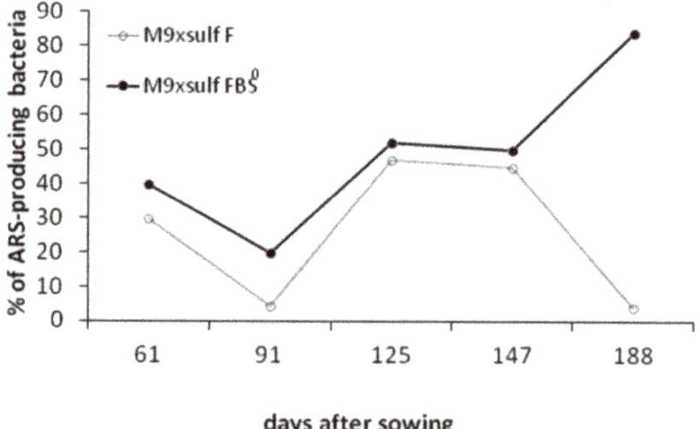


**Figure 5.** Abundance (percentage of ARS-producing bacteria in the total population of aerobic cultured bacteria growing in M9-Xsulf minimal media) in the F and $FBS^0$ crops during development.

### 3.4. *Diversity of ARS-Producing Bacteria*

The overall bacterial community was further characterized by sequencing the 16S rRNA gene of the 68 ARS-producing bacterial strains recovered from the examined phenological growth stages of wheat: 27 strains from the F crop and 41 strains from the $FBS^0$ crop. The ARS-producing bacterial isolates per sampling day and their characteristics from both the F and $FBS^0$ treatment are provided in Table S1. A data analysis revealed that the ARS-producing bacterial community associated with Rhs in both the F and $FBS^0$ crops was dominated by bacteria of the *Pseudomonas* genus, by 90% and 77%, respectively (Figure 6). Almost 80% of the identified ARS-producing bacterial isolates that were classified as *Pseudomonas* sp., *Pseudomonas koreensis*, or *P. fluorescens* showed fluorescent pigment production when they were tested on King medium B. The second most dominant group belonged to the *Bacillus* genus, whereas bacteria belonging to the *Cellulomonas*, *Acinetobacter*, *Stenotrophomonas*, and *Xanthomonas* genera were also found among the F- and $FBS^0$-treated wheat isolates (Figure 6). When two neighbor-joining trees were created, based on the 16S rRNA sequences from ARS-producing bacterial isolates (one for each treatment) (Figure 7), it was noticed that the identified ARS-producing bacterial isolates were classified among the clades of beneficial *Pseudomonas*, *Bacillus*, and other nonphytopathogenic antagonistic strains deposited in public databases.

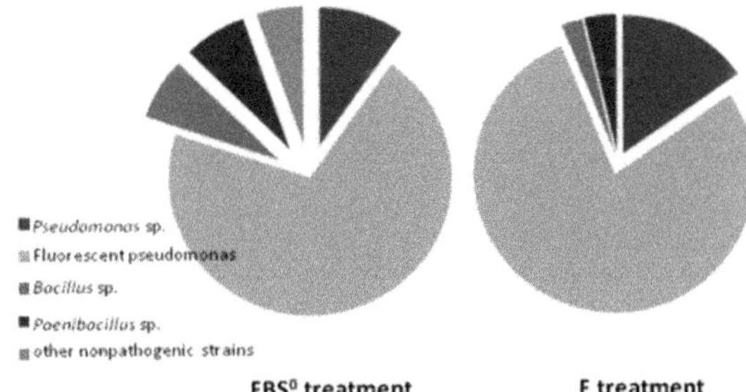


**Figure 6.** Relative abundance of ARS-producing bacterial isolates' phylotypes at the genus level in the FBS$^0$ versus F crop's rhizosoil. The phylogenetic position of the bacterial isolates was based on 16S rRNA sequences.

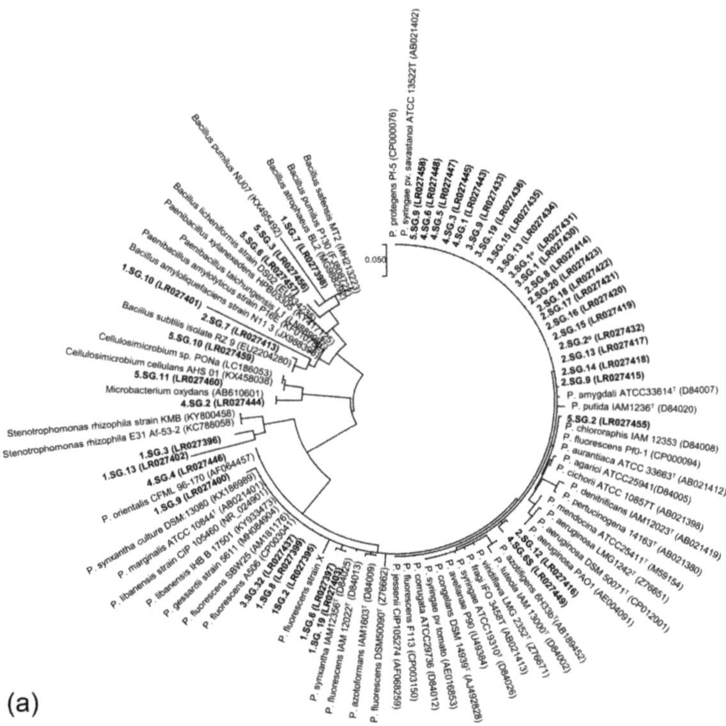


**Figure 7.** *Cont.*

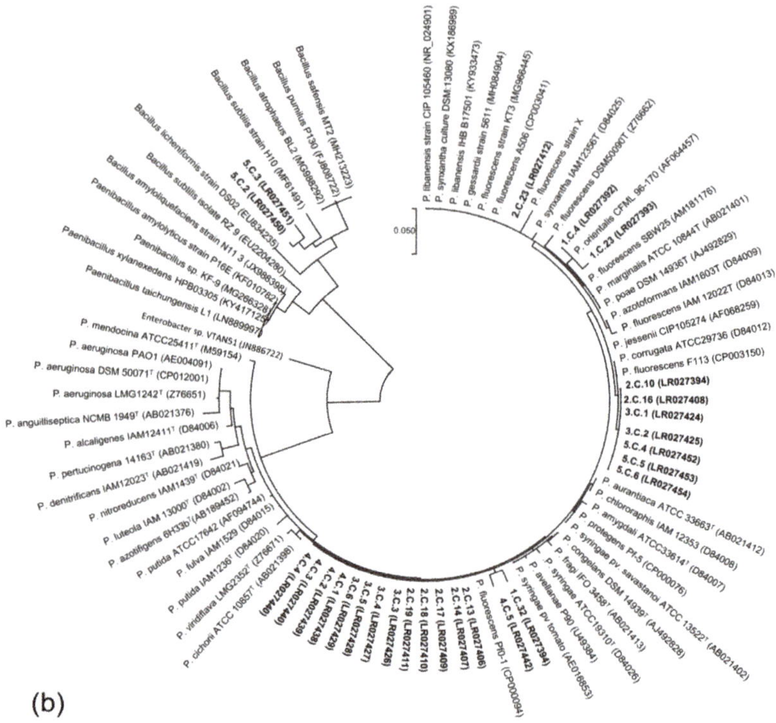

**Figure 7.** Phylogenetic position of ARS-producing bacterial isolates based on their 16S rRNA sequences. (**a**) FBS$^0$ treatment and (**b**) F treatment. List of ARS-producing bacterial isolates and their characteristics are presented in Table S1.

### *3.5. Phosphorus Dynamics in the Rhizosoil and the Aerial Part of the Plant*

During development, the Olsen P concentration of rhizosoil fluctuated around 10 mg kg$^{-1}$ in the FBS$^0$ crop and 18 mg kg$^{-1}$ in the F crop (Figure 8a). It is noticeable that the fluctuation pattern was the same in both treatments. In the aerial part of FBS$^0$ plants, the P concentration was 75% higher than in F plants at 125 DAS, i.e., after topdressing. Then, there was a tendency for a lower P concentration; however, it was not statistically significant (Figure 8b). Following this concentration pattern, P accumulation per plant presented the same sigmoid pattern as the F treatment. P concentration in the F crop increased up to 125 DAS and then decreased and stabilized to 60 µmol gDM$^{-1}$. The same pattern was followed by the FBS$^0$ treatment. P accumulation in the FBS$^0$ crop tended to be higher at 125 DAS; however, this difference was not statistically significant. Both curves reached a plateau at the same level (Figure 8c).

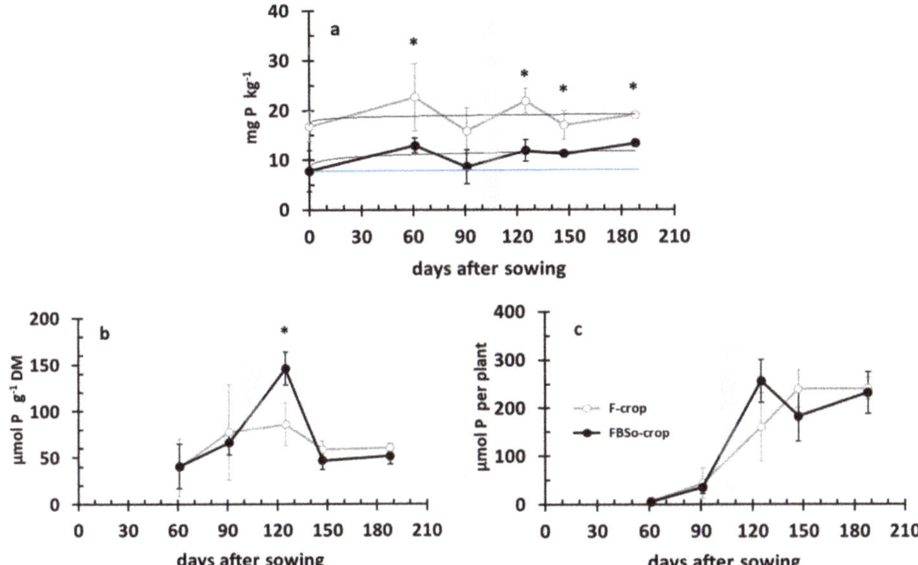


**Figure 8.** Phosphorus dynamics (**a**) in the rhizosoil (Olsen P concentration) and in the aerial part: (**b**) P concentration; (**c**) P accumulation, during crop development. Bars indicate standard deviation. An asterisk indicates a statistically significant difference ($p < 0.05$).

### *3.6. Iron Dynamics in the Rhizosoil and the Aerial Part of the Plant*

Crops started with 7.08 mg kg$^{-1}$ of Fe in the F crop's soil and 5.78 mg kg$^{-1}$ of Fe (i.e., 18.4% less) in the FBS$^0$ crop's one (Table 1). During development, Fe remained almost stable in the F crop's RS, up to 125 DAS, and then decreased to 5 mg kg$^{-1}$. Instead, the Fe concentration in the FBS$^0$ crop's RS increased up to 10 mg kg$^{-1}$ before topdressing and stabilized at 125 DAS. Then, a significant decrease to the F crop's levels took place (Figure 9a). In the aerial part of the plant, the Fe concentration decreased from 12 µmol gDM$^{-1}$ to 5 between 61 and 91 DAS, when it more or less stabilized (Figure 9b). FBS$^0$ treatment presented a differentiated pattern. After topdressing, the Fe concentration increased to the levels of 61 DAS and then decreased and stabilized at 3 mg kg$^{-1}$ (Figure 9b). Fe accumulation followed a sigmoid pattern in the F crop, while the FBS$^0$ crop presented a differentiated one. More specifically, after topdressing (125 DAS), Fe accumulation tended to be higher and then dropped and tended to remain lower towards the end of the crop cycle (Figure 9c).

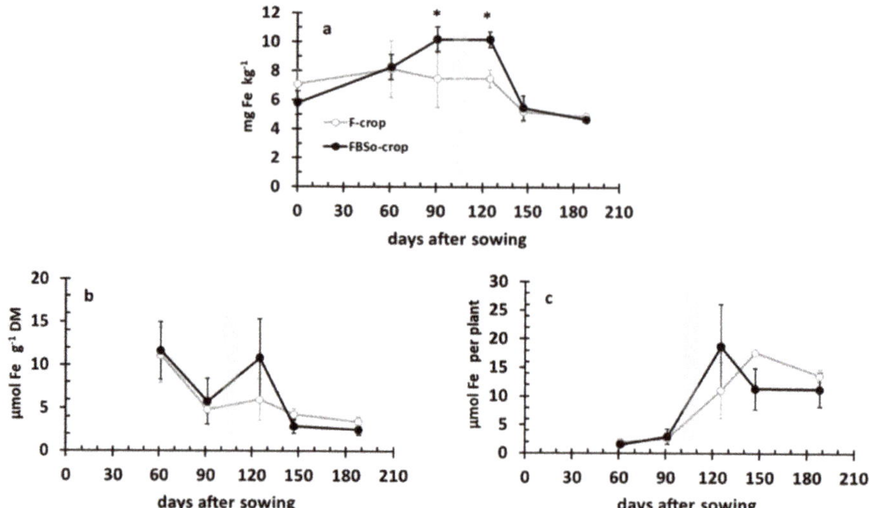


**Figure 9.** Iron dynamics in the rhizosoil (Fe-DTPA concentration (**a**) and in the aerial part (Fe concentration, (**b**); Fe accumulation (**c**) during crop development. Bars indicate standard deviation. An asterisk indicates a statistically significant difference ($p < 0.05$).

### 3.7. Correlations between $S_{org}$, Fe, P, and $N_{org}$ Crops' Nutritional Dynamics

$N_{org}$ and $S_{org}$ accumulations per plant in the F crop had a highly linear relationship ($R^2 = 0.9906$). The $FBS^0$ treatment presented a strong linear relationship, too ($R^2 = 0.9758$), with the same slope (Figure 10a), suggesting that the treatment did not disturb the relationship between $N_{org}$ and $S_{org}$ throughout crop development.

The correlation between $N_{org}$ and P accumulation per plant (Figure 10b) was strongly linear ($R^2 = 0.9037$) in the F crop. This held true in the $FBS^0$ crop ($R^2 = 0.8949$), too, with a tendency for a higher slope, suggesting enhanced accumulation of P in relation to $N_{org}$ after topdressing.

Comparing the accumulation of Fe with that of P or $S_{org}$ per plant in the F crop, strong linear relationships were obtained, with $R^2 = 0.95$ and 0.9251, respectively. The treatment disturbed these linear correlations (Figure 10c,d). By eliminating the point representing 125 DAS, the strong linear relationship was restored ($R^2 = 0.9382$ and 0.992, respectively; Figure 10e,f).

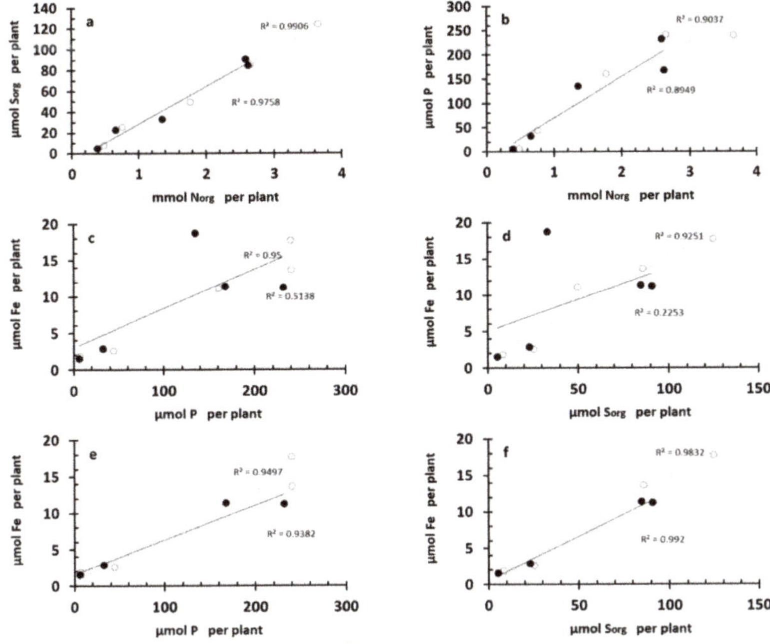


**Figure 10.** Correlations between the corresponding values of $N_{org}$ and $S_{org}$ (**a**), $N_{org}$ and P (**b**), P and Fe (**c**,**e**), $S_{org}$ and Fe (**d**,**f**) accumulations per plant. (**e**,**f**) were produced by eliminating the point representing 125 DAS (explanation in the text).

### *3.8. Phosphate Solubilization, and/or Siderophore-Producing Capacity, and/or Ureolytic Activity of the ARS-Producing Bacterial Isolates*

The ARS-producing bacterial isolates collected from both the F and $FBS^0$ crops in each developmental stage were further examined to see whether they possess additional PGP traits related to P solubilization, siderophore production, swarming motility, urease production, and antagonistic activity against phytopathogenic fungi (Figure 11). Interestingly, almost all isolates showed a swarming ability, whereas a significant proportion of the isolates collected from both the F and $FBS^0$ crops, 50% and 62% respectively, possessed both traits and involved P and Fe mobilization (Figure 11). The incorporation of low levels of $S^0$ had a noteworthy effect on the relative abundance of ureolytic bacterial isolates; a much larger proportion (66%) of isolates collected from the $FBS^0$ treatment exhibited ureolytic activity as compared to the F treatment (50%) (Figure 11). The proportion of ARS-producing bacteria possessing P or Fe mobilization or ureolytic activity changed during the development of both the F and $FBS^0$ crops (Figure 12). Interestingly, at 61 DAS, a small fraction (10%) of isolates collected from the $FBS^0$ crop possessed P-solubilizing capacity, whereas at 91 DAS this proportion rose to over 90%, while at 61 and 91 DAS the vast majority (90–100%) of isolates were found to produce siderophores (Figure 12a,b). In F-treated wheat plants at 61 DAS, a negligible fraction of ARS-producing bacteria possessed P or Fe mobilization activity, while at 91 DAS over 90% of the isolates possessed both traits. Taken together, these results suggest that at 61 and 188 DAS the growing wheat plants were selectively recruiting ARS-producing bacteria possessing both traits, and thus may enhance the levels of bioavailable phosphate and iron to the crop. At subsequent developmental stages of the plant, the abundance of ARS-producing bacteria possessing one or both traits were gradually decreasing, but not eliminated, suggesting that this group of rhizospheric bacteria may continuously provide bioavailable P and Fe to the crop plants, although at gradually reduced rates.

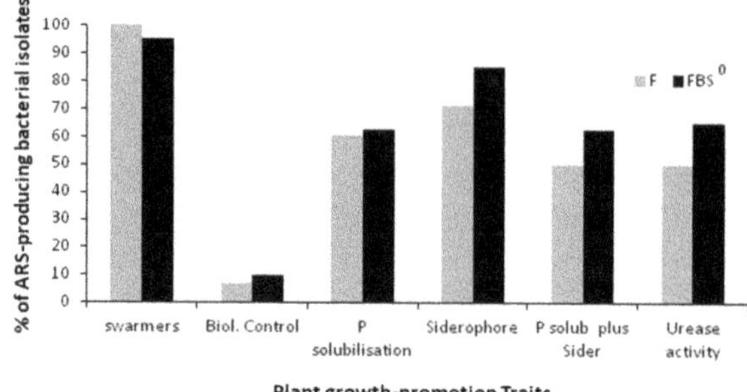


**Figure 11.** Percentage of 68 ARS-producing bacterial isolates' plant growth-promotion traits. The 68 bacterial isolates from F- and FBS$^0$-treated wheat rhizosphere were further tested for plant growth-promoting traits (swarming motility, biological control, phosphate solubilization, siderophore production, urea hydrolysis) in vitro. P solub + Sider represents the percentage of bacterial isolates possessing both siderophore production and phosphate solubilization activity.

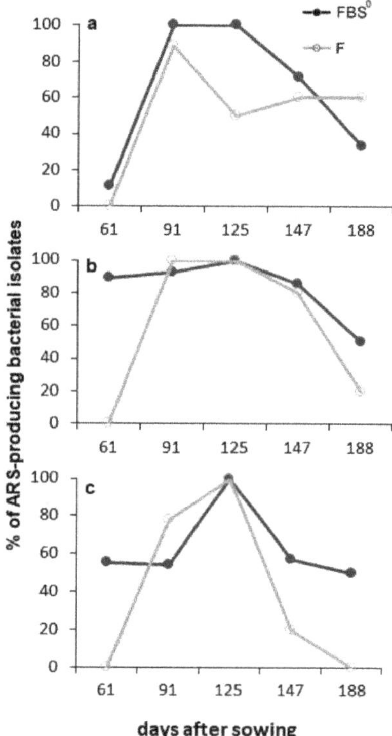


**Figure 12.** Proportions of 68 ARS-producing bacterial isolates possessing (**a**) P solubilization activity, (**b**) siderophore production, and (**c**) urease production in the total population of F- and FBS$^0$-treated wheat at each developmental stage.

The distribution of ureolytic bacteria also changed during the development of both the F and FBS$^0$ crops; at 61 DAS, none of the isolates collected from F-treated wheat showed ureolytic activity, whereas a large fraction (60%) of isolates collected from FBS$^0$ wheat were positive for ureolytic activity. After topdressing, at 125 DAS, all the isolates collected from both the F and FBS$^0$ crops showed positive ureolytic activity, but at subsequent developmental stages of the plant this gradually decreased (Figure 12c).

## 4. Discussion

The aforementioned data supported the working hypothesis: The $FBS^0$ crop provided a 3.4% lower yield compared to the F crop; therefore, the $FBS^0$ treatment significantly alleviated the potential yield loss due to the low amount of available phosphorus in the soil, clearly showing that the incorporation of 2% elemental sulfur in the conventional fertilizer granules helped the crop to effectively counteract the inferior quality of the rhizosoil. Moreover, the interactions between P, Fe, $S_{org}$, and $N_{org}$ were sustained, while the rhizospheric microbiome played a remarkable role in sustaining them. The interactions between iron nutrition and sulfur nutrition have received increasing attention in various plant species, with various approaches and at various levels (e.g., [37–41]). Significant interactions between plant Fe acquisition mechanisms and external sulfate supply have been reported [42]. Lower leaf concentrations have been observed in sulfur-deficient plants in relation to sufficient sulfur treatment [43]. The low availability of sulfate could affect the accumulation and release of phytosiderophores in Fe-deficient barley roots, associated with a possible impairment of Fe acquisition in these plants. These results suggest that the requirement of S may be higher when plants are experiencing Fe deficiency, and that plant responses to Fe deficiency are modified by S supply [42].

The nature of the $FBS^0$ fertilizer has been studied [27]. Briefly, the $FBS^0$ fertilizer used in this experiment contained sulfate as the accompanying anion of ammonium (i.e., directly available S), along with 2% $S^0$ (not directly available S). The ingredients of the binder (i.e., a mixture of molasses and glycerol) are both water-soluble and can be used by soil microorganisms, along with $S^0$, thus suggesting microbial action around the granules [26,27]. In this paper we show that the aforementioned interactions are highly affected by the rhizospheric microbiome under the circumstances.

### *4.1. The Addition of Elemental Sulfur to the Fertilization Scheme Promoted a Higher Density of ARS-Producing Bacterial Population in the Rhizosphere*

In the present study we demonstrated that the rhizosphere of $FBS^0$-treated wheat harbors a higher density of ARS-producing bacterial population, at all stages of plant development, as compared to conventional F-treated wheat. Therefore, we focused on the ARS-producing bacterial population in the rhizosphere and its phosphate and/or iron solubilization capacity, along with ureolytic activity, towards understanding the nutritional dynamics of the rhizosphere and aerial part of the plants with regard to organic sulfur, iron, phosphorus, and organic nitrogen. It has been established that PGPR, associated with the rhizosphere of cultivated plants, influence nutrient bioavailability in soil and nutrient uptake by the plant [44]. The abundance of ARS-producing bacteria appears to be primarily influenced by the incorporation of low levels of $S^0$ into conventional fertilizer, as well as the crop's developmental stage. The growth of ARS-producing bacteria is promoted by the presence of S in organic form in the soil and its activity may not be repressed by iNorganic sulfate [25]. Soil amending with plant residues, wheat, fescue, and mustard, exhibiting different C:S ratios, revealed that residues with high S content significantly stimulated the population density of ARS-producing bacteria as compared to those with low S content [45]. Similar findings were also reported in [46], demonstrating that rhizosphere soil arylsulfatase activity was influenced by the fertilization regime; the application of large amounts of manure and mineral fertilizer resulted in higher soil arylsulfatase activity than mineral fertilizers or the unfertilized control. The fact that S application to the soil has a positive effect when coupled with organic fertilization is pivotal for low-input or organic systems. This has been shown for durum wheat in a Mediterranean environment [47], where it has been showed that S application significantly increased grain yield under organic rather than mineral fertilization, and the present study provides a justification for this.

Furthermore, it has been demonstrated that enhancement of the population density of rhizobacteria that are able to transform organic sulfur into bioavailable forms promoted plant fortification with S of

*Lolium perenne* cultivated in acid soil [23]. The involvement of ARS-producing bacteria in S mobilization is further corroborated by two independent studies using plants cultivated in calcareous soils: Grecut et al. (2009) [22] demonstrated that S-demanding plants, such as rape, appear to recruit higher numbers of ARS-producing bacteria as compared to barley. Vong et al. (2007) [48] demonstrated that rape took up 2–3 times more S derived from fertilizer than barley, resulting in a significant reduction in soil S bioavailable pools at 63 DAS. Thus, ARS-producing bacteria represent a new class of potential biofertilizers that may have enhanced organic S mobilization and provide a sustained input of sulfate, irrespective of the sulfate added by the fertilizer, during crop development. In calcareous soils, S from added ammonium sulfate fertilizer probably reacts with $CaCO_3$ to precipitate $CaSO_4$ [49,50], and since $CaSO_4$ is slightly soluble in water [50], sulfate derived from fertilizers added to calcareous soils may become partly available for plant uptake, depending on the irrigation and precipitation. Thus, the action of ARS-producing bacteria may be important, in particular in unirrigated field crops, for maintaining a sustained input of bioavailable S to the plant.

### 4.2. A Large Fraction of the ARS-Producing Bacterial Isolates Possessed Phosphate Solubilization Capacity

The aforementioned data also demonstrated that a large fraction of the ARS-producing bacteria possessed P-mobilizing capacity, which may have enhanced P mobilization, leading to plant growth promotion. Most of the accumulated P applied to the soil remains in nonbioavailable forms, with P mobilization (measured as Olsen *p* values) ranging from 9 to 49 mg $Kg^{-1}$ [51], thus making the enhancement of bioavailable P a formidable task. Recent studies have shown that artificial enhancement of soil population density of calcareous soils with P-solubilizing bacteria (PSB) enhances the release of Olsen *p* values of natural calcareous soils or calcareous soils amended with mineral or organic fertilizers [10]. On the other hand, sewage sludge is an alternative P source to P fertilizer derived from rock phosphate [52]. Our data demonstrated that, regardless of the differences in Olsen *p* values of the $FBS^0$- and F-treated wheat RS, the P concentration in the $FBS^0$ crop at 125 DAS was significantly higher as compared to the F crop, suggesting a more efficient uptake of P by the $FBS^0$ crop after topdressing. The higher P concentration in the $FBS^0$-treated wheat may be a consequence of the high population density of PSB in the $FBS^0$ crop's Rhs; the high density of PSB at 91 and 125 DAS enhanced the P mobilization and provided a sustained input of bioavailable phosphate for uptake by the plant. These findings are in close agreement with recent studies wherein it has been demonstrated that the inoculation of wheat grown in different types of alkaline soils (with Olsen *p* values ranging from 4.8 to 8.7 mg $kg^1$) with P-solubilizing *B. subtilis* increased P uptake the more the Olsen *p* values in the soil decreased [13]. Furthermore, several studies have shown that inoculation with PSB increased growth and P nutrition in strawberries [53], broad beans [54], and *Zea mays* [55] grown in calcareous soil, rendering the exploitation of PSB a formidable task in improving agricultural productivity.

### 4.3. A Large Fraction of the ARS-Producing Bacterial Isolates Possessed Siderophore-Producing Capacity

The data from this study also demonstrated that a large fraction of the ARS-producing bacteria possessed Fe-mobilizing capacity, which may have enhanced Fe mobilization, leading to plant growth promotion. The Fe concentration of the $FBS^0$ crop's RS increased up to 125 DAS, while the F crop's one gradually decreased. However, the $FBS^0$ crop's rhizosphere was more abundantly populated by ARS-producing bacteria possessing siderophore-producing capacity as compared to the F-treated crop. The high population density of siderophore-producing bacteria may counteract the soil calcification, thus generating more bioavailable Fe in the $FSB^0$ as compared to the F crop's rhizosphere, which can explain the enhanced concentration of iron in $FBS^0$-crop at 125 DAS as compared to the F crop. The $FBS^0$ and F crops' RS Fe concentration was related to the population density of siderophore-producing bacteria in the Rhs during the crop's cycle; a drop in the population density of siderophore-producing bacteria at 149 and 191 DAS coincided with a decrease in the RS Fe concentration and a significant drop in the aerial part's Fe concentration, which was reflected in the overall Fe accumulation in wheat. This may point to the idea of artificial enhancement of a crop's RS population density at the proper time, as the addition of a mixture of selected ARS-producing bacterial isolates possessing siderophore-producing capacity to the RS can affect the final Fe uptake and accumulation. Artificial enhancement of the wheat Rhs population density by inoculation with siderophore-producing bacteria significantly increased the Fe content of wheat and rice [6,56]. Similarly, the inoculation of fertilized wheat with PGPR improved the growth, yield, and nutrient uptake [57]. The inoculation of peanuts cultivated in calcareous soils with bacteria producing siderophores alleviated the antagonistic effect of calcareous soils and increased the bioavailable content of Fe in soil, thus promoting iron nutrition [15].

### *4.4. A Large Fraction of the ARS-Producing Bacterial Isolates Possessed Ureolytic Activity*

Furthermore, our data demonstrated that the abundance of ARS-producing bacterial population possessing ureolytic activity was significantly affected by fertilization treatments; the $FBS^0$ fertilization scheme resulted in a higher population density of ureolytic bacteria as compared to the F one. The fate of urea mainly depends on bacterial urease activity, which is also considered an indicator of soil quality [58] d. As urease converts urea to ammonium, which is either taken up by the plant roots or further processed to nitrate by soil microorganisms, the high population of ureolytic bacteria in $FBS^0$-treated wheat may positively affect the N nutritional status of plant. The incorporation of the urease inhibitor NBPT along with 2% elemental sulfur as ingredients in the fertilization schemes resulted in even higher yields in similar experiments [27]. This supports the hypothesis that NBPT suppressed the ureolytic activity of the ARS-producing bacterial isolates, allowing for better ammonium utilization.

### *4.5. Towards a Sustainable Agronomic Biofortification Practice*

In addition to S assimilation and its effects on the yield and quality of wheat [59], the effect of S rate on durum wheat grain yield and quality has been studied under Mediterranean conditions [60]. In this experiment, a very small amount of $S^0$ (5.5 kg $ha^{-1}$) was added as an ingredient during fertilization at sowing and the practice had a positive impact on the rhizospheric bacteria of the calcareous soil. Obviously, this is the case with related experiments, too [26,27]. By extension, this fostered the idea that small amounts of $S^0$ fine powder could be applied alone in low-input or organic agricultural systems in order to trigger the same result. On the other hand, given that the $FBS^0$ crop is biofortified as regards the studied nutrients, these data demonstrate the merit of deploying inocula composed of the isolated strains from this experiment, or mixtures of these strains as a biofortification strategy for wheat, particularly under calcareous soil conditions. The compatibility issue with multi- versus single-strain plant growth promoting microbial inoculants has been discussed [61]. Taking into account the interactions between carbon, nitrogen, phosphorus, sulfur, and iron biogeochemical cycles, as discussed in [62], the application of such rates of $S^0$ by itself, or coupled with proper single- or multistrain inocula of ARS-producing bacteria, might effectively sustain these interactions, thus suggesting a sustainable agronomic biofortification approach that merits further investigation.

## 5. Conclusions

The incorporation of low levels (2% w/w) of $S^0$ as ingredient of fertilizer granuls in the fertilization scheme of durum wheat grown on a calcareous soil containing minimal available P increased the population density of the rhizospheric ARS-producing bacteria, a large fraction of which also possessed P- and Fe-mobilizing capacity, along with ureolytic activity. The changes induced in the rhizospheric microbiome promoted plant growth, fortified the crop's P, Fe, organic S, and organic N contents, and increased the grain yield.

**Supplementary Materials:** The following are available online at http://www.mdpi.com/2223-7747/8/10/379/s1. Table S1, List of ARS-producing bacterial isolates and their characteristics, isolated from F treatment (control) and $FBS^0$ treatment.

**Author Contributions:** D.L.B. and S.N.C. conceived and designed the experiment and carried out the chemical analyses; A.V. carried out the microbiological analyses; D.L.B., A.V., S.N.C, and P.K. elaborated the research questions, analyzed the data, and wrote the article. D.L.B. supervised all field work.

**Funding:** This research received no external funding.

**Acknowledgments:** The authors are grateful to the Greek fertilizer company Sulphur Hellas S.A. for providing the fertilizers and consumables for chemical and microbiological analyses; the farmer Pavlos Issaris for applying the $FBS^0$ products in his commercial crops; the agronomists Filippa Maniou and Dimitris Petrakos for their help with soil and plant sampling; the agronomist Haris Mavrogiannis for his help with supervising the various agronomic applications by the farmer during the experiment; and the reviewers for their stimulating suggestions. The samples used in this study were collected by the consent of the landowner.

**Conflicts of Interest:** The authors declare that they have no conflict of interest. Sulphur Hellas S.A. had no role in the design of the study; in the collection, analyses, or interpretation of data; in the writing of the manuscript, nor in the decision to publish the results.

## References

1. Marschner, P. *Marschner's Mineral Nutrition of Higher Plants*, 3rd ed.; Academic Press: London, UK, 2011.
2. Badri, D.V.; Vivanco, J.M. Regulation and function of root exudates. *Plant. Cell Environ.* **2009**, *32*, 666–681. [CrossRef] [PubMed]
3. Ahkamia, A.H.; White, A.R., III; Handakumburaa, P.P.; Janssona, C. Rhizosphere engineering: Enhancing sustainable plant ecosystem productivity. *Rhizosphere* **2017**, *3*, 233–243. [CrossRef]
4. Rana, A.; Joshi, M.; Prassara, R.; Shivay, Y.S.; Nain, L. Biofortification of wheat through inoculation of plant growth promoting rhizobacteria and cyanobacteria. *Eur. J. Soil Biol.* **2012**, *50*, 118–126. [CrossRef]
5. Vessey, J.K. Plant growth promoting rhizobacteria as biofertilizers. *Plant. Soil* **2003**, *255*, 571–586. [CrossRef]
6. Adnan, M.; Shah, Z.; Fahad, S.; Arif, M.; Alam, M.; Khan, I.A.; Mian, I.A.; Basir, A.; Ullah, H.; Arshad, M.; et al. Phosphate-solubilizing bacteria nullify the antagonistic effect of soil calcification on bioavailability of phosphorus in alkaline soils. *Sci. Rep.* **2017**, *7*, 16131. [CrossRef] [PubMed]
7. Shi, X.K.; Ma, J.J.; Liu, L.J. Effects of phosphate-solubilizing bacteria application on soil phosphorus availability in coal mining subsidence area in Shanxi. *J. Plant. Interact.* **2017**, *12*, 137–142. [CrossRef]
8. Oteino, N.; Lally, R.D.; Kiwanuka, S.; Lloyd, A.; Ryan, D.; Germaine, K.J.; Dowling, D. Plant growth promotion induced by phosphate solubilizing endophytic Pseudomonas isolates. *Front. Microbiol.* **2015**, *6*, 745. [CrossRef] [PubMed]
9. Zaida, A.; Khan, M.S.; Amil, M. Interactive effect of rhizotrophic microorganisms on yield and nutrient uptake of chickpea (*Cicer arietinum* L.). *Eur. J. Agron.* **2003**, *19*, 15–21. [CrossRef]
10. Hameeda, B.; Harini, G.; Rupela, O.P.; Wani, S.P.; Reddy, G. Growth promotion of maize by phosphate-solubilizing bacteria isolated from compost and macrofauna. *Microb. Res.* **2008**, *163*, 234–242. [CrossRef]
11. García-López, A.M.; Recena, R.; Avilés, M.; Delgado, A. Effect of Bacillus subtilis QST713 and Trichoderma asperellum T34 on P uptake by wheat and how it is modulated by soil properties. *J. Soils Sediments* **2018**, *18*, 727–773. [CrossRef]
12. Kumar, V.; Singh, P.; Jorquera, M.A.; Sangwan, P.; Kumar, P.; Verma, A.K.; Agrawal, S. Isolation of phytase-producing bacteria from Himalayan soils and their effect on growth and phosphorus uptake of Indian mustard (*Brassica juncea*). *World J. Microbiol. Biotechnol.* **2013**, *29*, 1361–1369. [CrossRef] [PubMed]
13. Rungin, S.; Indananda, C.; Suttiviriya, P.; Kruasuwan, W.; Jaemsaeng, R.; Thamchaipenet, A. Plant growth enhancing effects by a siderophore-producing endophytic streptomycete isolated from a Thai jasmine rice plant (*Oryza sativa* L. cv. KDML105). *Antonie Van Leeuwenhoek* **2012**, *102*, 463–472. [CrossRef] [PubMed]
14. Khalid, S.; Asghar, H.N.; Akhtar, M.J.; Aslam, A.; Zahir, Z.A. Biofortification of iron in chickpea by plant growth promoting rhizobacteria. *Pak. J. Bot.* **2015**, *47*, 1191–1194.
15. Liu, D.; Yang, Q.; Ge, K.; Hu, X.; Qi, G.; Du, B.; Liu, K.; Ding, Y. Promotion of iron nutrition and growth on peanut by *Paenibacillus illinoisensis* and *Bacillus sp.* strains in calcareous soil. *Braz. J. Microbiol.* **2017**, *48*, 656–670. [CrossRef] [PubMed]
16. De Santiago, A.; García-López, A.M.; Quintero, J.M.; Avilés, M.; Delgado, A. Effect of Trichoderma asperellum strain T34 and glucose addition on iron nutrition in cucumber grown on calcareous soils. *Soil Biol. Biochem.* **2013**, *5*, 598–605. [CrossRef]
17. Vejan, P.; Abdullah, R.; Khadiran, T.; Ismail, S.; Boyce, A.N. Role of plant growth promoting rhizobacteria in agricultural sustainability—A review. *Molecules* **2016**, *21*, 573. [CrossRef]

18. Pii, Y.; Mimmo, T.; Tomasi, N.; Terzano, R.; Cesco, S.; Crecchio, C. Microbial interactions in the rhizosphere: Beneficial influences of plant growth promoting rhizobacteria on nutrient acquisition process. A review. *Biol. Fertil. Soils* **2015**, *51*, 403–415. [CrossRef]
19. Kertesz, M.A.; Fellows, E.; Schmalenberger, A. Rhizobacteria and plant sulfur supply. *Adv. Appl. Microbiol.* **2007**, *62*, 235–268.
20. Gahan, J.; Schmalenberger, A. The role of bacteria and mycorrhiza in plant sulfur supply. *Front. Plant. Sci.* **2014**. [CrossRef]
21. Fox, A.; Kwapinski, W.; Grifffiths, B.S.; Schmalenberger, A. The role of sulphur- and phosphorus mobilizing bacteria in biochar-induced growth promotion of *Lolium perenne*. *FEMS Microbiol. Ecol.* **2014**, *90*, 78–91. [CrossRef]
22. Crégut, M.; Piutti, S.; Vong, P.-C.; Slezack-Deschaumes, S.; Crovisier, I.; Benizri, E. Density, structure, and diversity of the cultivable arylsulfatase-producing bacterial community in the rhizosphere of field-grown rape and barley. *Soil Biol. Biochem.* **2009**, *41*, 704–710. [CrossRef]
23. Crégut, M.; Piutti, S.; Slezack-Deschaumes, S.; Benizri, E. Compartmentalization and regulation of arylsulfatase activities in Streptomyces sp., Microbacterium sp. and Rhodococcus sp. soil isolates in response to iNorganic sulfate limitation. *Microbiol. Res.* **2013**, *168*, 12–21. [CrossRef] [PubMed]
24. Whalen, J.K.; Warman, P.R. Arylsulfatase activity in soil and soil extracts using natural and artificial substrates. *Biol. Fertil. Soils* **1996**, *22*, 373–378. [CrossRef]
25. Meldau, D.G.; Meldau, S.; Hoang, L.H.; Underberg, S.; Wunsche, H.; Baldwin, I.T. Dimethyl disulfide produced by the naturally associated bacterium *Bacillus* sp B55 promotes *Nicotia attenuata* growth by enhancing sulfur nutrition. *Plant. Cell* **2013**, *25*, 2731–2747. [PubMed]
26. Bouranis, D.L.; Chorianopoulou, S.N.; Margetis, M.; Saridis, G.I.; Sigalas, P.P. Effect of Elemental Sulfur as Fertilizer Ingredient on the Mobilization of Iron from the Iron Pools of a Calcareous Soil Cultivated with Durum Wheat and the Crop's Iron and Sulfur Nutrition. *Agriculture* **2018**, *8*, 20. [CrossRef]
27. Bouranis, D.L.; Gasparatos, D.; Zechmann, B.; Bouranis, L.D.; Chorianopoulou, S.N. The Effect of Granular Commercial Fertilizers Containing Elemental Sulfur on Wheat Yield under Mediterranean Conditions. *Plants* **2019**, *8*, 2. [CrossRef]
28. Jones, J.B., Jr. *Soil Analysis Handbook of Reference Methods*; CRC Press: Boca Raton, FL, USA, 1999; ISBN 9780849302053.
29. Sörbo, B. Sulfate: Turbidimetric and nephelometric methods. In *Methods in Enzymology: Sulfur and Sulfur Amino Acids*; Jakoby, W.B., Griffith, O.W., Eds.; Academic Press, Inc.: New York, NY, USA, 1987.
30. Miller, R.O. Extractable chloride, nitrate, orthophosphate, potassium, and sulfate sulfur in plant tissue: 2% acetic acid extraction. In *Handbook of Reference Methods for Plant Analysis*; Kalra, Y.P., Ed.; CRC Press LLC: Boca Raton, FL, USA, 1998; pp. 115–118.
31. Chorianopoulou, S.N.; Nikologiannis, S.; Gasparatos, D.; Bouranis, D.L. Relationships between iron, sulfur, nitrogen and phosphorus in lawns grown on a calcareous soil irrigated by slightly saline water. *Fresenius Environ. Bull.* **2017**, *26*, 1240–1246.
32. Pikovskaya, R. Mobilization of phosphorus in soil in connection with vital activity of some microbial species. *Mikrobiologiya* **1948**, *17*, 362–370.
33. Schwyn, B.; Neilands, J.B. Universal chemical assay for the detection and determination of siderophores. *Anal. Biochem.* **1987**, *160*, 47–56. [CrossRef]
34. Venieraki, A.; Tsalgatidou, P.C.; Georgakopoulos, D.G.; Dimou, M.; Katinakis, P. Swarming motility in plant-associated bacteria. *Hellenic Plant. Protect. J.* **2016**, *9*, 16–27. [CrossRef]
35. Weisburg, W.S.; Barns, S.M.; Pelletier, D.A.; Lane, D.I. 16S ribosomal DNA amplification for phylogenetic study. *J. Bacteriol.* **1991**, *173*, 697–703. [CrossRef] [PubMed]
36. Kumar, S.; Stecher, G.; Tamura, K. MEGA7: Molecular Evolutionary Genetics Analysis version 7.0 for bigger datasets. *Mol. Biol. Evolut.* **2016**, *33*, 1870–1874. [CrossRef] [PubMed]
37. Astolfi, S.; Cesco, S.; Zuchi, S.; Neumann, G.; Roemheld, V. Sulphur starvation reduces phytosiderophores release by Fe-deficient barley plants. *Soil Sci. Plant. Nutr.* **2006**, *52*, 80–85. [CrossRef]
38. Astolfi, S.; Zuchi, S.; Cesco, S.; Sanita di Toppi, L.; Pirazzi, D.; Badiani, M.; Varanini, Z.; Pinton, R. Fe deficiency induces sulphate uptake and modulates redistribution of reduced sulphur pool in barley plants. *Funct. Plant. Biol.* **2006**, *33*, 1055–1061. [CrossRef]

39. Chorianopoulou, S.N.; Saridis, Y.I.; Dimou, M.; Katinakis, P.; Bouranis, D.L. Arbuscular mycorrhizal symbiosis alters the expression patterns of three key iron homeostasis genes, ZmNAS1, ZmNAS3, and ZmYS1, in S deprived maize plants. *Front. Plant. Sci.* **2015**, *6*, 257. [CrossRef] [PubMed]
40. Saridis, G.I.; Chorianopoulou, S.N.; Katinakis, P.; Bouranis, D.L. Evidence for regulation of the iron uptake pathway by sulfate supply in S-deprived maize plants. In *Sulfur Metabolism in Higher Plants—Fundamental, Environmental and Agricultural Aspects, Proceedings of the International Plant Sulfur Workshop, Goslar, Germany, 1–4 September 2015*; De Kok, L.J., Hawkesford, M.J., Haneklaus, S.H., Schnug, E., Eds.; Springer: Berlin/Heidelberg, Germany, 2017; pp. 175–180.
41. Saridis, G.; Chorianopoulou, S.N.; Ventouris, Y.E.; Sigalas, P.P.; Bouranis, D.L. An Exploration of the Roles of Ferric Iron Chelation-Strategy Components in the Leaves and Roots of Maize Plants. *Plants* **2019**, *8*, 133. [CrossRef] [PubMed]
42. Astolfi, S.; Zuchi, S.; Neumann, G.; Cesco, S.; Sanita di Toppi, L.; Pinton, R. Response of barley plants to Fe deficiency and Cd contamination as affected by S starvation. *J. Exp. Bot.* **2012**, *63*, 1241–1250. [CrossRef] [PubMed]
43. Astolfi, S.; Zuchi, S.; Passera, C.; Cesco, S. Does the sulfur assimilation pathway play a role in the response to Fe deficiency in maize (*Zea mays* L.) plants? *J. Plant. Nutr.* **2003**, *26*, 2111–2121. [CrossRef]
44. Calvo, P.; Watts, D.B.; Kloepper, J.W.; Torbert, H.A. Effect of microbial-based inoculants on nutrient concentrations and early root morphology of corn (*Zea mays*). *J. Plant. Nutr. Soil Sci.* **2016**, *180*, 56–70. [CrossRef]
45. Piutti, S.; Slezack-Deschaumes, S.; Niknahad-Gharmakher, H.; Vong, P.C.; Recous, S.; Benizri, E. Relationships between the density and activity of microbial communities possessing arylsulfatase activity and soil sulfate dynamics during the decomposition of plant residues in soil. *Eur. J. Soil Biol.* **2015**, *70*, 88–96. [CrossRef]
46. Tang, H.; Xiao, X.; Sun, J.; Guo, L.; Wang, K.; Li, E.; Wenguang, T. Soil enzyme activities and soil microbe population as influenced by long-term fertilizer management during the barley growth in Hunan Province, China. *Afr. J. Microbiol. Res.* **2016**, *10*, 1720–1727.
47. Rossini, F.; Provenzano, M.E.; Sestili, F.; Ruggeri, R. Synergistic effect of sulfur and nitrogen in the organic and mineral fertilization of durum wheat: Grain yield and quality traits in the Mediterranean environment. *Agronomy* **2018**, *8*, 189. [CrossRef]
48. Vong, P.-C.; Nguyen, C.; Guckert, A. Fertilizer sulphur uptake and transformations in soil as affected by plant species and soil type. *Eur. J. Agron.* **2007**, *27*, 35–43. [CrossRef]
49. Hu, Z.Y.; Zhao, F.J.; McGrath, S.P. Sulphur fractionation in calcareous soils and bioavailability to plants. *Plant. Soil* **2005**, *268*, 103–109. [CrossRef]
50. Chien, S.H.; Gearhart, M.M.; Villagarcia, S. Comparison of ammonium sulfate with other nitrogen and sulfur fertilizers in increasing crop production and minimizing environmental impact: A review. *Soil Sci.* **2011**, *176*, 327–335. [CrossRef]
51. Recena, R.; Díaz, I.; Delgado, A. Estimation of total plant available phosphorus in representative soils from Mediterranean areas. *Geoderma* **2017**, *297*, 10–18. [CrossRef]
52. Houben, D.; Michel, E.; Nobile, C.; Lambers, H.; Kandeler, E.; Faucon, M.-P. Response of phosphorus dynamics to sewage sludge application in an agroecosystem in northern France. *Appl. Soil Ecol.* **2019**, *137*, 178–186. [CrossRef]
53. Ipek, M.; Pirlak, L.; Esitken, A.; Figen, M.; Dönmez, M.; Turan, M.; Sahin, F. Plant growth promoting rhizobacteria (PGPR) increase yield, growth and nutrition of Strawberry under high-calcareous soil conditions. *J. Plant. Nutr.* **2014**, *37*, 990–1001. [CrossRef]
54. Iqbal, S.; Khan, M.Y.; Asghar, H.N.; Akhtar, M.J. Combined use of phosphate solubilizing bacteria and poultry manure to enhance the growth and yield of mung bean in calcareous soil. *Soil Environ.* **2014**, *35*, 146.
55. Ibarra-Galeana, J.A.; Castro-Martínez, C.; Fierro-Coronado, R.A.; Armenta-Bojórquez, A.D.; Maldonado-Mendoza, I.E. Characterization of phosphate-solubilizing bacteria exhibiting the potential for growth promotion and phosphorus nutrition improvement in maize (*Zea mays* L.) in calcareous soils of Sinaloa, Mexico. *Ann. Microbiol.* **2017**, *67*, 801–807. [CrossRef]
56. Sharma, A.; Shankhdhar, D.; Shankhdhar, S.C. Enhancing grain iron content of rice by the applicationof plant growth promoting rhizobacteria. *Plant. Soil Environ.* **2013**, *59*, 89–94. [CrossRef]

57. Abbasi, M.K.; Sharif, S.; Kazmi, M.; Sultan, T.; Aslam, M. Isolation of plant growth promoting rhizobacteria from wheat rhizosphere and their effect on improving growth, yield and nutrient uptake of plants. *Plant. Biosyst.* **2011**, *145*, 159–168. [CrossRef]
58. Tscherko, D.; Kandeler, E.; Bardossy, A. Fuzzy classification of microbial biomass and enzyme activities in grassland soils. *Soil Biol. Biochem.* **2007**, *39*, 1799–1808. [CrossRef]
59. Zhao, F.J.; Hawkesford, M.J.; McGrath, S.P. Sulphur assimilation and effects on yield and quality of wheat. *J. Cereal. Sci.* **1999**, *30*, 1–17. [CrossRef]
60. Ercoli, L.; Lulli, L.; Arduini, I.; Mariotti, M.; Masoni, A. Durum wheat grain yield and quality as affected by S rate under Mediterranean conditions. *Eur. J. Agron.* **2011**, *35*, 63–70. [CrossRef]
61. Thomloudi, E.-E.; Tsalgatidou, P.C.; Douka, D.; Spantidos, T.-N.; Dimou, M.; Venieraki, A.; Katinakis, P. Multistrain versus single-strain plant growth promoting microbial inoculants—The compatibility issue. *Hellenic Plant. Protect. J.* **2019**, *12*, 61–77. [CrossRef]
62. Li, Y.; Yu, S.; Strong, J.; Wang, H. Are the biogeochemical cycles of carbon, nitrogen, sulfur, and phosphorus driven by the "FeIII–FeII redox wheel" in dynamic redox environments? *J. Soils Sediments* **2012**, *12*, 683–693. [CrossRef]

MDPI
St. Alban-Anlage 66
4052 Basel
Switzerland
Tel. +41 61 683 77 34
Fax +41 61 302 89 18
www.mdpi.com

*Plants* Editorial Office
E-mail: plants@mdpi.com
www.mdpi.com/journal/plants

www.ingramcontent.com/pod-product-compliance
Lightning Source LLC
LaVergne TN
LVHW070904170726
843515LV00005B/699

* 9 7 8 3 0 3 9 3 6 0 0 6 2 *